HEALTH MAINTENANCE OF CULTURED FISHES

PRINCIPAL MICROBIAL DISEASES

John A. Plumb, Ph.D.
*Department of Fisheries and Allied Aquacultures
and Alabama Agricultural Experiment Station
College of Agriculture
Auburn University
Auburn, Alabama*

CRC Press
Boca Raton   Ann Arbor   London   Tokyo

Library of Congress Cataloging-in-Publication Data

Plumb, John A.
   Health maintenance of cultured fishes : principal microbial
diseases / John A. Plumb.
      p.   cm.
   Includes bibliographical references and index.
   ISBN 0-8493-4614-2
   1. Fishes—Infections.   2. Fish—culture.   I. Title.
SH171.P66   1994
639.3—dc20   94-3864
                                                                      CIP

No claim to original U.S. Government works
International Standard Book Number 0-8493-4614-2
Library of Congress Card Number 94-3864
Printed in the United States of America   1 2 3 4 5 6 7 8 9 0
Printed on acid-free paper

PREFACE

Infectious diseases of cultured fish are among the most notable constraints on the expansion of aquaculture and the realization of its full potential. Viral, bacterial, and parasitic agents infect every species of cultured aquatic animal. Most pathogenic agents are endemic to natural waters where, under normal conditions, they cause no great problems. However, when these same viruses, bacteria, and parasites are present in aquacultural environments, they may cause significant disease and mortality. Fish are often held in environments to which they are not biologically accustomed, a circumstance that often makes them more susceptible to infectious disease. It is virtually impossible to separate the relationship of infectious disease from problems associated with environmental quality. In the following pages, the major objective is to emphasize the salient points of host-pathogen-environment relationships, elucidate the important aspects of infectious diseases, and explore how management can be used to help reduce the effects of disease on aquaculture.

The text concentrates on the infectious viral and bacterial diseases that are most prevalent in aquaculture. Although much information has been derived from North American studies, important disease problems from other parts of the world are included. Also, where applicable, the influence of the various diseases on wild populations has been included. I have tried to emphasize diseases of warmwater fishes while not ignoring those in cool and cold environments.

The viral and bacterial diseases of fish are organized into groups or families that are most extensively cultured. Where disease affects members of more than one family, it is included with the family which is most commonly or severely affected by that disease. Although some viral and bacterial diseases occur in both marine and freshwater fishes, no distinction is made between the two environments.

It was not my intention to list all of the published papers on each disease or subject mentioned; only those publications pertinent to the discussion at hand have been cited. This book is intended for students and scientists who are interested in health maintenance of aquatic animals, aquatic pathobiology, and infectious diseases of fin fish. Hopefully, it will be used as a text for beginning fish pathologists and as a reference source for those of broader experience.

THE AUTHOR

John A. Plumb, Ph.D., is a Professor in the Department of Fisheries and Allied Aquacultures at Auburn University in Auburn, Alabama, where he has served on the staff and faculty since 1969. Dr. Plumb graduated in 1960 from Bridgewater College in Bridgewater, Virginia with a B.S. degree in General Physical Science. He obtained his M.S. degree in Zoology from Southern Illinois University in Carbondale in 1963. In 1972, he received his Ph.D. in Zoology (Fisheries) from Auburn University. From 1962 through 1969, Dr. Plumb was employed as a Hatchery Biologist with the U.S. Fish and Wildlife Service where he worked at cold- and warmwater fish hatcheries. During that time he attended the Warm Water Inservice Fish Culture School in Marion, Alabama and the Fish Disease Long Course in Leetown, West Virginia.

Dr. Plumb is a member of the American Fisheries Society and the Fish Health Section. He has served as Secretary, Treasurer, and President of the Fish Health Section. Dr. Plumb has also served on the Editorial Board of *Progressive Fish Culturist* and is currently co-editor of the *Journal of Aquatic Animal Health,* a member of the Editorial Board of the *Journal of Applied Aquaculture,* and an Editorial Advisor for *Diseases of Aquatic Organisms.* His awards include the Distinguished Service Award from the Catfish Farmers of America and the S.F. Snieszko Distinguished Service Award from the Fish Health Section of the American Fisheries Society. Dr. Plumb has acquired an international reputation as an expert on fish health and has served short-term technical and advisory tours in South America, Southeast Asia, and Kuwait.

Dr. Plumb has received several research grants from the U.S. Department of Agriculture and the National Science Foundation, as well as numerous grants from private industry and other sources. He has served on project and grant review panels and symposia for the National Science Foundation, the U.S. Department of Agriculture, the U.S. Environmental Protection Agency, the Food and Agriculture Organization of the United Nations, and the Office of International Epizootiology. Dr. Plumb has been invited to present papers to numerous regional, national, and international scientific meetings, and has published 20 chapters in books on fish diseases and 84 papers in peer review journals and 13 other publications. His research includes investigations of viral and bacterial diseases of fish, immunology and vaccination of channel catfish, and epidemiology of infectious diseases of fish. Dr. Plumb presently teaches graduate courses on basic and advanced microbial diseases of fish and disease diagnosis at Auburn University.

ACKNOWLEDGMENTS

I wish to thank the many friends and colleagues who assisted in the preparation of this book and also those who so kindly allowed me to use their photographs. Without their help in reviewing the various chapters and providing constructive suggestions, it would not have been possible to complete the work.

I recognize and offer my appreciation to the following individuals who reviewed parts of the manuscript: Pierre de Kinkelin, Diane Elliott, Nikola Fijan, John I. Grover, John M. Grizzle, Ronald P. Hedrick, Brit Hjeltness, Richard A. Holt, John W. Jensen, Philip McAllister, Kathy Middleton, Takuo Sano, Mike Schiewe, Homer R. Schmittou, Emmett B. Shotts, and Y.-L. Song.

A special thanks is also extended to my wife, Peggy, for her invaluable assistance in proofing and editing and for her patience and understanding.

CONTENTS

PART ONE
HEALTH MAINTENANCE

Chapter I

Principles of Health Maintenance

"An ounce of prevention is worth a pound of cure" is a familiar phrase that describes one approach to the culture of food animal resources. Health maintenance actually refers to a concept in which animals are reared under conditions that optimize growth rate, feed conversion efficiency, reproduction, and survival while minimizing problems related to infectious, nutritional, and environmental diseases, all within an economical context. Therefore, "health maintenance" encompasses the entire production management plan for food animals, whether they be swine, cattle, poultry, or fish, either in the public (governmental) or private sector.

Fish health management is not a new approach to aquaculture. In 1958 Snieszko[1] recognized the need for health maintenance in fish culture when he stated, "We are beginning to realize that among animals (fish) there are populations, strains, or individuals which are not susceptible all of the time, or even temporarily, to some of the infectious diseases." He proposed the theory that fish possessed a certain level of natural resistance to infectious diseases that could be enhanced through proper management. Another contributor to the health maintenance concept of aquatic animals was Klontz,[2] who established a course in fish health management at Texas A & M University in 1973. This course combined the studies of culture and infectious diseases of fish into a health management concept. Also, The Great Lakes Fishery Commission published the *Guide to Integrated Fish Health Management in the Great Lakes Basin*[3] which defined a regional approach to fish health management. These references deal with the improvement of aquatic animal health through management. However, probably the most useful contribution to maintaining the health of domesticated (cultured) animals was made by Schnurrenberger and Sharman[4] when they set forth 24 principles for animal health maintenance that apply, in a general sense, to all domestic food animals. Although targeting terrestrial animals, these principles apply to all types of animal production and in the following pages these will be applied to aquatic animals. In theory, if these principles were utilized in the daily, monthly, yearly, and long-term management of an aquatic culture facility, there would be fewer environmental and disease problems and optimum production would be more readily obtained. This chapter describes how many of the principles of health maintenance can be applied to aquaculture.

A. MAINTAINING HEALTH

In an aquatic environment, there is a profound and inverse relationship between environmental quality and disease status of fish. As environmental conditions deteriorate, severity of infectious diseases increases; therefore, sound health maintenance practices can play a major role in maintaining a suitable environment where healthy fish can be grown.

Maintaining health in a fish population is a positive concept that will result in more efficient production, rather than the mere prevention of disease. Disease prevention emphasizes the interruption of a disease cycle and deals only with one segment of health maintenance. Maintaining a healthy fish population should include the use of genetically improved fish and certified specific pathogen-free (SPF) stocks whenever these are available and/or feasible, environmental control; prophylactic therapy; pond, cage, raceway, and tank management; control of vegetation, aeration, and use of other water quality maintenance practices; nutrition; and a commitment by the manager to provide an optimum habitat in terms of water quality for the fish being cultured.

Health maintenance does not simply target infectious diseases, but is a management approach inclusive of water quality and environmental manipulation, disease prevention, nutritional considerations, and economic factors. Its goal is to improve the health and well-being of animals that appear to be generally healthy. If sound health maintenance principles are followed, production will be more efficient and will result in a healthier product. Obviously, all improvements must be based on sound economic criteria.

B. STRESS

Stress is a general physiological reaction to trauma or to a physical or physiological insult to the body. Numerous definitions have been used to describe stress. Wedemeyer[5] quotes Brett,[6] who defined stress as "a state produced by any environmental factor which extends the normal adaptive responses of an animal beyond the normal range, or that which disturbs the normal functioning to such an extent that the chances of survival are significantly reduced." Stress may also have a related physiological meaning — the nonspecific response of the body to a factor that is perceived as harmful. Grizzle[7] believes that stress is often used imprecisely to mean "any change in an animal that impairs any aspect of performance including death, disease, reduced reproduction, growth, and stamina." Pickering[8] states that the problem of understanding and defining the term *stress* is one of semantics; therefore, when the word *stress* is used it should be accompanied by a definition pertaining to its usage. For clarification purposes in this text, *stress, stressor,* and *stress response* will be defined as follows: stressor will be considered synonymous with stress to indicate a stimulus that precipitates a physiological response in the fish[9] and stress response will be used to designate the physiological changes that take place in the fish as a result of exposure to the stressor. The level of corticosteroids and glucose present in plasma is the usual quantitative measure for stress.[10]

Fish react to stress in different ways, depending on the severity and length of exposure to the stressor. Fish may die almost immediately from shock if the stressor is sufficiently severe or, at the other extreme, they may adapt to a mild or slow developing stressor and suffer no long-term effects. Fish may also respond to a stressor by altering their physiology to the point that natural resistance and immunity to disease is reduced and they become more susceptible to infectious diseases.

The response of fish to stressful conditions is more rapid and extensive than in most cultured animals. Because the aquatic environment is in a continuous state of flux, fish must continually adapt physiologically to these environmental changes. Inability to make these adjustments may not result in immediate death, but will be manifested in lower productivity, reduced weight gain, increased feed conversion, decreased immunity, reduced natural resistance to infectious disease, and lowered hardiness. All of these conditions will result in reduced profits for the fish farmer and diminished production and efficiency for public hatcheries.

Some commonly known stressors present in the aquatic environment are unionized ammonia, chronic exposure to low concentrations of pesticides or heavy metals, insufficient oxygen or high concentrations of carbon dioxide, rapidly changing or extremes in pH, and extreme or rapidly changing water temperatures. A successful and efficient health maintenance program for an aquacultural facility must include measures to reduce the stressful conditions on the fish population.

C. HAZARD REDUCTION THROUGH MANAGEMENT AND ENVIRONMENTAL INFLUENCE ON DISEASE

Fish culture experience has shown that a wide variety of bacterial, parasitic, and other diseases will cause mortality if cultured fish are held under unfavorable environmental conditions.[10-12] Whenever management changes are contemplated, their effect on the environment and health of the fish should always be considered. It is important to remember that health management and environmental management decisions are interdependent, and a change in one area should not be made without evaluating the effect it may have in another area. The most notable stressor-related diseases include furunculosis, enteric redmouth, motile *Aeromonas* septicemia, columnaris, vibriosis, bacterial gill disease, external fungal infections, and some protozoan parasites (Table 1). These diseases are usually the culmination of many factors including management decisions and biological conditions (Figure 1).

Environmental stress on fish increases geometrically when environmental conditions approach the tolerance limits of the host.[13] For example, if the water temperature is critically high and oxygen concentration is adequate, fish may survive; if oxygen is critically low and the water temperature is normal, fish may adjust and survive. However, if both parameters are at stressful levels but not individually lethal, the problem is compounded exponentially and the fish will not be able to adapt and may become exhausted and die. Another example of an environmental stressor and its effect on fish involves the relationship between dissolved oxygen (DO) and carbon dioxide (CO_2) concentrations. Channel catfish *(Ictalurus punctatus)* can adapt to an elevated level of CO_2 (20 to 30 mg/l) if the DO

Table 1   **Microbial fish diseases commonly considered stress mediated**[16,43-45]

Disease	Predisposing environmental factors
Bacterial gill disease	Crowding, unfavorable environmental condition, and presence of causative bacteria, elevated ammonia, particulate material in water
Columnaris	Crowding, handling, seining, adverse temperature, other infectious diseases
Cold water disease	Temperature decrease from 10 to 15°C to 7 to 13°C
Enteric redmouth	High stocking density, elevated water temperature, excessive metabolites, handling, transport
Furunculosis	Low oxygen, handling when *A. salmonicida* is endemic
Motile *Aeromonas* septicemia	Injury to skin, fins, or gills, hauling, improper handling, temperature stress, low oxygen, other disease organisms, pesticides, seasonal changes, improper nutrition, crowding
Spring viremia of carp	Handling after over-wintering at low temperatures
Ulcer disease of goldfish	Handling and stocking in late winter or early spring
Vibriosis	Handling, poor environmental conditions, migration of fish from fresh water to sea water

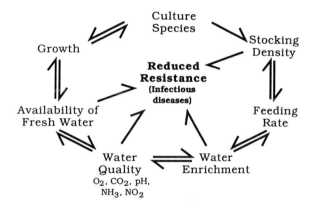

Figure 1   The relationship of environmental conditions and other biological factors in aquaculture that influence infectious diseases of fish.

concentration is optimal. However, if the CO_2 level is critically high and the DO is critically low, the fish cannot adapt. Although fish may survive these environmental stressors, their resistance can be so compromised that they become diseased.

Snieszko[13] applied a theory of host/pathogen/environment relationship to fish, with regard to the development of infectious diseases (Figure 2). This theory follows the premise that to have an infectious disease in fish, in addition to a host and pathogen, an unfavorable environmental condition must act as a trigger or stressor for the disease to develop. Often, especially in warmwater fish, potential infectious disease organisms are endemic in the environment and only environmental conditions, and/or the host's natural resistance can dictate the beginning of the disease process. Snieszko[13] described the interaction of these factors in a disease production equation as follows:

$$H(A + S^2) = D$$

where H = species or strain of the host (including age and inherited susceptibility), A = etiological agent, S = environmental stressors, and D = disease.

Environmental stressors are squared because as fish approach adaptation limits, effects of the stressors increase geometrically. Also, when more than one stressor is involved (oxygen, ammonia, carbon dioxide, temperature, etc.) detrimental effects are magnified.

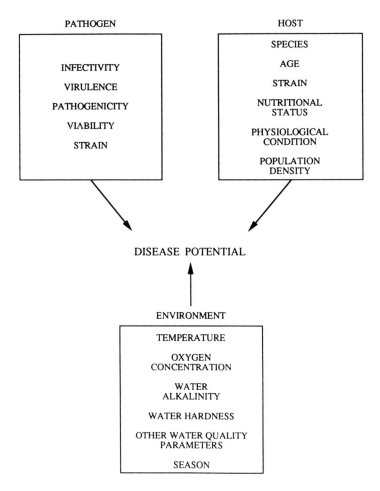

Figure 2 Some variables of the infectious agent, host, and the environment that influence the potential for disease occurrence of fish. (Adapted from Snieszko[13] and Schnurrenberger.[49])

The relationship between water quality deterioration, environmental stresses, and bacterial infection was shown in a fortuitous study by Plumb et al.[14] The sudden die-off of an algae bloom in a channel catfish pond was followed by a loss of DO, a decrease in pH, and increases in CO_2 and NH_4 levels. These water quality changes in the pond resulted in an ''oxygen depletion'' and a fish kill (Figure 3), a phenomenon that has since been described by Schmittou[15] as ''low dissolved oxygen syndrome'' (LODOS). When fish first began to die, no bacteria or other significant pathogens were found during necropsy. However, 4 d post oxygen depletion, channel catfish with hemorrhaged and depigmented lesions in the skin and muscle were seen (Figure 4). When first observed, the lesions were aseptic and no bacteria were isolated. On day 2 following the appearance of lesions and for several days thereafter, *Aeromonas hydrophila* was isolated from muscle and skin lesions and from internal organs. When fresh water was added to the pond and remedial aeration provided, mortality ceased and clinical signs of infectious disease abated. It was theorized that during the hypoxic condition of the water, some areas of muscle in the fish became anoxic which led to tissue necrosis, hemorrhaging, and skin depigmentation. When the integrity of protective epithelium was lost, *A. hydrophila* that was present in the water invaded any unprotected areas and established a systemic infection. Walters and Plumb[16] demonstrated that either low oxygen, low pH, high ammonia, or high CO_2 alone were not particularly stressful and did not lead to bacterial disease. However, if one or more of these adverse environmental conditions occurred simultaneously, infection was much more likely to occur (Figure 5).

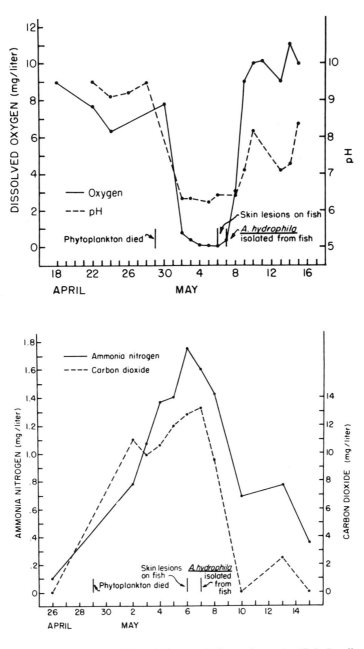

Figure 3 Water quality parameters before, during, and after a channel catfish die off and subsequent *Aeromonas hydrophila* infection. (From Plumb, J. A. et al., *J. Wildl. Dis.,* 12, 247, 1976. With permission.)

Currently, there is strong support for developing vaccines and new chemotherapeutics to control or prevent diseases of fish. The development of these therapeutic treatments is important and should be pursued. However, because the protection afforded by vaccination is relative, vaccines must be used in conjunction with other sound management practices. In time, bacteria may develop a resistance to antibiotics, and at best, chemoprophylactics are only sporadically effective. Therefore, disease control and prevention in a fish population is more economically achieved and maintained by proper management and effective environmental manipulation than solely by chemotherapeutic treatment.

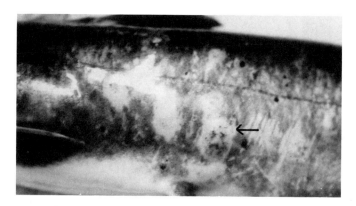

Figure 4   Channel catfish with depigmented, hemorrhaged, and necrotic lesions which appeared 6 d after oxygen depletion. (From Plumb, J. A. et al., *J. Wildl. Dis.,* 12, 247, 1976. With permission.)

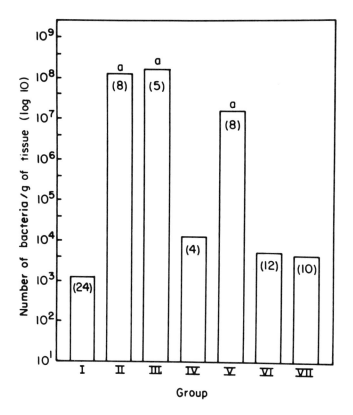

Figure 5   Number of bacteria (all species) per gram of trunk kidney from channel catfish held in various environmental treatments. I. Low dissolved oxygen (DO) only; II. Low DO and injected with *Aeromonas hydrophila;* III. Low DO, injected with *A. hydrophila* and added NH_3; IV. Low DO, injected with *A. hydrophila,* added NH_3 and CO_2; V. Low DO, injected with *A. hydropphila* and added CO_2; VI. Reaeration and injected with *A. hydrophila;* VII. Noninjected controls held in reaerated water. The number of fish samples in parentheses and significantly higher numbers of bacteria are designated by a. (From Walters, G. R. and Plumb, J. A., *J. Fish Dis.,* 17, 177, 1980. Printed by permission of The Fisheries Society of The British Isles.)

Table 2 **Range of temperature tolerance, optimum temperature for growth, and spawning temperature of selected cultured fish[43,46-48]**

Species	Temperature (°C)		
	Range	Optimum[a]	Spawning
Salmonids			
Atlantic salmon	1–24	10–17	7–10
Brook trout	1–22	8–13	8–13
Brown trout	1–25	9–17	9–13
Chinook salmon	1–25	10–14	8–13
Coho salmon	1–25	9–14	8–13
Lake trout	1–21	7–14	9–11
Rainbow trout	1–25	10–17	10–13
Sockeye salmon	1–21	10–15	8–12
Catfishes			
Channel catfish	1–35	21–30	22–27
Walking catfish	13–38	20–30	20–30
Cyprinids			
Common carp	1–35	23–30	13–27
Chinese carp[b]	1–35	22–30	22–27
Miscellaneous			
Eel		25–	NA
Largemouth bass	1–35	13–27	16–20
Milkfish		25–	NA
Pike (northern)	1–27	4–18	4–9
Striped bass	2–32	13–24	13–22
Walleye	1–27	8–16	9–13
Tilapia	15–35	23–32	23–32

[a]Temperature above or below the optimum range could be considered stressful. [b]Grass carp and silver carp.

Intensively raised fish provide unique, but manageable, cultural problems.[11,17] Fish require an adequate supply of water which must be maintained at a suitable temperature and oxygen concentration level for proper growth and reproduction. Water quality requirements vary from species to species (Table 2). For example, cultured channel catfish require a water temperature of 20 to 33°C for growth (optimum of 28 to 30°C), however, they spawn at 27°C and prefer oxygen concentrations above 5 mg/l. Channel catfish will survive at 1 to 5 mg/l of oxygen, but prolonged exposure below 1 mg/l is lethal. Rainbow trout (*Oncorhynchus mykiss*) require water temperatures of 8 to 18°C for growth, they spawn at a temperature of 6°C, and require an oxygen concentration above 5 mg/l. The walking catfish *(Clarias batrachus)* of Southeast Asia, can survive in water temperatures in the mid 30°C range or higher, and can survive very low concentrations of oxygen. However, these fish have an auxiliary gill that allows them to extract oxygen directly from the air. Environmental manipulation to alleviate stress on a species population may be necessary when deviations from optimum homeostasis occur.

Beneficial management changes in an aquatic environment can be made regardless of whether the environment is a natural fish community in an open, natural body (i.e., lake or river), or a confined, concentrated population in a closed or semiclosed, artificial body of water (i.e., culture pond, raceway, tank, or aquarium). However, the environmental manipulation of open natural waters is much more difficult to achieve because of the numerous physical, economical, legal, and social factors involved. Also, lack of control over the watershed complicates efforts to control pollution and other man-made environmental disorders and may have to be dealt with on a larger scale (i.e., regional, sectional, or national) or on a political level. These problems can apply to any culture situation where the watershed cannot be controlled.

Fish rearing environments are highly susceptible to management manipulation. For example, under natural conditions, a given body of water will support a certain standing crop of fish, but from an aquaculturist's point of view, this crop is too sparse to be economically profitable. By adding nutrients, the farmer can increase the standing crop of fish to a level that would be economically profitable. Unfortunately, the nutrient input proportionally increases environmental instability, and may cause extreme fluctuations in critical water quality factors. Consequently, a change in one management technique (e.g., nutrients) must be matched by another appropriate management technique (e.g., aeration). The carrying capacity of most minimally managed farm ponds is usually 25 to 150 kg of fish per hectare. Under these conditions, there is sufficient natural food, oxygen, and other life-sustaining substances to adequately support the fish population. However, a channel catfish culture pond which is stocked with fingerlings to yield as much as 5000 kg/ha represents the potential for a deteriorating environment, due to the amount of feed required to sustain that weight of fish. Uneaten feed and fecal wastes produce an enormous load of organic matter in the water which, in turn, provides an abundance of nutrients to support plant and bacterial growth. Also, surplus organic matter, including living and decomposing dead plants, consume oxygen that is essential for fish survival. If several consecutive cloudy days occur, photosynthesis may be reduced, or if plants grow so rapidly that they utilize all of a limiting nutrient, they may die. In either case, additional oxygen will be removed from the water as a result of decomposition and this can result in poor growth, disease, and even death of the fish population.

In some culture systems (e.g., trout), high-density problems are overcome by increasing water flow volume, but in pond culture systems (e.g., channel catfish) water volume is prescribed with little or no freshwater exchange. In an effort to increase productivity, some farmers increase stocking densities from approximately 10,000 fish per hectare (4500/acre) to approximately 20,000 fish per hectare (9000/acre) per year. Tucker et al.[18] showed that catfish production ponds stocked at rates of 11,000 fish per hectare produced 5800 kg/year, compared to stocking rates of 19,500 fish per hectare which produced 7900 kg over the same period. Although the higher stocking density resulted in greater weight of fish, these fish did not grow as well, there was greater variation in fish size, survival was reduced, feed conversions were higher, and the need for aeration increased. The presence of additional nutrients caused a deterioration in water quality and increased the potential for LODOS to occur, which leads to a lower disease resistance and a higher mortality rate in the overstocked fish ponds.

Problems concerning overstocking can be corrected, to a large degree, by manipulative management. Immediate corrective measures should include limiting quantity of feed, reducing standing crop, and increasing aeration. In the long term, routine and periodic applications of herbicides such as copper sulfate, aquazine, or other U.S. Food and Drug Administration (FDA)- and Environmental Protection Agency (EPA)-approved compounds can prevent massive blooms of phytoplankton or filamentous algae. Herbivorous fish, such as the grass carp (Ctenopharyngodon idella), that eat vascular plants and convert the nutrients to fish flesh may also be stocked. Silver carp (Hypophthalmichthys molitrix), or other filter feeders, will also help reduce phytoplankton while improving water quality. The use of herbivorous fish to control vegetation is generally more satisfactory, longer lasting, more economical than use of chemicals, less hazardous to the environment, and actually helps to improve water quality. Precautions must be taken, however, to avoid the escape of these fish into natural waters where they could be ecologically detrimental. Also, the use of exotic carps is prohibited by law in some states because of their potential damage to the aquatic environment. In most instances, these states will allow sterile, triploid grass carp to be stocked.

Reduced feeding rates will temporarily relieve environmental pressures due to over-fertility, but will also reduce weight gain. To help avoid oxygen depletion, periodic selective harvesting, or cropping, of marketable-size fish is advisable. This will reduce the organic load in the water as a result of reduced food requirements.

Because large amounts of feed are required to maintain adequate growth in heavily stocked ponds, oxygen levels should be carefully monitored during warm weather. If it drops to a critical, minimal level for the species of fish being cultured, immediate corrective measures should be taken. Because natural oxygenation is provided by photosynthesis and is driven by sunlight, oxygen levels fluctuate diurnally, with highest levels occurring in the afternoon and lowest levels occurring during the night and just before dawn. The need for supplemental aeration usually occurs between midnight and dawn,

Figure 6   Routine and emergency aeration with: (A) water pump powered by power takeoff; (B) paddlewheel in a catfish pond; and (C) aeration by flowing water over a column of perforated plates.

when oxygen levels are at their lowest. Aeration of water can be achieved by several methods: compressed air; pure or liquid oxygen can be bubbled into the water; or various mechanical water agitators, such as paddlewheels or sprayers can be used (Figure 6).

Disease agents are transmitted more easily from individual to individual in crowded populations, especially when environmental conditions are less than optimal. Channel catfish virus disease (CCVD), a severe infection of young cultured channel catfish, is a good example of such a disease. When young fish, that have been exposed to channel catfish virus (CCV), are held in overcrowded conditions and in water temperatures above 25°C, or the fish are otherwise stressed, an acute infection associated with a high mortality rate will develop. If precautions are taken to reduce stress, the losses due to CCVD are not nearly so dramatic or acute.

Tempering is a process of slowly changing the water temperature of the present environment to the same water temperature as the future environment. Considering this, one can understand why it is important to "temper" fish to reduce stress when moving them from one environment to another. Some species of fish (e.g., trout) require several hours to be properly tempered, while other species (e.g., carp) require less time. Fish generally can be tempered downward more quickly than upward, and can also tolerate a greater temperature differential downward than upward.

In geographic areas where water temperatures normally fall below the optimal level for growth and good health, water from warm underground aquifers or heated water discharge from power plants may be used to increase the water temperature in the rearing system. Heat exchange units or solar heating can also be used to warm the water if the amount of water required is relatively small, but this process is too expensive for use by large facilities.

Extreme deviations from optimal environmental conditions will result in stress manifested by reduced feeding, higher feed conversions, poor growth, disease, or death. High or low temperatures may result in greater susceptibility of the fish to viral, bacterial, and parasitic diseases.[19] There are numerous examples of diseases in fish that are directly associated with poor environmental conditions.

In order for the manager, biologist, or diagnostician to understand infectious diseases of fish, he or she must also understand how and to what extent the environment affects the host and the disease. It is not enough to simply culture the pathogen or to identify the parasite; it is important to understand what stressors may have occurred in the environment to predispose the diseased condition. When stressors are known, corrective and preventive measures can often be initiated to help prevent or minimize reoccurrence of the stressful condition. Therefore, in the aquatic world, maintenance of an optimal environment is essential to good animal health.

D. LOCATION, SOIL, AND WATER

Choosing the proper site for an aquaculture facility is paramount to its success. Topography of the land, quality of the soil, water quality and abundance, and a proximity to the market that will be served are primary factors that must be considered when choosing the location for an aquaculture facility. To maintain healthy populations of fish, the soil, water, and species of fish to be grown must be compatible. Sometimes, environmental modifications can be made to accommodate a specific species not indigenous to an area, but the modification must be cost-effective. For example, channel catfish are not grown in northern latitudes of the U.S. because of a lack of warm water and short growing season. However, if some type of heated water, i.e., from power plants or geothermal supplies, is available they can be grown successfully.

Soil requirements for culture ponds are very specific. The soil must be capable of holding water, thus a high clay content is essential. Leaky ponds are unstable, cannot be adequately fertilized, do not hold a suitable plankton bloom, and require continuous replenishment of precious water. Commercial aquaculture cannot afford these problems.

Certain regions of the world, particularly coastal areas, have acid-sulfate soils containing high levels of iron pyrite.[20] As long as these soils are submerged, the iron pyrite is stable and usually causes no problems. However, when the pond is drained and the pond bottom is exposed to air, iron pyrite is oxidized, and upon dehydration sulfuric acid is produced. When the pond is refilled, the water becomes highly acid (acidity may be as low as pH 3.5), rendering the environment unproductive. Generally, such areas should be avoided for aquaculture sites, but if this is impossible, the level of acidity can be corrected to some degree through intensive, laborious management.

It is very important that the soil is free of chemical and pesticide pollutants. If noxious substances are present, these toxicants may leach into the water and kill fish. One must be concerned not only with pond soil, but also with the soil in the watershed area if surface runoff will enter the culture system. Prior to making any commitment to pond construction, watersheds should be inspected for toxic substances that may contaminate the water supply.

The most important ingredient in successful fish health management is water quality and availability. Water quality, as described by Boyd[21] varies, but in reality most waters are suitable for aquaculture, except under unusual circumstances. Water quality parameters, such as hardness, alkalinity, pH, presence of toxins or undesirable gases, are affected by the source in which the water originates. Before a fish farm is built, the water quality, volume, and reliability of the water source to be used must be determined. Water chemistry criteria for aquaculture are summarized in Table 3.

Water sources for aquaculture may be lakes, rivers, springs, pumped or artesian wells, or strictly runoff. From a fish health management standpoint, spring- or well water is preferred because these water sources are free of wild fish that may be carriers of infectious disease agents. However, from a water quality standpoint these sources may not be directly usable. They may be low in oxygen, acidic, and high in carbon dioxide or nitrogen gases which must be removed. These waters may be very soft (less than 50 mg/l $CaCO_3$) or contain high concentrations of iron, sulfur, or manganese. Although well and spring water may have a constant temperature, it may not be optimum, thus requiring heating, or in the case of geothermal sources, cooling. If a pumping system is necessary to utilize a water source, it can be an expensive procedure prone to mechanical breakdowns and interrupted water flow. However, these water quality problems can be overcome by proper planning and management.

Table 3 **Suggested water quality criteria for optimum fish health management of warmwater and coldwater species in milligrams per liter (mg/l)**[21,42-44]

Characteristic	Trout	Warmwater
Acidity	pH 6.5–8	pH 6.5–9
Alkalinity (as $CaCO_3$)	10–400	50–400
Aluminum	0.1	
Ammonia	0.0125	0.02
Arsenic	0.7	
Cadmium (soft water)	0.0004	0.0003
Cadmium (hard water)	0.003	0.003
Carbon dioxide	0–10	0–15
Chromium	0.03	
Copper (soft water)	0.006	0.006
Copper (hard water)	0.03	0.03
Hydrogen sulfide	0.002	0.002
Iron (total)	0.03	0.03
Lead	0.03	0.03
Manganese	0.01	0.01
Mercury	0.005	0.005
Nitrate	3.0	3.0
Nitrogen (gas)	100% saturation	100%
Zinc	0.03	0.03
Total solids	80 mg/l	

When permanent surface water (stream or reservoir) is used to supply an aquaculture facility, it is important to be aware of certain inherent problems involved. These water sources may contain wild fish that harbor parasites, pathogenic bacteria, or viruses. Also, they may be prone to seasonal influence of toxicants, wide seasonal temperature fluctuations, variable oxygen levels, silt loads, increased organic loads, and volume fluctuations. They can also be a source of wild fish, including fry and eggs that would contaminate monoculture ponds. Installation of sand and gravel filters or use of porous (saran) socks on inlet pipes are among the options the manager can use to correct fish contamination.

If surface runoff is used as a water source, the greatest disadvantages are seasonal availability and annual fluctuations in volume. Also, these waters will reflect good or bad characteristics of the soil in the watershed.

E. AVOIDING EXPOSURE

The ideal way to control infectious diseases in fish would be to prevent their exposure to pathogenic agents, thus preventing most devastating fish health problems. However, when dealing with the aquatic environment, it is nearly impossible to define all disease-causing agents and to keep them isolated from the fish hosts. Water provides an excellent medium for transfer of many communicable agents from fish to fish or from locality to locality. This contamination by pathogens can be controlled to some extent by filtration or disinfection with ultraviolet light or ozone, but these methods are applicable only under limited situations. Moreover, many disease-causing organisms are normal inhabitants of the aquatic environment and are opportunistic, facultative pathogens that remain viable under various conditions. In spite of numerous problems involved in separating fish host and pathogen, it is possible to avoid exposure in some closed systems, and the concept should be used whenever feasible.

Avoidance of exposure to certain infectious agents has been an active practice in the U.S. since the mid 1960s, especially within federal and state fish hatcheries and, to a lesser degree, in the private aquaculture sector. Several avoidance methods are disease-free certification, quarantine, water disinfection, and destruction of populations infected with specific disease organisms. The pros and cons of the latter approach should be carefully weighed before a decision is made. To paraphrase a statement by Snieszko, *a disease management practice should not destroy more than it saves.*

The earliest attempts to systematically prevent exposure of fish to infectious pathogens involved trout. Stocks of trout that were known to be positive for infectious pancreatic necrosis virus (IPNV) were not used for egg production. The practice of verifying that fish stocks are negative for a specific disease is termed disease-free certification, or specific pathogenfree (SPF), and has now been expanded to involve other viral agents as well as several bacterial and parasitic organisms. It is impossible to declare a group or population of fish disease free, implying that they are free of all disease agents; therefore, the term disease free must be limited to a specific disease agent. Disease-free certification is currently the best way to prevent the introduction of a disease into a clean facility, but at the same time it must be understood that testing procedures are not infallible.

Many states have established fish health protection programs and specify that incoming fish must be accompanied by a document certifying them to be free of specific disease agents. California has one of the oldest and most rigorous individual state fish health regulatory programs in existence. Regional fish health plans have also been established to prevent the introduction of unwanted pathogens into geographically defined fish populations. For example, the Colorado River Basin Council has a strong inspection program to prevent the distribution of fish that are infected with certain disease agents into natural and hatchery waters. The Great Lakes Fisheries Commission also has a regional fish health protection program for the states and Canadian provinces contiguous to the Great Lakes. The major problem with state and/or regional fish regulatory plans is the inconsistency with which they are implemented. Some plans are stringently enforced while others are not.

International fish disease control concepts are also gaining strength to keep pace with the growing worldwide aquacultural industry. Many countries now require health certificates of some type before fish can be imported. These certificates generally require verification that eggs from which fish were hatched have been inspected and have been declared to be free of specific diseases. While not foolproof, these methods have generally been successful in inhibiting the spread of a number of infectious disease agents. Rohovec[22] reviewed international regulations prior to 1979 and reported that Great Britain, Europe, Canada, and the U.S. all had regulations limiting the movement of fish with emphasis on fish health. Other countries and regions have, or are presently putting into place, regulations to protect their natural aquatic resources from nonendemic diseases.

Quarantine is another approach to disease prevention through avoidance. When fish are moved from one area to another, they can be confined in isolated facilities for a period of time before they come in contact with the resident population. In this way, if a disease develops in the newly arrived animals, it can be dealt with more effectively and without exposing the entire stock.

Sometimes drastic measures, such as eradication of an entire population of fish, are necessary to avoid exposure of healthy fish to a highly infectious, exotic, or noninfectious disease agents. The practice of total eradication of an infected animal population has been a long-standing practice in poultry and cattle husbandry, and can sometimes be justified when dealing with diseased fish populations. A case in point was the attempted containment of viral hemorrhagic septicemia virus (VHSV) in 1989,[23] when salmonids at two hatchery sites in the U.S. Pacific Northwest were destroyed. This was the first reported occurrence of VHSV outside of Europe, so these drastic measures were felt to be justified. However, in subsequent years the virus was detected in different areas of the northwestern U.S. and the problem was more complicated than could be solved by simple eradication of a specific group of infected animals.[24] On the other hand, an incident when total eradication was probably not justified occurred in Michigan in the mid 1970s when an attempt was made to eliminate the disease *Myxobolus (Myxosoma) cerebralis* in trout. This particular disease-causing parasite had been present in the U.S. since the early 1960s,[25] thus the destruction of the fish in Michigan did nothing to prevent the establishment of the parasite. However, if these control efforts had been attempted when the disease was first identified in Pennsylvania, destruction of the diseased fish population would have been warranted.

Snieszko's theory on the application of fish health management practices comes to mind when one considers the destruction, in 1990, of valuable trout brood stock at the Jackson National Fish Hatchery (Wyoming)[26] and at the White Sulfur Spring National Fish Hatchery (West Virginia).[27] The administrative decision to kill a large number of valuable cutthroat *(Oncorhynchus clarkii)*, lake *(Salvelinus nemayacush)*, rainbow *(O. mykiss)*, and brook *(S. fontinalis)* trout was based on a highly sensitive method for detecting very low numbers of *Renibacterium salmoninarum* (bacterial kidney disease — BKD). The fish were killed in spite of the fact that neither facility demonstrated any evidence of BKD, and it is a ubiquitous organism. Because these two facilities are major trout egg suppliers for many federal and state agencies, the question must be asked — was the cost and consequence of implementation greater than what was saved?

Another example of a zealous, and in the author's opinion, ill-advised destruction of fish occurred in California when a shipment of 5000 Florida strain largemouth bass *(Micropterus salmoides)* finger-lings, from a commercial farm in Alabama, were killed by California biologists because they carried a light infestation of *Posthodisplostomum minimum,* the ubiquitous white grub.

While prevention of infectious diseases is best accomplished through avoidance, in some cases one must look beyond the host to find the disease source. Some disease agents, helminths for example, have complex life cycles that involve fish-eating birds, snails, or copepods. In these instances, the life cycle must be broken by controlling nonfish vectors, a nearly impossible task. Another excellent example of a disease agent that may have a nonfish vector is *Edwardsiella ictaluri,* the causative agent of enteric septicemia of catfish. This organism was found in the intestines of 53% of cormorants and other fish-eating birds examined,[28] and it would be extremely difficult to prevent exposure of warmwater fish to these sources of contamination. Furthermore, it is not fully known whether or not birds actually contribute to the dissemination of the disease.

Each disease outbreak must be considered individually, and rarely will a general policy be completely applicable. Health maintenance decisions must therefore be made using biological facts, common sense, and with an eye to the overall good of the aquatic environment.

F. THE EXPOSING DOSE

Total avoidance of an infectious agent is the best way to prevent fish disease; however, this is not always possible or even practical. Many organisms that are pathogenic to fish are normally free-living, facultative, and basically opportunistic pathogens. The fish and pathogens often coexist without incident, unless the fish's immune system or other defensive mechanism is compromised.

The bacterium *Aeromonas hydrophila,* the causative agent of motile *Aeromonas* septicemia (MAS), is an ubiquitous organism that occurs naturally in most fresh waters of the world. It is capable of living and proliferating in any water that has organic enrichment, usually causing disease in fish only after the fishes' resistance has been compromised by an environmental stressor or the fish have suffered some mechanical or biophysical injury. Thus, the best way of preventing MAS is to reduce bacterial numbers in the water, lower organic load, and keep the fish at a high state of resistance through maintenance of environmental quality. Prophylactic treatments, after fish are handled, will also aid in the healing of superficial wounds and in reducing bacterial populations on the skin. In a normal, healthy fish population a certain percentage of fish may have systemic *A. hydrophila.* However, only when bacterial numbers overwhelm the fishes' resistance does disease occur.

Most fish, wild or cultured, normally have some parasites present on their gills or skin. As long as these parasites, either protozoans or monogenetic trematodes, are present in low numbers there generally is no health hazard. If parasites become abundant, or if water quality deteriorates to the point that the fish are stressed (i.e., LODOS), the fish may then be adversely affected. The population of parasites can be reduced with prophylactic chemotherapy and fish health reestablished by restoring environmental quality.

Metazoan parasites (e.g., helminths) follow a similar scenario in fish health. These parasites usually have a complex life cycle that involves copepods, snails, or forage fish as an intermediate host. The adult worms live in the digestive tract of fish-eating birds or predatory fish. As long as the larval worms in the fish are present in low numbers, they usually cause no problem, and even when worm populations become more numerous in the flesh, internal organs, or intestines, the fish rarely become debilitated. It is usually impossible to totally eliminate metazoan parasites in natural waters, but their numbers can be reduced by initiating snail control and barring birds access to culture units. Management of fish populations by culling old brood stock and using younger fish for this purpose, before they become overburdened with parasites, can also reduce the effects of metazoan parasites. The use of young brood stock has been successful in the southeastern U.S., where there has been a problem with the bass tapeworm larvae infecting ovaries of adult fish and causing a reduction in egg production.[29]

Most infectious agents require specific levels of infective units before they inflict injury to, or adversely affect, the health status of the host. Therefore, it stands to reason that if these organisms can be maintained at levels below the disease threshold, losses due to disease will be reduced.

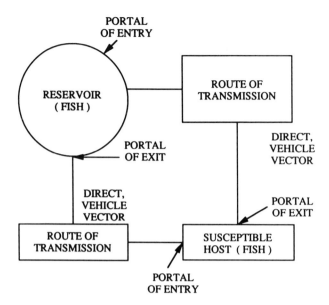

Figure 7   Typical transmission cycle for communicable disease agents (adapted from Swan[50]).

G. EXTENT OF CONTACT

Diseases that are strictly opportunistic (i.e., facultative) are not considered in this particular discussion. The following remarks pertain to fish diseases that are associated with obligate pathogens (i.e., viruses, some bacteria, and a few parasites). In managing such diseases, initial consideration must be given to the pathogen source as it exists in the culture system.

Pathogens must have a route of transmission to a susceptible host (Figure 7). Transmission of pathogens may be by direct contact from fish to fish; through the water; on nets, buckets, and other gear; or via reproductive products. After a susceptible host is sufficiently exposed to a pathogen, it may become infected. Fish that survive an epizootic may become carriers and finally, reservoirs for the disease organism. There are a number of examples of fish diseases (particularly viruses) where epizootics are associated with a reservoir host. The pathogen must have a way of escaping the reservoir host; in fish this can be via the shedding of pathogens across the gill membrane, through the mucus on the skin, and in feces or urine. Bacterial and viral agents can be shed through these portals, but internal metazoan parasites often must wait until the fish is eaten by an intermediate or final host, or until infected fish die and decay, releasing the parasite.

Some pathogens are released from the reservoir host during spawning or in the feces. For example, viruses may be released from carrier fish in eggs, ovarian fluid, or milt. In some viral diseases e.g., infectious hematopoietic necrosis (IHN), the number of virus-shedding fish increases during spawning time.[30] This virus characteristic enhances the transmission of infective units to other fish. IPN virus is shed in the feces, as are some bacteria, *Yersinia ruckeri* for example.

Bacteria and parasites can be transmitted to cultured fish populations by feeding them raw wild fish flesh. Marine fish are an especially good reservoir source for these pathogens. Mycobacteria (acid-fast staining organisms) can be easily transmitted to salmonids by feeding them uncooked flesh or viscera. This transmission route was common in Pacific salmon hatcheries where it was a practice to feed raw, ground marine fish carcasses to hatchery stocks. Another example of parasitic organisms being transmitted by feeding contaminated fish flesh occurred in Florida. Chopped Atlantic menhaden (*Brevortia tyrannus*) infected with *Goezia* (a nematode that inhabits the stomach and intestinal wall), were fed to striped bass (*Morone saxatilis*) fingerlings. The infected striped bass were stocked into inland lakes in Florida, where the parasite was transmitted to resident tilapia (*Tilapia* spp.) and largemouth bass (*Micropterus salmoides*).[29] Although the transmission of viral disease as a result of feeding raw fish has not been reported, most virulent fish viruses have been transmitted orally under laboratory conditions.

While extensive infectious disease-induced mortalities in wild fish populations are occasionally reported, high mortalities are much more common in cultured fish populations. Like other animals, fish are more susceptible to infection when crowded. As fish density increases, the rate of pathogen transmission and extent of contact, increases geometrically. Therefore, in high-density fish populations common to aquaculture facilities, many epizootics have the potential for reaching catastrophic proportions.

H. PROTECTION THROUGH SEGREGATION

Segregating animals according to species and age can reduce the transmission of fish disease-producing agents. Disease-producing organisms are not equally pathogenic to all species of fish, all strains within a species, or to all ages of fish within a susceptible species. A fishery manager can utilize this information to help prevent and/or control disease in his stocks.

Some fish disease-producing organisms are not host specific, while others are more-or-less host specific. The nonhost specific organisms are generally facultative or free-living, opportunistic parasites or bacteria. Some diseases, such as furunculosis *(Aeromonas salmonicida),* and most viral agents, are generally host specific. Brook trout are considered highly susceptible to *A. salmonicida* and should be reared completely segregated from infected populations. Rainbow trout and brown trout *(Salmo trutta)* are also susceptible to *A. salmonicida,* but to a lesser degree, and may be considered for culture where this disease is established. Rainbow trout and some species of Pacific salmon are highly susceptible to IHNV, therefore, culture of a less susceptible species of salmonid should be considered where the virus is endemic.

Warmwater fish pose a somewhat different susceptibility problem than do the salmonids. In the case of CCVD, channel catfish is the most susceptible species. However, within channel catfish stocks, some strains are less susceptible to the virus than others. Inbred strains are more susceptible to the virus than outbred strains.[31] In areas where CCVD is endemic, less susceptible species of catfish, such as the blue catfish *(I. furcatus),* could be cultured.

Generally, young fish are more susceptible to pathogens than older fish, particularly with regard to viral diseases and some parasites. In the case of IPNV and IHNV of trout and CCV of ictalurids, fish less than 3 months of age are highly susceptible. Susceptibility and effects of the disease diminish with increasing age. If these species of fish can be reared in noncontaminated water until they are past the critical age of susceptibility, severe losses to these diseases can be avoided. The same is true with *Myxobolus cerebralis* (Whirling disease of trout), a parasite which infects very young trout when soft cartilage is still present in the skeletal tissues. Young, susceptible fish should be reared in well water or protected spring water that is free of wild fish or contamination. After fish have grown past the susceptible size or age, it is safe to transfer them into less-protected waters.

Water temperature affects the disease-causing ability of some fish pathogens. IHNV is more devastating at a temperature below 17°C; thus, if fish can be reared in water above 17°C, the disease can be avoided. However, at elevated temperatures fish may be more susceptible to other disease organisms and their growth rate will be reduced. CCV is most severe at temperatures above 23°C. If catfish can be held at cooler temperatures, CCV can be avoided, but growth rate will be reduced. It is obvious that when one problem is avoided, others may have to be addressed and all options must be carefully weighed. Unfortunately in most instances, optimum temperature for the pathogen is also optimum temperature for the host.

I. THE PROBLEM OF NEW ARRIVALS

Newly arrived fish at a facility create a unique problem. They may bring pathogenic organisms with them, either in an active or carrier state, to which the resident population has not been exposed. Therefore, no new animals should be introduced into an existing population until it is reasonably certain that they will not be detrimental. This rule applies to home aquaria, large aquaculture facilities, movement of fish from farm to farm, and to interstate and international shipments of fish.

The farmer or aquarium owner can take several precautions to reduce the potential of introducing disease with new arrivals. They should be familiar with the health history of any fish to be introduced, whether any serious disease problems occurred, and if so, what type treatment was given. If not treated before transport, newly arrived fish or eggs should be treated prophylactically with appropriate chemo-therapeutics to remove any external pathogens. If possible, new fish should be segregated (i.e., quarantined) from the endemic population until it is reasonably certain that they are disease-free.

Fish farmers must be extremely careful when replenishing their brood stocks with fish from outside sources. There is always the risk of introducing a pathogen that has not previously existed in their facility. This danger is especially prevalent if fish from wild populations are used to replace brood stocks. For example, epithelial epizootic virus disease of lake trout only occurs if wild brood stocks are used.[32]

Disease-free certification of fish or eggs, while not 100% reliable, is applicable to IPNV, IHNV, VHSV, and some bacterial diseases of salmonids, but is less applicable for warmwater fish diseases. For example, there is no definitive way, at the present time, to certify channel catfish free of channel catfish virus.

Some state, federal, and foreign governments require disease inspections before they will issue permits to transport fish across jurisdictional lines. Simply stated, in order to stop the movement of disease agents, one must stop the movement of infected animals. The prudent fish farmer must take every precaution available to him when acquiring new stock.

J. BREEDING AND CULLING

Domestication of wild animals has been the foundation of successful animal husbandry throughout the ages. Selection, culling, cross-breeding, and hybridization have been vital in developing today's herds and flocks of domesticated animals. Culturing fish for food has been practiced for thousands of years in some areas of the world, especially in China. However, comparatively few studies have dealt with development of disease-resistant fish through selective breeding and genetic manipulation. When disease resistance has occurred, it has usually been by happenstance or as a byproduct of research for other desirable characteristics. This is especially true with warmwater species. In most experiments involving fish, the objectives have been to increase growth rates and fecundity, or to improve feed conversions.

The lack of available, scientifically developed, disease-resistant brood fish should not prevent the aquaculturist from trying to improve his stocks. He can gradually accomplish this by selecting fish that are less affected by specific disease outbreaks. Also, if a scientific research facility is available to help with brood fish selection, that expertise should be used. The routine culling of inferior individual brood fish will also improve performance. If a particular lot of brood fish routinely produces offspring that develop a specific disease year after year (e.g., CCV), replacing those brood fish should be strongly considered.

There is a great deal of interest at the present time in genetically improving fish for aquacultural purposes. This includes gene manipulation, hybridization, and genetic engineering. Such studies may provide some valuable culture animals in the future, but in the meantime, the aquaculturist should continue to improve present stocks by culling and selective breeding.

K. THE NUTRITIONAL BASIS OF HEALTH MAINTENANCE

Proper nutrition is essential for the survival, growth, and reproduction of any animal species.[33] This applies to both wild or cultured fish populations, even though nutritional management in the two systems is entirely different. Nutrition of wild fish populations is dependent upon the natural food chain and the availability of primary nutrients. Nutrients can be supplemented in wild populations by addition of organic or inorganic fertilizers to stimulate growth of macroflora, microflora (i.e., phytoplankton), and zooplankton. Forage fish utilize these organisms as food and, in turn, are fed upon by predator species. The health of a fish population, at any given level in the food chain pyramid, depends on the available food source.

Recently, the role nutrition plays in infectious disease susceptibility of fish has been addressed. Nutrient requirements of fish for growth, reproduction, and normal physiological functions are similar to those of other animals. Manufactured fish feeds contain proteins, fats, carbohydrates, fiber, vitamins, amino acids, and minerals. A deficiency of any of these nutrients may result in nutritional disease or lowered resistance to infectious disease. The amount of each nutrient needed for proper growth depends on the species and age of the fish being fed.

The aquaculturist should always use the best feed available to meet the needs of the culture system. For example, in lightly stocked ponds where significant natural food is available, a supplemental feed will probably be adequate. At the other end of the spectrum, when intensive systems with maximum stocking densities are used, a nutritionally complete feed is required.

It has been shown that certain nutritional deficiencies or excesses can result in specific pathological conditions.[34] Severe injury to the spinal column of channel catfish, (i.e., broken-back syndrome) results from a vitamin C deficiency. Lack of sufficient riboflavin can result in eye cataracts in rainbow trout and possibly channel catfish. A niacin deficiency increases sensitivity to UV irradiation and "sunburn". Nutritional gill disease of trout is caused by a pantothenic acid deficiency, and can progress into bacterial gill disease if the deficiency is not corrected. Aflatoxins produced by *Aspergillus* sp., at doses less than 1 μg/kg of diet, will cause liver tumors (hepatomas) in rainbow trout.[35] Other species of fish, (e.g., channel catfish) may not be as sensitive to the fungal toxins as are trout.

Several researchers have shown that a deficiency of specific elements in fish diets can lead to increased disease susceptibility, while megadoses of specific elements can lead to increased immune response. Durve and Lovell[36] and Li and Lovell[37] reported channel catfish that were fed a vitamin C-deficient diet were more susceptible to *Edwardsiella tarda* and *E. ictaluri*, while megadoses of vitamin C (5 times the recommended 60 mg/kg of diet per day) enhanced immune response. However, these studies were conducted in aquaria, and when the same principles were applied to pond-reared channel catfish, vitamin C-deficient diets did not affect susceptibility nor did megadoses of the vitamin enhance immunity.[38] This study probably indicates the availability of natural vitamin C sources in fish ponds.

More subtle effects of a nutritionally deficient diet are reduced growth, increased feed conversion, and reduced fecundity. All of these dietary factors have the potential to affect the overall efficiency, health status, and disease susceptibility of fish.[33]

L. ERADICATION, PREVENTION, OR CONTROL

Before discussing eradication, control, or prevention, these terms should be defined. *Eradication* is the complete elimination of a disease-causing agent from a facility or defined geographical region. However, "practical eradication" has a slightly different meaning; namely elimination of an agent from its reservoir of practical importance. *Control* is the reduction of a problem to a level that is economically and biologically manageable, or the confinement of a problem to a defined area. *Prevention* is avoidance of the introduction of a disease-producing agent into a region or farm, or stopping the disease process before it gets started.

Eradication of a fish disease from a farm, watershed, or region is highly desirable, but in most instances very difficult to accomplish. There are a few examples where total eradication of a disease has been accomplished in humans and other animals, especially in developed countries with sophisticated medical technology, but there are no reported examples of total eradication of a fish disease agent from a large geographical region. However, practical eradication of several diseases from individual fish hatcheries and farms has occurred. By following a strict set of criteria, some federal and state trout hatcheries have eliminated IPNV, furunculosis, and bacterial kidney disease from their facilities. All fish were eradicated, and the facilities were sterilized with a chlorine solution or formaldehyde gas. These select hatcheries also had closed water supplies with no indigenous fish populations. Elaborate precautions were then taken to prevent reintroduction of the disease through fish, eggs, or contaminated equipment, including fish-hauling trucks. When a commitment is made to eradicate a specific disease from an area or facility, an equal commitment must be made to keep the facility free of the disease.

A commitment to control a disease implies that the disease continues to be present and a decision has been made to accept the resulting losses. However, stringent control measures including sanitation, prophylactic treatments, vaccination, a good health management program, and chemotherapeutic treatments when disease does occur should be utilized to reduce these losses.

Biological and economic constraints usually dictate what approach will be used in dealing with a fish disease problem — eradication, control, or prevention. Eradication may be the best approach where isolated cases of obligate pathogens are involved if it can be accomplished within the criteria set forth. In any case, attempted eradication of a disease agent through destruction of entire populations of animals is going to be controversial. When contemplating eradication, several factors need to be considered: the economic and biological significance of the disease; if the disease is indigenous to the area; and whether or not the disease can adequately be controlled by chemotherapeutics. In other words, is the expense of eradication worth the savings in fish, and can the eradicated area be maintained free of that organism?

Control is the appropriate approach when nonobligate pathogens are involved or when effective chemotherapeutic and management practices are available. The objective of control is to reduce the pathogen level in the environment, so that acute disease will not occur. This approach dictates that some fish will be lost while every attempt is made to keep these losses at a minimum. Often, control of infectious diseases through chemotherapeutics is successful only when coupled with improved environmental conditions. However, chemotherapeutics should be used only after all management, environmental, and husbandry practices have been put in place and the disease still occurs.

Prevention of disease is primarily a farm management approach and should begin with construction of the aquaculture facility. It includes site selection, development of the water supply, and selection of fish to be stocked. Original fish stocks, as well as any subsequent fish brought into the facility, should come from populations known to be free of serious obligate pathogens. As fish or eggs are moved within or brought into the facility, they should be prophylactically treated according to acceptable guidelines. When available, vaccination should be considered as a disease prevention tool. Sanitation practices at the facility are very important and disinfection of nets, buckets, boots, and other equipment should be a routine practice between uses on different groups of fish. This precaution will also help prevent the spread of diseases within a facility, when they do occur.

Eradication vs. control should be determined on a case by case basis, with an eye to the economic consequences involved. Sometimes these decisions must be made on a regional basis by professional fishery scientists. Individual fish farmers should strive to eradicate diseases at their facility when feasible, but only if the facility, after sterilization, can be maintained free of that specific disease agent.

M. VARIABLE CAUSES REQUIRE VARIABLE SOLUTIONS

Epizootiology is the study of disease occurrence patterns in animals other than man. A basic epizootiological precept allows that the absence of overt disease is not necessarily indicative of the presence or absence of a disease-causing agent. As discussed earlier in this chapter, disease occurrences in fish are determined by complex interacting factors involving host, pathogen, and environment.

Although a host and pathogen must be present to cause a disease, environmental influences are often the trigger that stimulates actual development of the disease. It is relatively easy to identify and describe clinical signs of a pathogenic organism that is causing an acute infection, but the environmental influences that precipitated the disease are far more complex to identify. Therefore, each disease must be evaluated and dealt with on an individual basis.

When treating a fish disease problem, an evaluation of environmental conditions prior to, and at the time of, disease outbreak must be part of the diagnosis. Facts pertinent to making an accurate diagnosis are: Was there a recent oxygen depletion?; Did the water change color?; Did heavy rains occur recently?; and Did mortality occur suddenly or increase gradually? Answers to these questions will, in large part, dictate the treatment that will be used. If oxygen levels are low, formalin should not be used in a pond, because its use will cause an oxygen depletion in 30 to 36 h and supplemental aeration or fresh water will be required. Copper sulfate cannot be used in waters with an alkalinity below 25 mg/l, because at these low levels it becomes toxic to the fish. As the alkalinity increases, appropriate amounts of copper sulfate may be used with caution.

At the time of disease outbreak, it is imperative that a complete disease examination be performed to determine if a single bacterium, parasite, or virus is involved or if multiple pathogens are present. It must be emphasized that even after a pathogen has been found, the necropsy must continue to completion or other serious pathogens may go undetected. Different disease organisms, requiring different treatments, can produce very similar clinical signs. An incomplete necropsy may result in an incorrect diagnosis and improper treatment. A good case in point would be a catfish with dual infections of *E. ictaluri* (bacterium) and *Trichodina* sp. (protozoans). The *Trichodina* infection should be treated with a chemical bath (formalin), while the bacterial infection should be treated with medicated feed. The key to treating multiple infections is to first identify and treat the primary disease-causing agent and then deal with lesser disease agents. In this particular case, the bacterium represents the more critical problem and should be corrected first.

N. DON'T JUST CURE, PREVENT

When disease does occur, the manager of an aquaculture facility often expects an immediate diagnosis and treatment prescription from a fish disease specialist, and sometimes the diagnostician may be tempted to recommend a treatment that will correct the immediate problem while ignoring the long range prognosis. Often a chemical can be added to the water and/or an antibiotic incorporated into the feed, that will result in an immediate, positive response and a cessation of mortality. However, the positive effect of these treatments may be temporary and the disease may reoccur unless a commitment has been made by the aquaculture manager to eliminate all predisposing disease factors.

A health maintenance program should be in place when an aquacultural facility opens, and health management issues should be addressed on a daily basis, not just when acute disease outbreaks occur. Even though the fish farmer is more likely to accept chronic disease losses because they are not dramatic and may occur over a long period of time, these losses can be minimized through proper health maintenance. However, if an acute disease occurs and a great number of fish die in a few days, the loss is dramatic and the farmer will be more inclined to take preventive steps. The bacterium *Aeromonas hydrophila* causes chronic mortality, and over an extended period of time, 25 to 30% of the fish may die. This loss would in all likelihood be acceptable and, because the mortality pattern was not dramatic, no health maintenance program would be initiated. However, if an outbreak of CCV killed over 80% of fry or fingerlings in a few days, strong measures to prevent the reoccurrence of the disease would most likely be taken by the facility manager.

While preventive health maintenance is very important in helping to control fish disease, it is not a "cure all". Some diseases cannot be prevented through management or environmental manipulation, and there is simply no known prevention for some serious diseases that affect fish. While no health maintenance program is perfect, one that emphasizes disease prevention is well worth the effort economically, productively, and for the pure satisfaction of knowing that the best possible aquaculture techniques are being implemented at your facility.

O. THE LAW OF LIMITING FACTORS

The law of limiting factors plays an important role in our ecology as well as in the health management of fish. Schnurrenberger[39] likened the law of limiting factors to the idea that "a chain is no stronger than its weakest link." Also, each major link consists of its own set of limiting factors, and a weak spot within a subset of factors will cause failure in the overall chain. If one considers the fish production pond a chain, with one end being the spawning adult and the other end the harvestable-size fish, there are many potential weak links in the chain, any of which would be considered a limiting factor. Water is essential for fish production. Components of the water make it either suitable or unsuitable to support a healthy fish population. Temperature, pH, quality of water, and volume of water must all be compatible with the species of fish being cultured. For example, if adequate water is available, but the temperature is not suitable, temperature becomes a weak link in the water chain and water becomes a limiting factor in the production chain, because healthy fish cannot be grown at an optimal rate.

Food is usually the limiting factor in unmanaged fish ponds. With fertilization, the base of the food chain in these ponds can be extended to support an increased standing crop, but if culture procedures become too intensive, other limiting factors may come into play. As the standing crop increases, supplemental feeding is required so that food will not again become a limiting factor. Because supplemental feeding increases organic load in the water, pond fertility also increases as a result of uneaten feed and an accumulation of metabolic wastes (feces, urine, ammonia, etc.). This organic enrichment stimulates growth of algae and other plants. Periodically, decomposition of dead plants and uneaten feed remove oxygen from the water, which may cause oxygen to become a limiting factor for the fish. The accumulation of other substances, such as ammonia, nitrite, and CO_2, may also become limiting factors.

The law of limiting factors becomes more acute as the fish culture system becomes more intensive: open pond culture to cage culture, to raceway culture, and finally to closed recirculating systems. As fish density increases more feed is required; more water is necessary to remove, dilute, and neutralize metabolic wastes, possibly leading to LODOS; more efficient supplemental aeration is required; and stressors on the fish become more critical. At what point these limiting factors become detrimental to effective production must be determined by the facility manager, and when this point is reached, appropriate measures should be taken. Remember that a limiting factor does not always have to be biological, but may be financial or dependent on the manager's ability to properly operate a facility.

Each species of fish has its own set of factors that is necessary for it to reproduce, grow, and survive. The factors that allow survival of fish lie between minimum requirements and maximum tolerances. One must also consider the synergistic effect of one limiting factor on another; altering one factor to improve health may produce a previously unknown limiting factor.

P. STAYING ON TOP OF THE OPERATION

Many people go into fish farming without fully understanding what is required of them to initiate and maintain a successful aquaculture facility. Fish farming, especially of warmwater species, is possibly the most demanding of all agricultural enterprises during certain seasons of the year. During critical periods in the production cycle, the farmer must often maintain a 24-hour vigil in order to stay on top of the operation. It is as important for the fish farmer to routinely observe and evaluate his aquaculture procedures and facility, as it is for a cattle or swine farmer to inspect his herd daily.

Although casual scrutiny of ponds, tanks, or raceways can be helpful in detecting possible health changes, the manager must also look at specifics. Operational procedures should be observed in such detail that any departure from good health practices will be immediately detected.

The most obvious indicator of deteriorating health in a fish population is a change in feeding behavior. When fish are affected by adverse changes in environmental conditions and/or by infectious disease, it is usually manifested in reduced feeding activity. The individual who feeds the fish must be alert to any change in feeding patterns. Some fish, such as channel catfish, normally reduce their feed consumption in autumn as water temperatures drop, but at other times of the year, this could indicate an impending health problem. If one day the fish are eating actively and voraciously and the next day there is a noticeable reduction in feeding activity, the cause should be investigated. Water quality should be checked to determine dissolved oxygen, nitrite, or carbon dioxide concentrations and a sample of fish should be examined for the presence of infectious or parasitic disease.

Color changes in the water may indicate a potential water quality problem. When a pond turns from green (phytoplankton bloom) to a dingy brown, it could be an indication that algae are dying and oxygen depletion is imminent. It could also be an indicator of environmental problems such as pond "turnover". If the cause can be detected and corrected while it is occurring, stress on the fish can be minimized.

During warmer months, DO concentrations should be measured on a daily and nightly basis in all culture ponds. This is particularly true in heavily stocked ponds that are receiving large amounts of feed. The oxygen should be measured at nightfall, 9:00 p.m., at 12:00 midnight, and again at daybreak when it will be at the lowest level.[21] If over a period of days and nights a downward trend becomes evident, the oxygen may be ready to "crash" and the manager should initiate nightly aeration or at least have emergency aeration available to prevent LODOS. Other characteristics of a LODOS problem are the presence of increased amounts of carbon dioxide and ammonia, accompanied by a drop in pH.

Fish that are observed to flash, i.e., swim lethargically just beneath the surface of the water or in the shallows, should be examined for disease immediately. Erratic swimming or flashing can be caused by poor water quality, an infectious disease, or a parasitic infestation.

"Staying on top" can prevent catastrophes from occurring in a fish population, but it requires diligence on the part of employees and regularly scheduled visits to the entire culture facility. During critical times, if several days or nights are allowed to pass without checking water quality, heavy losses can occur in a very short period of time. Also, sublethal stressful conditions (e.g., LODOS) may develop that can lead to infectious diseases.

Q. EARLY DIAGNOSIS

Making an early and accurate diagnosis when disease does occur is a major factor in maintaining healthy fish. Therefore, it is imperative that an aquaculturist know what is normal and be able to recognize when something is wrong. It is important to be able to recognize clinical signs of disease and to have fish exhibiting these signs necropsied as soon as possible. It is also important to be able to distinguish signs of oxygen depletion and chemical toxicants from those of infectious disease.

Very few specific infectious fish diseases can be diagnosed solely upon clinical signs; therefore, whenever possible fish should be examined by a fish pathologist. In most cases of infectious disease, only a few fish die during the early stages of an outbreak followed by a gradual increase in the daily mortality rate. When the first few sick or dead fish appear, suitable specimens should be submitted for examination. After a manager gains experience, he may be able to recognize clinical signs of some diseases and be able to initiate treatments. However, samples of infected fish should still be examined by a fish pathologist to rule out the possibility of multiple infections.

Although an early and accurate diagnosis is important in nearly all infectious diseases of fish, it is absolutely crucial in the case of bacterial infections where the use of medicated feed is indicated. If an infection is allowed to progress to the point that the fish are no longer feeding, oral treatment of the disease is no longer an option. This is especially true when treating enteric septicemia of catfish (ESC). In any disease situation, the earlier the diagnosis is made the sooner treatment can begin, thus reducing losses.

R. A DYNAMIC TEAM EFFORT

A dynamic fish health maintenance program must be flexible and coordinated, not a mixture of unrelated, unchanging procedures. While objectives and principles for fish health maintenance programs can follow a general outline, each facility needs to design and implement a health plan that best meets the needs of the facility it serves. For example, chemical, physical, and biological factors that influence animals and their environment will seldom be the same between any two units or farms, which means that each health management package must be individualized.

It has been said that there are three principal situations when an animal food resource producer is ready to discuss the establishment of a health maintenance program:[40] (1) when experiencing a disease crisis, (2) when starting a new unit, and (3) when considering expansion with borrowed capital. The broad scope of basic knowledge and technology that is applicable to health maintenance of aquatic animals is usually beyond the expertise of any one individual. Therefore, a team approach which utilizes input from a variety of specialists is often indicated. In most geographical areas, State Cooperative Extension Services have qualified fishery specialists available to help farm managers develop a good health maintenance program that will meet their specific needs. The wise farm manager will utilize available expertise whenever possible.

S. KEEPING CURRENT

All health management decisions should be based on the most current and accurate information available. Major advances have been made in aquaculture in the past 25 years in breeding, genetics, nutrition, disease identification, treatment, and processing. However, during this time few new drugs or chemicals have been developed to aid in the treatment and/or prevention of diseases. Because of constraints imposed by the FDA, the fish farmer has actually seen a reduction in the number of drugs and chemicals that can be used on food fish. This reduction is due, in part, to advances that have been made in analytical technology and the identification of hazardous characteristics of some therapeutic chemicals. It is, therefore, imperative that a producer stay current on which drugs and chemicals are approved for use on food fish (Table 4). In the U.S., there are only four drugs or chemicals that can be used on food fish. In contrast, Aoki[41] lists 23 drugs that can legally be used on cultured fish with bacterial infections in Japan.

Prevention of disease is always the best health management technique. When moving, buying, or selling fish, acceptable prophylactic treatments should be used and stressful conditions should be avoided.

A fish farmer can stay current in aquaculture management techniques by being active in professional or trade organizations on a local, regional, state, and national level and by reading literature published by these organizations. Extension fishery specialists and agricultural agents have access to pamphlets, brochures, and newsletters with accurate and up to date aquaculture information. If modern fishery managers are to maximize their production facilities, they must keep pace with any technological advances made in the field of aquaculture.

Table 4 Drugs and chemicals approved[a-b] by the U.S. Food and Drug Administration and/or the U.S. Environmental Protection Agency for use on food fish[51-52]

Drug/chemical	Species	Indication	Dosage regimen	Limitations/comments
Oxytetracycline	Salmonids	Control ulcer disease, furunculosis, motile *Aeromonas* septicemia, and pseudomonas disease (*A. salmonicida*, *A. liquefaciens*, *Pseudomonas*)	2.5–3.75 g/100 lb/d for 10 d (50–100 mg/kg of fish per day)	In mixed ration; water temperature not below 48.2°F; 21-d withdrawal
	Catfish	Control motile *Aeromonas* septicemia and pseudomonas disease (*A. liquefaciens*, *Pseudomonas*)	2.5–3.75 g/100 lb/d for 10 d (50–100 mg/kg of fish per day)	In mixed ration; water temperature not below 62°F 21-d withdrawal
	Lobster	Control gaffkemia (*Aeromonas viridans*)	1 g/lb medicated feed for 5 d	In feed; 30-d withdrawal
Sulfadimethoxine ormetoprin (Romet)	Salmonids	Control furunculosis (*Aeromonas salmonicida*)	50 mg/kg/d for 5 d	In mixed ration; 42-d withdrawal
	Catfish	Control enteric septicemia	50 mg/kg/d for 5 d	In mixed ration; 3-d withdrawal
Tricaine methanesulfonate	Fish (Ictaluridae, Salmonidae, Esocidae,	Sedation/anesthesia	15–330 mg/l	Added to water; concentration depends on desired degree of anesthesia, species, size, water temperature and softness, stage

Drug	Species	Use	Dosage	Comments
(continued)	(Percidae)			of development (preliminary tests of solution should be made with a few fish); 21-d withdrawal time (fish); use when water temperature is over 50°F
Povidine-iodine	Eggs	Egg disinfectant	100–200 mg/l	Use during water hardening or when eyed
Formalin solution (Paracide-F)	Salmonids, catfish, largemouth bass, bluegill	Control protozoa and monogenetic trematodes (*Ichthyopthirius Ichthyobodo, Scyphidia, Epistylis, Trichodina* spp., and *Cleidodiscus, Gyrodacytlus, Dactylogyrus* spp.)	Tanks and raceways above 50°F: up to 170 µl/l, up to 1 h Below 50°F: up to 250 µl Earthen ponds: 15 to 25 µl/l indefinitely	Do not apply to ponds when water warmer than 80°F, heavy phytoplankton bloom, dissolved is less than 5 mg/l; ponds may be retreated in 5–10 d if needed
Potassium permanganate	All	Oxidizer and detoxifier	2 mg/l indefinite	Food fish, exempted from registration by EPA
Copper sulfate	All	Algicide	0.25–2 mg/l indefinite	Toxic to fish in waters of low hardness
Salt	All	Osmoregulator	0.5–3%	Exempt

[a]Approval applies to the specific drug that is the subject of a new animal drug application (NADA); bulk drug from a chemical company or similar compounds made by companies other than those specified in the NADA are **not** approved new animal drugs. [b]Approval applies only to use of the drug for the indications and manner specified on the label. An INAD exemption is not the same as approval; it merely temporarily permits research (under specific conditions) on an unapproved compound.

T. MAINTAINING A CLEAN ENVIRONMENT

Everyone recognizes the fact that an accumulation of trash, excrement, decaying animals, etc. in a terrestrial environment is associated with unhealthy conditions. This is also true in the aquatic environment. Generally speaking, when visiting private and governmental aquaculture facilities around the world, it is possible to identify facilities that have the most disease problems by their appearance. This is not to say that a "spit and polish" facility will never have a disease problem, but they do tend to have fewer and less severe outbreaks.

Water is an excellent medium for transmitting disease agents, therefore sick or dead fish should be removed to reduce pathogen transmission. Maintaining clean areas around ponds and controlling vegetation at the water's edge will not allow sick or dead fish to go unnoticed. If dead fish are allowed to remain in a pond, there is also the possibility that scavengers will carry diseased carcasses to other ponds, thus spreading the disease. Accumulation of feces, uneaten feed, and other organic detritus is a major problem in heavily stocked ponds, raceways, and recirculating systems. Removal of excreta from water will help improve and maintain water quality and will reduce areas where pathogenic organisms can multiply. These problems can be avoided and/or corrected by regular cleaning and periodic draining, drying, and when necessary, disinfection of holding facilities.

Equipment sanitation, including the cleaning and disinfection of seines, nets, buckets, tanks, and other equipment, will reduce transmission of many pathogens. A classic example is *Ichthyophthirius multifiliis* being transmitted from one infected catfish pond to another on a seine that had not been properly disinfected or air-dried between uses.[29] Maintaining a clean aquaculture facility will pay dividends by providing a safer work place, reduced incidence of disease, increased production, and generally healthier fish, all of which will translate into financial profit.

U. THE HIGH-RISK CONCEPT

Aquaculture, by its very nature, is a high-risk endeavor because of its close relationship to environmental change. Some aspects of fish culture present higher risk factors than do others, and the pond manager is wise to concentrate on the high-risk areas. Water quality is the most important factor in successfully producing a healthy crop of fish. If water quality is not suitable, fish will not feed or grow well, regardless of feed quality, and potential for disease will be more prevalent. Consequently, major emphasis must be placed on water quality in any aquaculture health maintenance program.

Infectious disease must also be classified as a major risk factor, but these risks will differ from farm to farm depending on water type and quality, species and strain of fish being reared, and general management practices, etc. Disease problems also change from year to year.

A good aquaculturist will always be aware of the high-risk factors at his facility and will practice preventive health maintenance in these areas, while never losing sight of lesser risk factors and the potential they have for creating problems.

V. RECORD KEEPING AND COST ANALYSIS

A well organized set of records is essential to a successful aquaculture business. In addition to production records which should include stocking rates, feeding rates, feed conversion, and harvest totals, each fish production facility should maintain a set of health records. These records should include mortality patterns and dates, infectious disease incidences, treatments used, and results of those treatments. The size of each production unit, volume of water, and water quality characteristics should also be included in the health records of that unit. After several growing seasons, the manager may be able to ascertain from these records whether or not a certain disease is likely to occur and what conditions predispose fish to a specific disease. By using this information, the manager may be able to prevent disease outbreaks by timely prophylactic treatments or remedial actions. Detail and extent of record keeping may vary with size and intensity of an operation; however, records should be kept by every facility no matter its size.

Every health maintenance procedure should be evaluated for economic feasibility and cost-effectiveness, and without accurate records this is impossible. To be practical, every disease prevention or corrective measure must be cost-effective. When considering economic feasibility, all alternatives —

cost, efficacy, and benefits — must be considered and the most cost-effective, not necessarily the cheapest, procedure selected.

When initiating fish health management practices, a long range, cost-effective analysis should be made. The initial financial outlay may be high, but over a period of time, good health management practices will pay for themselves in reduced mortalities and in the maintenance of optimal growth. For example, setting up a disinfection station for all nets, seines, and other equipment; establishing a quarantine facility; or routinely applying prophylactic treatments may be initially expensive and time consuming, but over a period of years they will be cost-effective.

Occasionally, a fish farmer will be forced to initiate procedures that are not economically feasible, such as total eradication of a fish population due to the presence of a highly virulent disease. If the pathogen can be eliminated from the facility by total destruction of the infected fish population and recontamination can actually be prevented, eradication is probably a practical solution, even if not an economic one.

A cost to benefit ratio (C:B) should be calculated for each health management procedure. The benefits must outweigh the cost, except under unusual circumstances. The C:B ratio can also be used to compare effectiveness of different procedures that have the same results.

REFERENCES

1. **Snieszko, S. F.,** Natural resistance and susceptibility to infections, *Prog. Fish. Cult.,* 20, 133, 1958.
2. **Klontz, G. W.,** *Fish Health Management,* Texas A & M Sea Grant Program, Texas A & M University, College Station, 1973.
3. **Meyer, F. P., Warren, J. W., and Carey, T. G.,** *A Guide to Integrated Fish Health Management in the Great Lakes Basin,* Great Lakes Fishery Commission, Special Publication 83-2, 1983.
4. **Schnurrenberger, P. R. and Sharman, R. S., Eds.,** *Principles of Health Maintenance,* Praeger Publishers, NY, 1983.
5. **Wedemeyer, G.,** The role of stress in the disease resistance of fishes, in *A Symposium on Diseases of Fishes and Shellfishes,* Snieszko, S. F., Ed., American Fisheries Society, Washington, D.C., Publ. No. 5, 1970, 30.
6. **Brett, J. R.,** Implications and assessments of environmental stress, in *Investigations of Fish-Power Problems,* Larkin, P. A. and MacMillan, H. R., Eds., Lectures in Fisheries, University of British Columbia, 1958, 69.
7. **Grizzle, J. M.,** Department of Fisheries and Allied Aquacultures, Auburn University, AL, personal communication, 1992.
8. **Pickering, A. D.,** Introduction: The concept of biological stress, in *Stress and Fish,* Pickering, A. D., Ed., Academic Press, London, 1981, 1.
9. **Seley, H.,** *Stress and Health and Disease,* Butterworth, Boston, 1950, 1256.
10. **Wedemeyer, G. and Wood, J. W.,** Stress as a predisposing factor in fish diseases, U.S. Department of the Interior, Fish and Wildlife Service, Washington, D.C., Fish Disease Leaflet No. 38, 1974.
11. **Wedemeyer, G., Meyer, F., and Smith, L. S.,** *Environmental Stress and Fish Diseases,* TFH Publications, Neptune City, NJ, 1977, 192.
12. **Barton, B. E., Peter, R. E., and Paulencu, C. R.,** Plasma cortisol levels of fingerling rainbow trout *(Salmo gairdneri)* at rest, and subject to handling, confinement, transport and stocking, *Can. J. Fish. Aquat. Sci.,* 37, 805, 1980.
13. **Snieszko, S. F.,** Recent advances of scientific knowledge and development pertaining to diseases of fishes, *Adv. Vet. Sci. Comp. Med.,* 17, 291, 1973.
14. **Plumb, J. A., Grizzle, J. M., and deFigueiredo, J.,** Necrosis and bacterial infection in channel catfish *(Ictalurus punctatus)* following hypoxia, *J. Wildl. Dis.,* 12, 247, 1976.
15. **Schmittou, H. R.,** Department of Fisheries and Allied Aquacultures, Auburn University, AL, personal communication, 1993.
16. **Walters, G. R. and Plumb, J. A.,** Environmental stress and bacterial infection in channel catfish, *Ictalurus punctatus* Rafinesque, *J. Fish Dis.,* 17, 177, 1980.

17. **Meyer, F. P.,** Seasonal fluctuations in the incidence of disease on fish farms, in *A Symposium on Diseases of Fishes and Shellfishes,* Snieszko, S. F., Ed., American Fisheries Society Publ. No. 5, 1970, 21.

18. **Tucker, C. S., Steeby, J. A., Waldrop, J. W., and Garrard, A. B.,** Effects of cropping system and stocking density on production of channel catfish in ponds, Bulletin 988, Mississippi Agricultural & Forestry Experiment Station, Mississippi State, MS, 1992, 25.

19. **Austin, B. and Austin, D. A.,** *Bacterial Fish Pathogens: Diseases in Farmed and Wild Fish,* Ellis Horwood Limited, Chichester, 1987.

20. **van Breemen, N.,** Dissolved aluminum in acid soils in acid mine waters, *Soil Sci. Soc. Proc.,* 37, 694, 1973.

21. **Boyd, C. E.,** *Water Quality in Ponds for Aquaculture,* Alabama Agricultural Experiment Station, Auburn University, AL, 1990.

22. **Rohovec, J.,** Review of international regulations concerning fish health, in *Proc. from a Conf. on Disease Inspection and Certification of Fish and Fish Eggs,* Oregon State University Sea Grant College Program, Publ. No. ORESU-W-79-001, Oregon State University, Corvallis, 1979, 7.

23. **Hooper, K.,** The isolation of VHSV from chinook salmon at Glennwood Springs, Orcas Island, Washington, *Fish Health Sect./Am. Fish. Soc. News Lett.,* 17(2), 1, 1989.

24. **Winton, J. R., Batts, W. N., Deering, R. E., Brunson, R., Hooper, K., Nischzoua, T., and Stehr, C.,** Characteristics of the first North American isolates of viral hemorrhagic septicemia virus, *Proc. Second Int. Symp. Viruses of Lower Vertebrates,* Oregon State University Press, Corvallis, OR, 1991, 43.

25. **Hoffman, G. L., Dunbar, C. E., and Bradford, A.,** Whirling disease of trouts caused by *Myxosoma cerebralis* in the United States, Special Scientific Report Fisheries No. 427, U.S. Department of the Interior, Fish and Wildlife Service, Washington, D.C., 1962.

26. **Anderson, D. E.,** BK episode (epitaph), Jackson National Fish Hatchery, A 1988 case study, in *Proc. 16th Ann. Eastern Fish Health Workshop* (Abstract), Martinsburg, WV, June 11–13, 1991.

27. **Cipriano, R. C., Starliper, C. E., and Teska, J. D.,** Case study on the investigation of salmonids at the White Sulfur Springs National Fish Hatchery for *Renibacterium salmoninarum,* in *Proc. 16th Ann. Eastern Fish Health Workshop* (Abstract), Martinsburg, WV, June 11–13, 1991.

28. **Taylor, P. W.,** Fish-eating birds as potential vectors of *Edwardsiella ictaluri, J. Aquat. Anim. Health,* 4, 240, 1992.

29. **Rogers, W. A.,** Department of Fisheries and Allied Aquacultures, Auburn University, AL, personal communication, 1993.

30. **Mulcahy, D. and Pascho, R. J.,** Vertical transmission of infectious hematopoietic necrosis virus in sockeye salmon *Oncorhynchus nerka* (Walbaum): isolation of virus from dead eggs and fry, *J. Fish Dis.,* 8, 393, 1985.

31. **Plumb, J. A., Green, O. L., Smitherman, R. O., and Pardue, G. B.,** Channel catfish virus experiments with different strains of channel catfish, *Trans. Am. Fish. Soc.,* 104, 140, 1975.

32. **Horner, R.,** Illinois Department of Natural Resources, Matawan, IL, personal communication, 1991.

33. **Lovell, T.,** *Nutrition and Feeding of Fish,* AVI Publishing/Van Nostrand Reinhold, NY, 1988.

34. **Halver, J.,** *Fish Nutrition,* Academic Press, NY, 1972.

35. **Halver, J. E.,** Crystalline afflatoxin and other vectors for trout hepatoma, in Trout Hepatoma Research Conference Papers, Res. Rep. 70, Halver, J. E. and Mitchell, I. A., Eds., Bureau of Sport Fisheries and Wildlife, Government Printing Office, Washington, D.C., 1967, 78.

36. **Durve, V. S. and Lovell, R. T.,** Vitamin C and disease resistance in channel catfish *(Ictalurus punctatus), Can. J. Fish Aquat. Sci.,* 39, 948, 1982.

37. **Li, Y. and Lovell, R. T.,** Elevated levels of dietary ascorbic acid increase immune response in channel catfish, *J. Nutr.,* 115, 123, 1984.

38. **Liu, P. R., Plumb, J. A., Guerin, M., and Lovell, R. T.,** Effect of megalevels of dietary vitamin C on the immune response of channel catfish *Ictalurus punctatus* in ponds, *Dis. Aquat. Org.,* 7, 191, 1989.

39. **Schnurrenberger, P. R.,** The law of limiting factors, in *Principles of Health Maintenance,* Schnurrenberger, P. R. and Sharman, R. S., Eds., Praeger Publishers, New York, 1983, Chap. 3.

40. **Hudson, R. S.,** A dynamic, coordinated entity, in *Principles of Health Maintenance,* Schnurrenberger, P. R. and Sharman, R. S., Eds., Praeger Press, NY, 1983.

41. **Aoki, T.,** Chemotherapy and drug resistance in fish farms in Japan, in *Diseases in Asian Aquaculture,* Shariff, I. M., Subasinghe, R. P. and Arthur, J. R., Eds., Fish Health Section, Asian Fisheries Society, Manila, 1992, 519.

42. **Post, G.,** *Textbook of Fish Health,* TFH Publications, Neptune, NJ, 1987.

43. **Piper, R. G., McElwain, I. B., Orme, L. E., McCraren, J. P., Gowler, L. G., and Leonard, J. R.,** *Fish Hatchery Management,* U.S. Department of the Interior, Fish and Wildlife Service, Washington, D.C., 1982.

44. **Wedemeyer, G. and Wood, J. W.,** Stress as a predisposing factor in fish diseases, Fish Disease Leaflet No. 38, U.S. Department of the Interior, Fish and Wildlife Service, Washington, D.C., 1974.

45. **Roberts, R. J.,** *Fish Pathology,* 2nd ed., Bailliere Tindal, London, 1989, 467.

46. **Rothbard, S.** Induced reproduction in cultivated cyprinids — The common carp and the group of Chinese carps. I. The technique of induction, spawning and hatching, *Bamidgeh,* 33, 103, 1981.

47. **Ney, J. J.,** A synoptic review of yellow perch and walleye biology, in *Selected Coolwater Fishes of North America,* Spec. Publ. No. 11, Kendall, R. L., Ed., American Fisheries Society, Bethesda, MD, 1978, 1.

48. **Jhingran, V. G. and Pullin, R. S. Y.,** *A Hatchery Manual for the Common, Chinese and Indian Carps,* Asian Development Bank, Manila.

49. **Schnurrenberger, P. R.,** Variable causes require variable solutions, in *Principles of Health Maintenance,* Schnurrenberger, P. R. and Sharman, R. S., Eds., Praeger, NY, 1983, Chap. 6.

50. **Swan, A. I.,** The extent of contact, in *Principles of Health Maintenance,* Schnurrenberger, P. R. and Sharman, R. S., Eds., Praeger Publishers, New York, 1983, Chap. 4.

51. **Schnick, R. A., Meyer, F. P., and Gray, D. L.,** A guide to approved chemicals in fish production and fishery resource management, University of Arkansas and U.S. Fish and Wildlife Service, National Fisheries Research Center, LaCrosse, WI, MP 241, 1989, 27.

52. **Anon.,** Aquaculture drug use: answers to commonly asked questions, FDA Workshop — Requirements for Investigational New Animal Drugs, Eastern Fish Health Section, Center for Veterinary Medicine, U.S. Food and Drug Administration, Auburn, AL, June 19, 1992.

Chapter II

Epizootiology of Fish Diseases

To understand the many problems involved in the management of infectious diseases of fish, one must first understand the aquatic environment and the role it plays in fish health. Fish respond quickly to environmental changes which can affect their disease susceptibility and overall general health. A good fish health maintenance program, therefore, must give careful attention to the interaction between a fish population and its environment.

A. EPIZOOTIOLOGY

Epizootiology in a broad sense is the study of disease: the spread of pathogens, their mode of infection, and the control of disease in animals other than humans. When discussing epizootiology, one must define and understand the following terms:

1. *Disease* — a deviation from normal or good health, not necessarily implying the cause of the deviation
2. *Communicable disease* — a disease that results from the multiplication, replication, or reproduction of the causative organism in a host and the likelihood that the organism can be transmitted (communicated) to other hosts
3. *Epizootic (epidemic when discussing human diseases)* — a disease outbreak among animals, other than humans, in a specified and localized area (to be classified as an epizootic, an arbitrary number of animals must be infected)
4. *Enzootic (endemic)* — the continuing presence of a causative disease organism in a localized area, such as a pond, raceway, hatchery, river, or a fish population, not necessarily expressed as disease
5. *Infection* — presence of a pathogenic organism in a host without necessarily causing disease

It is important to note the distinction between disease and infection, as the latter simply implies the presence of a pathogenic organism in the host. Infection is a common occurrence in fish and does not always manifest itself in disease. Many fish disease organisms are enzootic in an aquatic environment, but will only cause disease when conditions are favorable.

There are basically two types of organisms involved in communicable disease: (1) *obligate pathogens* and (2) *facultative or nonobligate pathogens*. Obligate pathogens must have a host or intermediate host to survive indefinitely in nature. Examples of obligate pathogens in fish include all viruses, some bacterial pathogens, such as *Mycobacterium marinus* (Mycobacteriosis) and *Renibacterium salmoninarum*, and the parasitic protozoan *Ichthyophthirius multifiliis* ("Ich"). Facultative or nonobligate pathogens can live and multiply in a host or in the environment independent of the host. They have the ability to derive needed nutrition from organic matter in water and mud or from a living host. *Aeromonas hydrophila* (motile *Aeromonas* septicemia) and the fungus, *Saprolegnia* spp., are examples of facultative pathogens.

B. SEASONAL TRENDS OF FISH DISEASES

The incidence of fish disease is generally seasonal.[1-3] Disease occurrence tends to fluctuate with changes in temperature, the presence of young, susceptible fish in a population, and environmental conditions that affect immunity and natural resistance. In tropical climates, only minor differences are evident in the number of fish disease outbreaks throughout the year, however, in temperate and colder climates, a seasonal fluctuation can be noted, particularly in warmwater species (Figure 1). In the southern U.S., the occurrence of disease is low in cultured and wild fish populations from November through February, but increases during March through June as waters begin to warm. The correlation between disease and increased water temperatures was noted by Snieszko.[4]

Disease outbreaks are more prevalent in spring and early summer because: (1) spring is the time of year when spawning occurs and there is an infusion of highly susceptible young fish into the population;

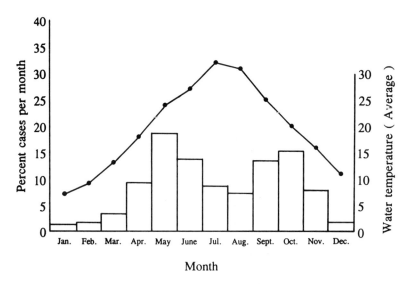

Figure 1 Monthly distribution of 3781 bacterial disease outbreaks in Alabama from 1980 through 1990. Data compiled from diagnostic records at the Southeastern Cooperative Fish Disease Laboratory, Auburn University, Alabama and the Alabama Fish Farming Center, Greensboro, Alabama.

(2) fish have just emerged from over-wintering conditions and their immunity and natural resistance tends to be low. Also, at this time naturally occurring antimicrobial substances in the blood are low[4] making fish more susceptible to infection by facultative disease-causing organisms present in the water; and (3) the water temperature in spring and early summer normally ranges from 18 to 28°C, which is the optimum growth temperature for many fish pathogens. As water temperatures rise in July and August, disease decreases before increasing again in September and October as water temperatures begin to moderate. A similar pattern prevails as one moves northward, only the period of increased disease incidence occurs later in the spring and earlier in the fall.

C. FACTORS IN DISEASE DEVELOPMENT

When an obligate pathogen-caused fish disease outbreak occurs, predisposing factors that dictate the seriousness of the disease are (1) source of infection, (2) mode of transmission, (3) portal of entry, (4) virulence of the pathogenic organism, and (5) resistance of the host. If this chain can be broken or altered through management, a particular disease may be prevented or its severity greatly reduced.

1. SOURCE OF INFECTION
Every infectious disease requires a pathogen source. The source may be infected fish (dead or sick), carrier fish that show no clinical signs of disease, contaminated eggs from infected brood stock, or infected water supplies. Contaminated feed, especially if it contains uncooked fish flesh or viscera, may be the source of some bacterial or parasitic infections. Humans can also transmit infection to fish populations by mechanically transporting infective organisms on nets, boots, and other equipment. The perpetuation of infection can be interrupted through good management practices such as disinfection of equipment, filtration of water, manipulation of stocking procedures in infected waters, removal of dead and moribund fish from infected populations, utilization of seed fish from certified disease-free brood stocks, and proper preparation of feed.

2. MODE OF TRANSMISSION
The mode of fish disease transmission is closely related to, and sometimes inseparable from, the source of infection. Water, because it contains fish metabolites and waste products, often provides an ideal, high nutrient environment for potential fish pathogens to grow and proliferate and thus serves as a source of pathogens as well as a primary mode for their transmission. Chances that an infection will be transmitted from one animal to another increases with the density of the fish population.

Eggs produced by disease-carrying brood stock are often responsible for transmission of viral and some bacterial agents from one generation to another. Also, intermediate hosts, such as birds and snails, are important in life cycles of some fish parasites and serve as vehicles for disease. Good management practices can help disrupt the perpetuation of disease by using only eggs from certified disease-free sources when applicable, use of prophylactic treatments, control of intermediate hosts, and utilization of filtered water or water from a fish-free source.

3. PORTAL OF ENTRY

Each disease organism has an optimal point of entry to the host and when these entry points are known, management measures can be taken to prevent introduction of the pathogen. Common points of entry are the intestine, gills, and skin. The mucus layer that covers the epithelium of fish provides protection against some pathogens, however, when this mucus layer is disrupted, the underlying epithelium is exposed to bacteria or parasites that are present in the water and as a result the skin, gills, or intestines become portals of entry for disease organisms. Also, some parasites may penetrate the layer of mucus and epithelium and enhance the invasion of bacteria. Proper handling of fish and the use of prophylactic treatments will often curtail or prevent infections associated with the epithelium.

4. VIRULENCE OF THE PATHOGENIC ORGANISM

Virulence is a measure of a pathogenic organism's ability to cause disease. The disease-causing ability of a pathogen may range from low to high, and all levels between the two extremes. Some organisms, especially facultative bacteria, can vary greatly in virulence from location to location. *A. hydrophila*, a facultative bacterium in water, may be highly virulent in one area, while an isolate from another body of water may be avirulent (i.e., not pathogenic). Also, when *A. hydrophila* is isolated from diseased fish, it is usually more virulent than when isolated from water.[5] When highly virulent strains of pathogens are present, it is more likely that a severe disease outbreak will occur than when a low virulence organism is present. Although it is difficult to manipulate the virulence of pathogenic organisms, management can help prevent severe disease outbreaks by ensuring that environmental conditions are more favorable to the host than to the pathogen.

5. NATURAL RESISTANCE OF THE HOST

Natural resistance means that the host has an inherent ability to subdue a pathogen to the point that it will not cause disease. The natural resistance of a fish host to infectious disease is one of the most important factors in deterring infection.[6] Factors that influence the natural resistance of fish to disease are phagocytic activity of leukocytes and macrophages, tissue integrity, nonspecific serum components (interferon, complement, etc.), nutritional well-being, age, species, and environmental conditions. Natural resistance of adult fish is often reduced during the spawning season when most of the fish's energy goes into reproductive activities. In temperate zones, older fish tend to have reduced resistance in the spring due to over-wintering, and the young fish often have not yet acquired their natural resistance leaving them more susceptible to infections. Natural resistance is an integral part of fish health management and can be enhanced by the farm manager, if disease-resistant strains of fish are selected for brood stock.

D. HOST/PATHOGEN RELATIONSHIP

That a basic host/pathogen relationship exists is a concept fundamental to disease susceptibility of fish. A disease organism may be present in the environment or host, but as long as the host remains in good physical condition and its natural resistance remains high, disease is less likely to occur. Disease organisms often invade a fish population when there is a host/pathogen imbalance. Humans, through poor management practices, often contribute to this imbalance by placing stressors on the fish which compromise their natural resistance. Poor nutrition or deterioration of the aquatic environment may reduce the fish's resistance and allow "infection" to progress to "disease". Through good management, a proper host/pathogen balance can be maintained and disease outbreaks kept to a minimum, which will result in better growth and higher fecundity.

E. DEGREE OF INFECTION

Fish diseases are categorized according to severity of infection, rate of morbidity, and presence of clinical signs and are described as either acute, chronic, or latent. Mortality pattern is related to degree of infection and is helpful in determining whether cause of death is due to infectious agents, toxins, nutritional, or adverse water quality (Figure 2).

1. ACUTE

An acute disease outbreak is marked by a high rate of mortality with a large number of fish being affected in a short time. Overt clinical signs (ulcerative lesions, severe hemorrhage, etc.) may be minimal or totally lacking. Some highly virulent organisms which are capable of causing acute mortality include channel catfish virus (CCV), infectious hematopoietic necrosis virus (IHNV), and infectious pancreatic necrosis virus (IPNV). On rare occasions, bacterial pathogens (e.g., *Flexibacter columnaris*) and some parasites (e.g., *I. multifiliis*) may cause acute mortality under ideal conditions. Also, some noninfectious conditions, such as oxygen depletion and chemical toxicants, can cause acute mortality.

2. CHRONIC

Many infectious diseases of fish exhibit chronic characteristics. In these infections, the mortality rate is gradual and it is difficult to detect a peak period of morbidity or mortality. In chronic disease outbreaks, severe clinical signs may be present and may persist for weeks, sometimes resulting in large numbers of deaths. Examples of chronic infections in fish are motile *Aeromonas* septicemia (*Aeromonas* spp.), columnaris, furunculosis *(A. salmonicida)*, bacterial kidney disease *(R. salmoninarum)*, and most protozoan parasitic infections.

3. LATENT

In a latent condition, a disease organism is present, but the host exhibits no overt clinical signs of disease and there is little or no mortality. Latent infections are usually associated with viral infections (IPN, IHN, etc.); obligate bacterial pathogens such as *A. salmonicida, R. salmoninarum;* and the protozoans *Myxobolus cerebralis* (whirling disease) and *I. multifiliis.*

Diseases in fish may result from a *primary infection* or from a *secondary infection* caused by a secondary pathogen. Primary infections are those caused by primary pathogens, usually obligate pathogens, and result in morbidity or death without the presence of additional factors or outside influences. Primary pathogens are able to cause disease in healthy fish, even under ideal environmental conditions. However, any adverse environmental factor can synergize the infection to a more serious level.[7] Channel catfish virus and the bacterium *Edwardsiella ictaluri* are examples of primary pathogens that infect fish.

Generally, a secondary pathogen is a free-living, facultative, opportunistic organism that infects hosts when their defenses are compromised by other pathogens or stressors and results in a secondary infection in the host. Secondary infections usually occur as a result of some stressful condition or disease that has left the host in a weakened state. Poor management practices or environmental stressors, such as low oxygen levels, temperature shock, and excessive or improper handling, may precipitate a secondary infection. There are many secondary pathogens that affect fish, including *A. hydrophila, F. columnaris, Saprolegnia* spp., and others.

It is also important to remember that an environmental stressor can cause a minor disease condition to progress into a more severe one and that every disease has the potential to become more serious if good management procedures are not practiced during disease outbreaks. For example, if diseased pond fish are moved to a confined holding facility for treatment, the act of moving the fish will probably be more detrimental and stressful than any benefit derived from the treatment.

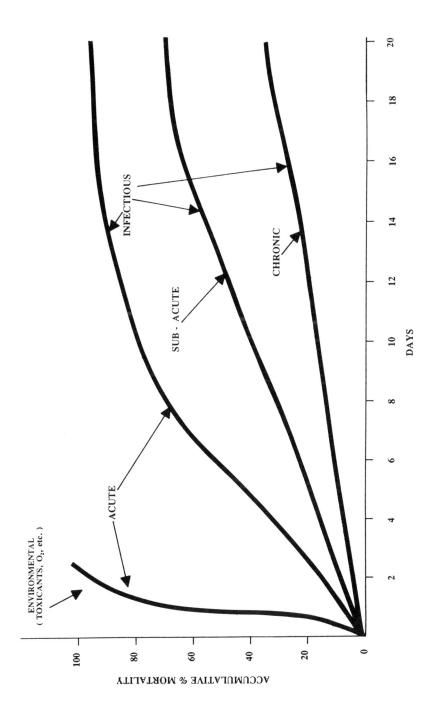

Figure 2 Theoretical cumulative mortality patterns of fish affected by environmental, inanimate, etiological agents and infectious, animate, etiological agents. The curves may change to steeper or flatter depending on the disease, species of fish, water temperature, and other environmental or biological factors.

REFERENCES

1. **Meyer, F. P.,** Seasonal fluctuations in the incidence of disease in fish farms, in *A Symposium on Diseases of Fishes and Shellfishes,* Snieszko, S. F., Ed. American Fisheries Society, Publ. No. 5, Washington, D.C., 1970, 21.
2. **Plumb, J. A.,** An 11-year summary of fish disease cases at the Southeastern Cooperative Fish Disease Laboratory, *Proc. Annu. Conf. Southeastern Assoc. Game and Fish Comm.,* 29, 254, 1976.
3. **Warren, J. W.,** *Diseases of Hatchery Fish,* 6th ed., U.S. Fish and Wildlife Service, Washington, D.C., 1991.
4. **Snieszko, S. F.,** Natural resistance and susceptibility to infections, *Prog. Fish. Cult.,* 20, 133, 1958.
5. **de Figueiredo, J. and Plumb, J. A.,** Virulence of different isolates of *Aeromonas hydrophila* in channel catfish, *Aquaculture,* 11, 349, 1977.
6. **Chevassus, B. and Dorson, M.,** Genetics of resistance to disease in fishes, *Aquaculture,* 85, 83, 1990.
7. **Wedemeyer, G.,** The role of stress in the disease resistance of fishes, in *A Symposium on Diseases of Fishes and Shellfishes,* Snieszko, S. F., Ed., American Fisheries Society, Publ. No. 5, Washington, D.C., 1970, 30.

Chapter III

Pathology

Pathology is the study of disease that includes functional and morphological alterations and reactions that develop in a living organism as the result of injurious agents, nutritional deficiencies, or inherited conditions. Fish pathology, as a specific science, has not developed to the point that it has terminology and clinical characteristics specific to that group of vertebrates, but principles of vertebrate pathology, as described by Slauson and Cooper[1] and Thomson,[2] generally do apply to fish. Ferguson,[3] in his text and atlas of systematic pathology of fish, states that lesion description and interpretation are essentially the same in fish as in higher vertebrates. The terminology of general pathology has been adapted for fish, even though all of the pathological conditions observed in homothermic animals do not occur in fish. The subject of fish pathology is too vast to be covered in one chapter. It is the objective of this chapter, therefore, to introduce the reader to some basic pathologic terms and to present examples of the most common pathological changes that occur in fish.

A. PATHOLOGICAL TERMS

One must be familiar with the following terms when describing a disease or considering the disease process.

1. *Etiology* — the study of how to determine the cause of disease
2. *Etiological agent* — the agent that causes a particular disease (for example, channel catfish virus, CCV, is the etiological agent of channel catfish virus disease, CCVD)
3. *Pathogenesis* — the sequence of events by which a disease develops, including method of infection, establishment of the agent in the host, and injury inflicted upon the host
4. *Pathogenicity* — refers to those characteristics of a microorganism that enable it to cause disease and its ability to gain access to susceptible tissue, become established, proliferate, avoid host defense mechanisms, and cause injury to the host
5. *Virulence* — refers to degree of pathogenicity [avirulent (not virulent), low, medium or highly virulent] and usually relates to a specific group, species, or strain of pathogenic organism
6. *Lesion* — any gross, microscopic, or biochemical tissue abnormality associated with disease (it may be either functional or morphological)
7. *Clinical signs* — characteristics or conditions associated with a disease that can be seen during an external examination of the affected animal (clinical signs include behavioral and gross morphological changes, but do not include microscopic, biochemical, or internal changes)
8. *Histology* — study of tissues on a microscopic level (representative species from several important groups of fishes have been studied, described, and documented on a microscopic level, including channel catfish, *Ictalurus punctatus*,[4] rainbow trout, *Oncorhynchus mykiss*,[5] striped bass, *Morone saxatilis*,[6] and walking catfish, *Clarias batrachus*)[7]
9. *Histolopathology* — the microscopic study of injuries to tissue on a cellular level (most examples given in this text will be related to gross pathology, i.e., seen with the naked eye)

B. CAUSE OF DISEASE

A disease can be caused by an etiological (specific cause) or nonetiological (contributing cause) agent. Etiological agents are specific causes of disease and can be classified as either inanimate or animate agents. Inanimate etiological agents are factors without life of their own and can originate within a host (endogenous) or outside of a host (exogenous). Endogenous, inanimate agents are factors associated with the genetics and/or metabolic disorders of the host. Exogenous, inanimate agents include trauma such as handling, temperature shock, electrical shock, chemical toxicity, and dietary deficiencies. If etiological agents are severe enough, they can become sublethal stressors which may predispose fish to infectious disease. Animate etiologies are living, communicable infectious agents, including viruses, bacteria, fungi, protozoa, helminths, and copepods.

Nonetiological causes of disease can be characterized as extrinsic (from outside the body) or intrinsic (from within the body). Extrinsic diseases are usually associated with environmental conditions or dietary problems while intrinsic factors that influence disease include age, sex, heredity, and species or strain of fish. Species or strain of fish is important because they are not all equally susceptible to a specific disease organism. Feed quality (e.g., protein and vitamin content), water quality factors (e.g., inadequate oxygen concentration, excessive carbon dioxide or ammonia), and water temperature extremes can be classified as either etiological or nonetiological, extrinsic factors and can contribute significantly to infectious disease.

C. PATHOLOGICAL CHANGE

Pathological changes in a fish population can aid in recognition and identification of disease. The following brief discussion describes some pathological changes that can be recognized in diseased fish using definitions taken from Slauson and Cooper,[1] Thomson,[2] and Hopps.[8] For further information on fish pathology, *Fish Pathology* by Roberts[9] is recommended.

1. CIRCULATORY DISTURBANCES

Circulatory disturbances are those abnormalities (lesions) that reflect injury to the vascular system. The most common types of circulatory disturbances in fish are hemorrhage, edema, hyperemia, congestion, emboli, and telangiectasis.

a. Hemorrhage

Hemorrhage is the escape of blood from blood vessels and may occur in the skin, mucus membranes, within serous cavities, or between cells of any tissues or organs. Blood may leave the vessel because of a hole in the wall or may pass through an intact vessel wall because of increased porosity. Hemorrhages are usually classified according to degree of area affected. *Petechiae* (in gross appearance) are those hemorrhages that measure less than 2 mm in diameter. They are common in bacterial and viral infections and typically occur in the epithelium or on the surface of visceral organs. *Ecchymotic* hemorrhages are larger and more diffuse. They measure 2 to 3 mm or larger in diameter. *Paint-brush* hemorrhage refers to larger affected areas that appear as though they were splashed with red paint. *Hematomas* (seldom seen in fish) are localized hemorrhages that result in a blood-filled swelling or lump.

Hemorrhages can be caused by trauma, rupture of the blood vessels, or increased porosity resulting from the presence of infectious bacteria, viruses, or toxicants. In fish, the overall effect of hemorrhaging is mild if the process is gradual because hematopoiesis is able to compensate for loss of blood cells. However, if hemorrhaging is acute, as often is the case in infectious disease, anemia can occur and will be characterized by pale gills and internal organs.

b. Edema

Edema is an abnormal accumulation of fluids in body cavities or in the interstitial spaces of tissues and organs which causes swelling to occur. Edema is characterized by the presence of yellow fluid in the abdominal cavity or watery, gelatinous material in tissues. There are four types of edema that occur in fish: (1) *ascites* edema is the accumulation of fluid in the body cavity caused by dysfunction of the heart, liver, or kidney and gives affected fish a "pot-bellied" appearance. (2) *Pitting* edema is the presence of fluid in connective tissues thus giving the fish a swollen appearance. When pressure is applied on the skin, the indentation will remain. (3) *Lepidorthosis* edema is associated with pitting edema. The accumulation of fluid in the scale pockets causes the scales to be pushed away from the body surface. (4) *Exophthalmia* results from the accumulation of fluid in the eye socket which causes protrusion of the eyeball and gives the fish a "popeyed" appearance.

Edema indicates an imbalance of hydrostatic pressure or improper, osmotic pressure of the blood, increased permeability of capillaries, lymphatic obstruction, or kidney dysfunction. These conditions are normally associated with chemical toxicants; viral, bacterial, and parasitic diseases; or mechanical injury. Edema caused either by mechanical injury or disease may predispose the fish to further infection, as edematous fluids provide a good medium for bacterial growth. The effects of edema can be harmful and long lasting.

c. Hyperemia and Congestion

Hyperemia and congestion both involve an excessive volume of blood within the vessels; however, the mechanisms of development differ. Hyperemia is an active engorgement of the vascular bed with blood and is often associated with inflammation. In contrast, congestion is a passive accumulation of blood in vessels caused by an impairment of blood flow. Hyperemia and congestion cause affected tissue to appear red with possible swelling and will cause blood vessels in the fins and mesenteries to become more prominent.

d. Telangiectasis

Telangiectasis is the bulging of a blood vessel in gills of fish and can be equated to an aneurysm in higher vertebrates. However, an aneurysm is the permanent swelling of an artery, while telangiectasis is a passive and reversible condition. Telangiectasis may be caused by mechanical injury, toxicants, viruses, bacteria, bacterial toxins, parasites, and in some cases, nutritional deficiencies.

e. Embolism

An embolism is an obstruction of some part of the vascular system that interrupts the flow of blood. One of the most common types of fish emboli is the formation of ''gas bubbles'' in blood vessels caused by a supersaturation of the water by a gas, usually oxygen or nitrogen.[10-11] When a gas in the water exceeds 100% saturation for an extended period of time, gas bubbles will form in the blood of fish producing a condition known as ''gas bubble diseases''. This disease is similar to the bends that afflict divers. Fish affected by gas bubble disease may have microscopic or macroscopic gas bubbles in the opercle, gills, around the mouth, in the eyes, or in the fins. In cultured fishes, the most common occurrences of gas bubble disease are caused when deep wells are used as a water supply or when cool water is heated just before it flows into the fish culture unit. Wild populations of fish are usually affected by gas bubble disease just below high dams as a result of the water tumbling over the spillway and the subsequent entrapment of gases in the plunge pool.

2. CELLULAR DEGENERATION

Degeneration is a broad term which refers to a retrogressive process in which cells or tissues deteriorate, usually with a corresponding degree of functional inhibition. In the process of cellular degeneration, cells go through biochemical alterations and functional abnormalities that finally result in morphologic change. The degree and progress of the degeneration varies with the type and number of cells affected, the nature of the injurious agent, and the quantity (intensity) and duration of the injury. Degeneration of cells and tissues can be reversible and can change from a retrogressive to a progressive process with affected cells being restored to normal, if the injurious cause is removed before necrosis (death) of the cells occurs.

Degeneration may result from: (1) deficiency of a critical material (oxygen or a vital nutrient), (2) lack of a source of energy that interferes with metabolism, (3) mechanical, thermal, or electrical injury, or (4) accumulation of an abnormal substance in the cell caused by viral, bacterial, or parasitic pathogens and their toxins, or by toxic chemicals, nutritional imbalance, mild irritants, and others. Types of cellular degeneration that occur in fish include cloudy swelling, fatty change, and mucoid degeneration.

a. Cloudy Swelling

Cloudy or cellular swelling is often the result of bacterial toxemia, the earliest morphological (microscopic) indication of cellular degeneration. Cells affected with cloudy swelling are enlarged and their cytoplasm has a homogeneous, hazy appearance. This condition can only be identified by microscopy.

b. Fatty Degeneration

Fatty degeneration (or fatty change) is the result of an accumulation of lipids, primarily in the liver. This change is usually caused by an infectious disease, nutritional imbalance, hypoxia, anemia, and possibly some toxicants. In fatty degeneration, the liver is characterized by a pale yellowish to gray appearance. Although macroscopic evidence of fatty degeneration may eventually appear, early indications are microscopic.

3. NECROSIS
Necrosis is the death of cells or tissues following cell degeneration in a *living* animal and is the final stage in irreversible degeneration. Necrosis should not be confused with post-mortem change which is cellular deterioration after death of an animal. Characteristics of necrotic tissue can include paler than normal color; loss of tensile strength (tissue becomes friable and easily torn); may have a cheesy or pasty consistency; and an unpleasant odor may be present.

Necrosis can be caused by trauma, biological agents (e.g., viruses, bacteria, fungi, and parasites), chemical agents, or an interrupted blood supply to a specific area. Three types of necrosis are (1) liquefaction necrosis, (2) coagulation necrosis, and (3) caseation necrosis. When a cell dies, enzymes within the cell will destroy it. This liquefying effect by enzymes is called *liquefaction necrosis* and suggests a semisolid or fluid mass that has undergone self-destruction due to its own enzymes. When this occurs in the epithelium or exposed muscle of fish, the necrotized tissue is sloughed into the water leaving an open, ulcerative lesion that can be invaded by facultative pathogens present in the water. Liquefaction necrosis is the most common type of necrosis in fish. *Coagulation necrosis* refers to an area of necrosis in which the gross and microscopic nature of the tissues are still recognizable. It is associated with injury caused by several types of toxicants. *Caseation necrosis* results when a pathogenic organism produces cheesy, whitish material in a lesion. It is not common in diseased fish.

4. DISTURBANCES OF DEVELOPMENT AND GROWTH
Several growth changes may occur in an animal in response to an infection, toxicant, or other irritant. These changes can involve excessive growth, deficient growth, or an abnormal growth pattern in a tissue or organ. Generally, disturbances of growth involve either an abnormal number of cells, cells of abnormal size in an organ or tissue, or a combination of the two.

a. Atrophy
Atrophy refers to a decrease in size of a mature body part or organ due to a decrease in size or number of cells present. Atrophy is a slow process and can result from starvation or malnutrition (most common cause), lack of adequate blood supply, or chronic infection.

b. Hypertrophy
Hypertrophy is an increase in the size or volume of a body part or organ due to an increase in the size of individual cells. Hypertrophy usually results from an increased demand for function, but can also be initiated by an infectious agent such as lymphocystis virus which causes infected cells to become greatly enlarged.

c. Hyperplasia
Hyperplasia refers to an increase in the size of a body part or organ due to an increase in the number of cells present. One form of hyperplasia is characterized by an increased thickness of gill lamellae epithelium due to infection or exposure to a continuous mild irritation. Hyperplasia can result from water pollutants as well as from some fish viruses that cause formation of hyperplastic lesions, particularly of the integument. (Enlargement of a thyroid gland to form a goiter in humans is an example of hyperplasia.) Numerous examples of hyperplastic growth on the head, mouth, skin, and fins of fish have been documented.

5. INFLAMMATION
Inflammation is the aggressive vascular and cellular response of living animal tissue to a sublethal injury and is one of the most important defensive reactions an animal possesses. It may be acute or chronic. The function of inflammation is to help minimize the effect of an irritant or pathogen on the injured tissue. When injury occurs to the body, the primary response to the injury is an accumulation of fluids from the vascular system and the migration of lymphocytes, neutrophils, macrophages, and other blood components to the injured area. The accumulated fluids and cells dilute, localize, destroy and remove the irritant or infectious agent, and stimulate replacement and repair of injured tissue. Inflammation can be caused by viruses, bacteria, parasites, trauma, heat, irradiation, and toxicants. The "cardinal signs" of inflammation in higher vertebrates are (1) redness (hyperemia), (2) swelling, (3) heat, (4) pain, and (5) loss of function. How these apply to fish is not clearly understood.

Inflammation is classified according to the type of exudate involved: serous, fibrinous, hemorrhagic, catarrhal, purulent, or granulomatous. The exudates are named according to gross morphological appearance and all of these types of inflammation are seen in fish in varying degrees.

a. Serous
Serous inflammation is the exudation of clear fluid from the vascular system and mucosal surfaces, in response to the presence of a mild irritant or infection. Serous inflammation is usually acute and reflects injury to the vascular system. A blister is a good example of serous inflammation although it is seldom seen in fish.

b. Fibrinous
Fibrinous inflammation occurs when large amounts of fibrin escape from blood vessels and result in a clear clot, especially when exposed to air. This type of inflammation usually occurs on the intestinal mucosa and peritoneum of fish and may be associated with hyperemia and hemorrhage. Fibrinous inflammation, along with hemorrhage, has been observed in the body cavities of tilapia that were systemically infected with amoebas.[13]

c. Purulent
Purulent inflammation indicates that pus containing a combination of necrotic cells, neutrophils, and proteolytic enzymes is the primary feature of the exudate. Generally, fish do not produce pus, although, in some bacterial diseases, focal infections do produce purulent exudate. Purulent inflammation does occur in furunculosis of trout and has occasionally been found in *A. hydrophila* infections.

d. Catarrhal
Catarrhal inflammation occurs on the mucus-producing surfaces of fish epithelium of the digestive tract, skin, and gills and is characterized by an excessive production of mucus. The exudate may be clear, cloudy, or pink and has a consistency of fluid to mucoid. The fecal cast of trout infected with IHNV contains catarrhal exudate. Catarrhal exudate will form on the skin of fish infected with *I. multifiliis*.

e. Hemorrhagic
Hemorrhagic inflammation is characterized by the presence of large numbers of erythrocytes and other blood components on the surface of organs or in the exudate. Hemorrhagic inflammation will be generally diffused on serous or mucus membranes. It can be caused by viruses, bacteria, parasites, or toxicants. Erythemia, a skin condition, is usually associated with hemorrhagic inflammation.

f. Granulomatous Inflammation
Granulomatous inflammation is a chronic condition in which macrophages or related cell types are predominant. This type of inflammation is associated with several microbial infections of fish. *E. ictaluri* causes granulomatous inflammation, but does not form a typical granuloma, although the spongy, soft tissue surrounding the lesion in the skull of infected fish is granulomatous.[12] Other diseases, e.g., *Mycobacterium* spp. (bacteria) or *Ichthyophonus hoferi* (fungus), cause granulomatous inflammation, but these infections progress into macroscopic granulomas. Acute inflammation occurs initially in internal organs and muscle, but once this has disappeared there remains a central necrotic area that retains the causative agent. Macrophages and other inflammatory cells form a layer resembling epithelium around the irritant. As the lesion ages, increased numbers of fibroblasts and lymphocytes appear. The resulting granulomas are white to yellow in color and have a cheesy to hard consistency. In some cases, several granulomas may be connected by fibrous connective tissue to form a large, encapsulated, hard nodule.

D. RECOGNITION OF INFECTIOUS DISEASES OF FISH

To properly diagnose an infectious fish disease, one needs to know the case history of the fish involved, mortality patterns, clinical signs, and necropsy results.

1. CASE HISTORY

The case history of a fish population exhibiting morbidity or mortality can be acquired from the pond owner, farm manager, or biologist. The history should include the species, size and number, type of fish involved, type of holding facility (pond, cage, raceway, tank, aquarium, etc.), how long fish have shown clinical signs of illness, behavior and feeding patterns, water quality parameters, and type of treatment administered if any.

2. MORTALITY PATTERN

In a fish population, the mortality pattern may indicate whether the cause of death was due to an infectious agent, poor water quality (e.g., LODOS), or the presence of a water toxicant. Generally, infectious disease outbreaks are marked by a gradual acceleration in morbidity and mortality unless an extremely virulent pathogen is involved. Mortalities due to nutritional deficiencies are usually very protracted. Theoretical mortality curves representing deaths from toxicants or severe environmental disorders (oxygen depletion, etc.) and acute, subacute, and chronic infections are illustrated in Chapter 2 (Figure 2). Mortality rates of 30 to 40% or less are common during an infectious disease and they seldom reach 100%. However, when highly virulent agents and highly susceptible fish are involved, 90 to 100% mortalities can occur. When the majority of a fish population dies overnight or in a 24-hour period, oxygen depletion (LODOS) and any water quality changes that occurred in conjunction with this event (Chapter 1, Figure 3), chemical toxicants, or other environmental causes should be considered. Infectious disease agents seldom, if ever, result in "overnight" mass mortality.

3. CLINICAL SIGNS

The term "symptom" is not used in fish health because it refers to a phenomenon of physical or mental disorder which can only be described by the patient. The proper terminology used to describe what is found upon examination of animals other than humans, is clinical signs. Clinical signs describe behavioral, external, physical, or gross pathological changes. These clinical signs, combined with gross internal pathological changes, are very helpful in diagnosing fish diseases. However, in most cases, it is impossible to diagnose a specific disease, or to determine a specific etiological agent based solely on these signs because very few clinical signs are specific (i.e., pathognomonic) for a particular disease (Table 1). For our purposes, emphasis will be placed on the recognition of clinical signs that are readily apparent, overt, or obvious by gross inspection.

a. Behavior

The initial reaction of fish to many diseases is a cessation of feeding activity; therefore, any sudden drop in feed consumption should be quickly and carefully investigated for its cause. Stress of any type can cause fish to "go off feed". In addition to not feeding, diseased fish may swim into shallow water, swim lethargically at the surface, lie listlessly on the pond or tank bottom, float downstream, swim erratically, or scrape against underwater structures as if trying to eliminate an irritant from their skin. Some disease agents cause afflicted fish to swim in a longitudinal spiral or "tail chase" in a circle. Crowding around a water inlet to take advantage of oxygenated water may be a sign of diseased gills, particularly if water quality is good. Sick fish may gather in a tight school or distribute themselves evenly throughout the culture unit.

b. External Clinical Signs

In fish, the most obvious external clinical signs are inflammation and/or hemorrhage of fins, skin, or head; frayed fins; open necrotic and ulcerative lesions at any location on the body; lepidorthosis of scales; and excessive mucus production. A total lack of mucus, edema, an enlarged abdomen, growths on the body surface, presence of yellow or black spots on the skin, prolapsed anus, and exophthalmia are all clinical signs of disease. Varying degrees of erythemia, hyperemia, or hemorrhage may also be observed in the skin and fins of diseased fish.

Gills are adversely affected by viruses, bacteria, and parasites. They can become frayed and necrotic (gray or white), pale (anemic), swollen as a result of hyperplasia or can produce excessive amounts of mucus. Excessive mucus interferes with oxygen absorption from the water and causes affected fish to crowd around water inflow, a behavior simulating the response to a lack of dissolved oxygen in the water. Excessive mucus production on the skin tends to give the fish a grayish appearance.

Table 1 **Summary of behavioral and external clinical signs and internal lesions of fish and types of diseases that could be associated with them**

Clinical sign or lesion	Type of disease
Behavior	
Reduce or stop feeding	Viral, bacterial, parasitic diseases, environmental factor
Lethargic swimming	Viral, bacterial, parasitic, fungus, low oxygen
Crowding to fresh water	Bacteria or parasites on gill, low oxygen
Spinning, corkscrew or erratic swimming	Viral, parasitic, toxicants
External	
Gills necrotic	Bacterial, parasitic
Gills with excess mucus	Bacterial, parasitic, environmental
Skin with excess mucus	Parasitic, environmental
Depigmented areas of skin	Bacterial, parasites on skin or in muscle under skin
Hemorrhage, erythemia	Viral, bacterial, parasitic, toxicants in water
Frayed, eroded, erythemic fins	Bacterial, parasitic, mechanical injury
Exophthalmia, hemorrhaged opague eye	Viral, bacterial, parasitic, gas supersaturation
Ulcerative, necrotic lesions on skin	Bacterial, parasitic
Hydropsy	Bacterial, viral
Lepidorthosis	Bacterial, viral
Enlarged abdomen	Viral, bacterial, parasitic
Growths or nodules on skin or fins	Viral, parasitic, fungal, neoplasms
Internal Pathological Lesions	
Clear, yellow fluid in peritoneal cavity	Viral, rarely bacterial
Cloudy, bloody fluid in peritoneal cavity	Bacterial
Viscera is generally hyperemic	Viral, bacterial
Liver pale, friable	Viral, bacterial, nutritional imbalance
Liver mottled, hyperemic or petechiae, friable	Viral, bacterial
Spleen dark red, swollen	Viral, bacterial
Kidney dark red, pale or soft	Viral, bacterial
Intestine, erythemic, flaccid, void of food	Viral, bacterial
Kidney with large white, irregular shaped nodules of various size	Bacterial, fungal
Internal organs rough with small white spots	Parasitic, bacterial
Muscle with petechiae, necrotic pustules	Viral, bacterial

Lesions on the skin appear as slightly raised or swollen, depigmented areas where the epithelium has not been totally necrotized or sloughed off. In advanced stages of disease or where pathogens severely affect the skin, the epithelium may be totally missing (necrotized and sloughed into the water) leaving the underlying musculature exposed. Deep, necrotic lesions accompanied by areas of deep ulceration may occur in muscles, on top of the head, or in the mouth. Margins of ulcerative lesions are often white (necrotic) and surrounded by hemorrhages or erythemia in the epithelium. In the center of such lesions, the muscle bundles are often distinguishable.

Scale pockets may show lepidorthosis or the body may appear swollen because of muscle edema or from the presence of ascitic fluid in the abdominal cavity.

Growths on the skin and fins of fish may appear as cotton-like material; solid tumors of various size, color and texture; or as parasites embedded in, or beneath, the skin. The embedded parasites will appear as yellow, white, or black spots.

In fish that have a systemic bacterial infection, the anus is often prolapsed, swollen, and red; however, this condition should not be confused with the swollen vent of a gravid female that is ready to spawn.

Protruding (exophthalmic), hemorrhaged, or opaque eyes may be associated with an accumulation of fluid (edema) behind the eye, helminth parasites, or a supersaturation of gas in the water (gas bubble). Opaqueness of the eye may also be due to a trematode or nematode infestation, a dietary deficiency, or a bacterial infection.

4. GROSS INTERNAL LESIONS (PATHOLOGY)

After behavioral and clinical signs of disease have been observed and recorded, the fish must then be necropsied (i.e., post-mortem examination). If during necropsy internal lesions are found, a study of the lesions may help the diagnostician determine whether the infection was caused by a virus, bacteria, or parasites. The presence of small white or yellow, uniform-sized cysts in internal organs is indicative of metazoan parasites. Clear, straw-colored fluid in the abdominal cavity usually indicates the presence of a viral infection. If the fluid is bloody and cloudy and organs have large white, pustular, or granulomatous lesions of variable sizes, a bacterial infection is usually indicated. Some bacterial infections will cause the visceral cavity to have an intense putrid odor, even if the fish is fresh. However, as in most diagnostic procedures, there are exceptions to these generalizations.

A generalized hyperemia or hemorrhage in the viscera is indicative of a viremia or bacterial septicemia, in which case, the spleen will usually be swollen and dark red; the liver may be friable or soft and pale or mottled with petechial hemorrhages. The kidney is often swollen and friable. Fish that have infectious disease usually have intestines that are devoid of food, may contain white or bloody mucoid material, and an intestinal wall that is often flaccid and erythemic.

E. DISEASE DIAGNOSIS

In diagnosing fish diseases, a small number of animals are examined and this information is used to determine the health status of an entire population; therefore, it is essential that the sample of fish examined be representative of the general population. It is imperative that fish used for necropsy display clinical signs of disease. Moribund specimens are best for examination purposes and yield the most dependable results; never use fish that have been dead more than a few minutes, unless they were iced immediately upon death, or apparently healthy fish because they may not yield accurate results.

The etiologies of a fish disease should never be determined from clinical signs alone. Any attempt to do so often leads to erroneous diagnosis, improper treatment, and a waste of time, money, and fish. Disease diagnostic procedures should include examination for external parasites, attempted isolation of possible viruses and/or bacteria, and pathogen identification. If a pathogen is found on initial examination, all phases of the necropsy should be completed because multiple infections involving different pathogens are very common among fish. Successful therapy will require that all pathogens be treated. It is also very important to collect as much environmental and water quality data as possible. These data should include water temperature; dissolved oxygen determinations; any change in water color or water flow; carbon dioxide, ammonia, and nitrite concentrations; stocking density; weather conditions just prior to mortality; agricultural activities in the area (spraying of pesticides or herbicides); and size of culture unit. Diagnostic information should also include species, age, and size of fish. All of these data may be vital in accurately diagnosing a disease, treating it, and more importantly, preventing its reoccurrence.

For more detailed diagnostic and fish necropsy procedures, the reader is referred to Plumb and Bowser[12] and Meyer and Barclay.[13]

REFERENCES

1. **Slauson, D. O. and Cooper, B. J.**, *Mechanisms of Disease: A Textbook of Comparative General Pathology*, Williams & Wilkins, Baltimore, 1982.
2. **Thomson, R. G.**, *General Veterinary Pathology*, W.B. Saunders, Philadelphia, 1984.
3. **Ferguson, H. W.**, *Systematic Pathology of Fish: A text and atlas of comparative tissue responses in diseases of teleosts*, Iowa State University Press, Ames, 1989.
4. **Grizzle, J. M. and Rogers, W. A.**, *Anatomy and Histology of the Channel Catfish*, Alabama Agricultural Experiment Station, Auburn University, AL, 1976.
5. **Anderson, B. G. and Mitchum, D. L.**, *Atlas of Trout Histology*, Wyoming Game and Fish Department, Cheyene, 1974.
6. **Groman, D. B.**, *Histology of the Striped Bass*, American Fisheries Society, Bethesda, 1982.
7. **Chinabut, S., Limsuwan, C., and Kitsawat, P.**, *Histology of the Walking Catfish, Clarias batrachus*, International Development Research Centre, Ottawa, Canada, 1991.
8. **Hopps, H. C.**, *Principles of Pathology*, 2nd ed., Appleton-Century-Crofts, NY, 1964.
9. **Roberts, R. J.**, *Fish Pathology*, 2nd. ed., Bailliere Tindall, London, 1989.
10. **Bouck, G. R.**, Etiology of gas bubble disease, *Trans. Am. Fish. Soc.*, 109, 1980, 703.
11. **Marking, L. L.**, Gas Saturation in Fisheries: Causes, Concerns, Cures, U.S. Fish and Wildlife Services, Leaflet 9, 1987.
12. **Plumb, J. A. and Bowser, P. R.**, *Microbial Fish Disease Laboratory Manual*, Alabama Agricultural Experiment Station, Auburn University, AL, 1983.
13. **Meyer, F. P. and Barclay, L. A., Eds.**, Field Manual for the Investigation of Fish Kills, U.S. Fish and Wildlife Service, Washington, D.C., Resource Publ. 177, 1990.

PART TWO
VIRAL DISEASES

The study of viral diseases of fishes is still in its infancy when compared to the study of bacterial and parasitic diseases of aquatic animals. Infectious pancreatic necrosis (IPN) of salmonids, described in the mid 1950s by Wood et al.,[1] with viral etiology later demonstrated by Wolf et al.[2-3] in 1959, was the first proven viral disease of fish. The first major compilation of literature on viral diseases of fish, "Viral Diseases of Poikilothermic Vertebrates", published in 1965 by the New York Academy of Sciences[4] included contributions by fisheries scientists from the U.S., Europe, and Asia. The most extensive work published on fish viruses to date, *Fish Viruses and Fish Viral Diseases* by Wolf,[5] is a very detailed documentation of all the viruses that were known to infect fish at time of publication.

By 1985, approximately 50 viruses or virus-like agents had been reported in marine and freshwater fish species with about 75% being from freshwater fishes.[7] In 1988, Wolf[5] listed 59 diseases of fish that were in some way associated with a viral agent. Today, several new viruses or viral diseases of fish are being described every year.

The majority of known fish viruses have been reported in cultured species that have high economic value, and it is in culture environments the full devastating potential of some viruses emerges. Fish viruses are now more readily identified because intensive aquaculture is conducive to the expression of viral infections, more and better trained fish disease diagnosticians are available, disease facilities are better equipped to detect and research fish viruses, and improved and more sensitive virus detection reagents are being used.

An important characteristic of viruses is that they must invade a living cell in order to replicate. Viruses cannot reproduce themselves, but rely upon the synthetic machinery of the cells to produce new virus. In infected fish, the virus may manifest itself in several ways, the most drastic being production of hemorrhagic inflammation and primary necrosis, often resulting in high mortality. Tumorous growths on the skin and fins may result from viral infections or there may be an absence of pathology. Viruses vary in virulence: several are capable of killing a high percentage of infected fish (e.g., infectious pancreatic necrosis virus of trout), some have relatively low virulence (e.g., golden shiner virus), while others have no known virulence (e.g., catfish reovirus). Also, virulent viruses can be carried for extended periods of time by apparently healthy fish without the host showing any overt signs of infection.

Viral infections can be confirmed either by isolation of the causative agent in cell or tissue cultures, or by observation of the virus in tissues by electron microscopy. The first fish cell cultures in which a virus was replicated *in vitro* was reported in 1962 by Wolf and Quimby,[6] and this technology has led to major advances in the field of fish virology during the last three decades.

Viruses infecting fish cells or tissue cultures often affect the cells in a particular way. This cell injury, called cytopathic effect (CPE), can be detected with relatively low magnification using light microscopy. The CPE produced by a virus in cell culture is commonly characteristic of a type or group of viruses, but is seldom specific enough to definitively identify the virus.

Many factors determine to what extent a virus affects a fish population; however, water temperature is among the most important. Typically, fish viruses have a temperature range in which the virus is most pathogenic. Channel catfish virus is a good example as severe mortality occurs in young-of-the-year fish when water temperatures are above 25°C, moderate mortality occurs when water temperatures are 21 to 24°C, and little, if any mortality occurs when water temperatures are below 18°C. Age also affects the susceptibility of fish to viral infection. Generally, younger fish suffer the greatest mortality. Although severe mortality is rare among fish that are older than 1 year, the virus may still be present in older fish which often become carriers. Stressful effects of shipping, handling, poor water quality, high stocking densities, and inadequate nutrition may also affect severity of a viral infection among fish populations, but are often difficult to measure. These factors will be discussed in more detail as they relate to a particular virus disease.

Virus diseases having the greatest impact, or potential for impact, on cultured fish populations are discussed in this text and are organized according to the primary family or group of fish they infect. Only minor attention is given to nonvirulent viruses or those that have been detected by electron microscopy only. No attempt has been made to include all viruses that affect fish populations, therefore, the reader is referred to *Fish Viruses and Fish Viral Diseases*.[5]

REFERENCES

1. **Wood, E. M., Snieszko, S. F., and Yasutake, W. T.,** Infectious pancreatic necrosis in trout, *Am. Med. Assoc. Arch. Path.,* 60, 26, 1955.
2. **Wolf, K., Snieszko, S. F., and Dunbar, C. E.,** Infectious pancreatic necrosis, a virus-caused disease of fish, *Excer. Med.,* 13, 228, 1959.
3. **Wolf, K., Dunbar, C. E., and Snieszko, S. F.,** Infectious pancreatic necrosis of trout, a tissue-culture study, *Prog. Fish Cult.,* 22, 64, 1960.
4. **Whipple, H. E.,** Viral diseases of poikilothermic vertebrates, *Ann. N.Y. Acad. Sci.,* 126, 680, 1965.
5. **Wolf, K.,** *Fish Viruses and Fish Viral Diseases,* Cornell University Press, Ithaca, NY, 1988, 476.
6. **Wolf, K. and Quimby, M. C.,** Established erythermic line of fish cells *in vitro, Science,* 135, 1065, 1962.
7. **Ahne, W.,** Virusinfektionen bei fischen: Atiologie, diagnosis und bekampfung, *Z. Vet. Med.,* 32, 237, 1985.

Chapter IV

Catfish (Ictaluridae)

A. CHANNEL CATFISH VIRUS DISEASE

Channel catfish virus disease (CCVD) is an acute, highly communicable infection of cultured fry and fingerling channel catfish *(Ictalurus punctatus)* that was first reported by Fijan[1-2] in 1968. The etiological agent of CCVD is channel catfish virus (CCV), a member of the family Herpesviridae, but Wolf et al.[3] suggested *Herpesvirus ictaluri* as the specific epithet.

1. GEOGRAPHICAL RANGE AND SPECIES SUSCEPTIBILITY

Channel catfish virus has been isolated from channel catfish in most southern states and from channel catfish in culture systems in many other areas of the U.S. The virus was also isolated in Honduras, Central America in 1972 from a group of channel catfish fry shipped from the U.S. These fish were totally destroyed and the virus apparently failed to become established. There have been unconfirmed reports of diseases resembling CCVD in channel catfish populations in other parts of the world where this species has been introduced.

The channel catfish appears to be the most susceptible species of catfish affected by CCV. Experimental infection was induced in fingerling blue catfish *(I. furcatus)* and in channel catfish × blue catfish hybrids by injection with the virus, but no infection occurred when CCV was introduced orally or by cohabitation with virus-infected channel catfish fingerlings.[4] Contrary to experimental refractiveness of blue catfish to CCV, the virus was isolated from diseased fingerlings of this species from Missouri in 1988.[5] Brown bullheads *(I. nebulosus)* and yellow bullheads *(I. natalis)* cannot be infected experimentally with CCV by injection or by feeding. Plumb et al.[6] demonstrated that the European catfish *(Silurus glanis)* is also resistant to CCV.

A study where CCV was fed to different strains of channel catfish fry indicated that there was a variation in susceptibility among strains of the species.[7] Young fish that were the result of cross-breeding between different strains of channel catfish were more resistant to CCV than were the inbred, parental strains suggesting hybrid vigor.

2. CLINICAL SIGNS AND FINDINGS

Clinical signs of CCV vary among diseased fish and some or all of the following may be present (Figure 1): distension of the abdomen due to the accumulation of a clear straw-colored fluid in the peritoneal cavity, exophthalmia, pale or hemorrhagic gills, hemorrhage or erythemic areas at the base of fins and throughout the skin (particularly on the ventral surface).[2] Infected fish swim erratically or convulsively, sometimes rotating about their longitudinal axis. Moribund fish sink to the bottom, become quiescent, respire weakly but rapidly, and then die.

Internal lesions include a general hyperemia throughout the visceral cavity although the liver and kidneys may be pale. The spleen is generally dark red and enlarged; the stomach and intestine are void of food, but contain a mucoid secretion.

3. DIAGNOSIS AND VIRUS CHARACTERISTICS

Any time a sudden increase in morbidity occurs among young channel catfish during the summer CCV should be suspected, and they should be necropsied for the virus. Homogenized viscera or whole fish (fry) samples from virus-suspect populations should be decontaminated by filtration or addition of antibiotics and a 10^{-2} dilution inoculated into the cells. Brown bullhead (BB) cells are susceptible to CCV, but channel catfish ovary (CCO) cells are the cell line of choice because of their higher sensitivity to the virus and greater viral productivity.[8] CPE will often be visible in 12 to 24 h at 30°C in lower dilutions, if an active CCV infection is in progress. Focal CPE is first indicated by pyknotic cells which then coalesce into a multinucleated syncytium that is connected to surrounding normal cells by protoplasmic bridges resembling irregular spokes of a wheel (Figure 2). As CPE progresses, the cells are released from the surface and form a loose network. The appearance of syncytia in susceptible cell

Figure 1   A 10-cm channel catfish naturally infected with channel catfish virus. Note the abdominal distension and exophthalmia. The upper lobe of the caudal fin has a columnaris lesion.

cultures is presumptive identification of CCV; positive identification can be made by serum neutralization using CCV antiserum produced in a rabbit, goat, or fish.

CCV has been recognized as a herpesvirus since Wolf and Darlington[3] suggested *H. ictaluri*. Davison et al.[9] using DNA sequencing, determined that CCV is genetically dissimilar from herpesviruses of higher vertebrates. They concluded that CCV is related morphologically to the other herpesviruses, but is unrelated in their evolution. It is suggested that CCV is the only known member of a subfamily in the Herpesviridae. Channel catfish virus is an enveloped, icosahedral virus with DNA genome and a nucleocapsid diameter of 95 to 105 nm. Enveloped virions have a diameter of 175 to 200 nm. Infectivity is inactivated by either 20% ether or by 5% chloroform. CCV is heat-labile at 60°C for 1 h, unstable in sea water, and sensitive to ultraviolet light, requiring 20 to 40 min for inactivation.[10] The virus survives for less than 24 h on dried fish netting or glass cover slips, it retains infectivity in pond water for about 2 d at 25°C and for 28 d at 4°C. In dechlorinated tap water, infectivity is retained for 11 d at 25°C and for over 2 months at 4°C. Under experimental conditions when the virus is introduced into pond-bottom mud, infectivity is immediately neutralized. Infectious CCV can not be isolated from decomposing infected fish at 22°C 48 h after death; however, it is recoverable for up to 14 d from viscera of whole iced fish, for 162 d from fish frozen at −20°C, and for 210 d from fish frozen at −80°C.[11]

Intranuclear CCV replication occurs in CCO cells at 15 to 35°C with 30 to 35°C being optimum.[8] Cells derived from the walking catfish kidney (K1K) are also susceptible to CCV,[12] and Lewis developed a tilapia heart cell line that is susceptible to CCV.[13] Fernandez et al.[14] also reported that cell lines developed from two species and one hybrid of marine fishes (Japanese striped knife jaw, *Oplegnathus faciatus;* purplish amberjack, *Seriola dumerilli;* and the hybrid of kelp and spotted grouper, *Epinephelus moora* X *E. akaara*) produced CPE when infected with CCV. Other poikilothermic and mammalian cell systems are refractive to CCV.[3]

4. EPIZOOTIOLOGY

Outbreaks of CCVD on channel catfish farms have been diagnosed during June through October when water temperatures are above 25°C (Figure 3). In Alabama, 84% of the CCVD cases between 1980 and 1990 occurred during June through August, with none being diagnosed from November through May. Epizootics on catfish farms occur most frequently during years of high water temperature and in heavily stocked fingerling ponds, but many of these outbreaks are preceded by handling and transport. There is evidence that survivors of CCV epizootics may be stunted.[15]

Fry and fingerling channel catfish can transmit the virus horizontally, as long as an active infection is in progress. Young-of-the-year fish, less than 4-months-old, are the most susceptible to CCV and the younger or smaller the fish, the higher the mortality.

Under experimental conditions, it is possible to transmit the virus from infected moribund or dead fish to healthy fish through the water. Modes of virus transmission, other than by water, are intramuscular or intraperitoneal injection; by incorporating it into feed; or by swabbing the gills with a saline solution containing virus. Infectivity can be experimentally induced in fish weighing up to 10 g by waterborne exposure, but injection is required to infect larger fish. Occasionally, 1-year-old fish suffer outbreaks

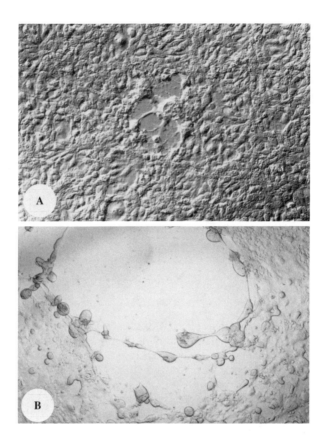

Figure 2   Channel catfish ovary cells infected with channel catfish virus and incubated at 30°C. (A) Focal CPE 18 h after inoculation surrounded by normal cells; (B) total CPE 24 to 36 h after inoculation.

of CCVD, but this usually follows stressful transportation of the fish. Often, CCVD occurs in conjunction with a secondary *Flexibacter columnaris* infection which may prolong the disease.

The incubation period, between exposure of fish to virus and the appearance of clinical signs and morbidity, is inversely related to water temperature.[16] Experimental infection at 30°C is followed by clinical signs in 32 to 42 h, with the first deaths occurring several hours later.[17] At 20°C, the incubation period is 10 d. Under field conditions at 25 to 30°C, healthy channel catfish fingerlings develop the disease within 72 to 78 h after exposure, and up to 100% die within 6 d. In 1 documented instance a group of fry, probably infected with CCV from the brood fish, were held at 28°C in troughs receiving well water.[18] The fish developed clinical CCVD when 21 d old and 72 h later, 100% were dead. A sample of fish assayed for CCV when these fish were 14-days-old was negative. The disease developed rapidly, even though the infected fish were not exposed to a known source of CCV between days 14 and 21.

Channel catfish virus is communicable from the time fish show clinical signs until soon after their death. In experimentally infected juveniles, the virus begins to disappear in surviving fish 120 h after infection. It is difficult, or impossible, to isolate virus by routine procedures once the clinical signs of enlarged abdomen, exophthalmia, and hemorrhage have passed. However, Bowser et al.[19] isolated CCV from apparently normal adult fish during winter.

Adult fish are considered a possible source of CCV infection by vertical transmission of the virus to offspring via the reproductive products, but, to date, this has not been conclusively demonstrated. While active CCV infections are relatively easy to diagnose by isolating the virus in cell cultures, the detection of covert CCV carrier fish in a population is a difficult problem. Plumb[20] demonstrated that CCV antibodies could be detected in channel catfish that had been exposed to the virus and suggested that this may be a way of separating adult fish that had been exposed to CCV from those that had not.

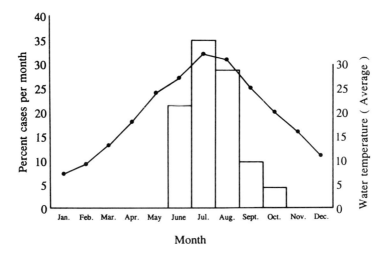

Figure 3   Monthly distribution of 137 cases of channel catfish virus diagnosed in Alabama between 1980 and 1990.

Plumb and Jezek[21] and Amend and McDowell[22] independently used serum neutralization tests on adult catfish to separate possible carrier populations from noncarrier populations. The latter demonstrated that 7 of 17 major brood populations tested showed positive CCV serum neutralization, and CCV was subsequently isolated from young-of-the-year fish at 2 of the 7 CCV antibody-positive farms. Hedrick et al.[23] showed that CCV-neutralizing antibody persisted in adult channel catfish for 2 years following the last known exposure to the virus. Adult fish, not exposed to CCV as juveniles, became infected and one adult died as result of the virus. Survivors produced anti-CCV antibody for 6 months although the titers were low. These authors suggested that CCV does establish latent infections in epizootic survivors and that expression of certain viral antigens, or periodic virus reactivation, may be a stimulus for continued anti-CCV antibody production.

Other studies have been made which strongly support the idea that adult fish may carry CCV and can possibly develop active infections. Plumb et al.[24] demonstrated that CCV nucleic acid could be detected in gonads of adult fish by immunofluorescence, and Bowser et al.[19] isolated CCV from adult channel catfish. Their studies showed that older fish can develop active infections. Wise and Boyle[25] and Wise et al.[26] utilized a CCV-specific DNA probe to clearly demonstrate the presence of CCV nucleic acid in livers of adult channel catfish, including those adults from which virus had been isolated. It was further shown that by using the CCV-nucleic acid probe, CCV genetic material was present in adults and it was also present in offspring of these fish.[27] The results further indicate vertical transmission of the virus. Bird et al.[28] used molecular cloning of fragments of the CCV genome to suggest that, in channel catfish, CCV can persist in a dormant or transcriptionally active state, without causing clinical signs.

5. PATHOLOGICAL MANIFESTATIONS

Histopathological changes are similar in both natural and experimental CCV infections.[29-31] Renal hematopoietic tissue is edematous and has extensive necrosis and cellular dissolution, coupled with an increase in macrophages. The liver develops regional edema, necrosis, and hemorrhage, and hepatic cells have eosinophilic intracytoplasmic inclusions. Pancreatic acinar cells are necrotic. The submucosa of the digestive tract is edematous and has focal areas of macrophage concentrations and hemorrhage. The spleen becomes congested with erythrocytes and this change is coupled with extensive reduction of lymphoid tissue. Virus particles have been seen in electron micrographs of the liver, kidneys, and spleen of infected fish (Figure 4).

A generalized viremia is established within 24 h after experimental infection. The kidneys, liver, spleen, and intestine become active sites of virus replication 24 to 48 h after infection, and virus can sometimes be isolated from brain tissue after 48 h. Peak virus titers occur in the kidney and intestine 72 h after infection and in the spleen, brain, and liver after 96 h. Virus titers in the muscle remain comparatively low.

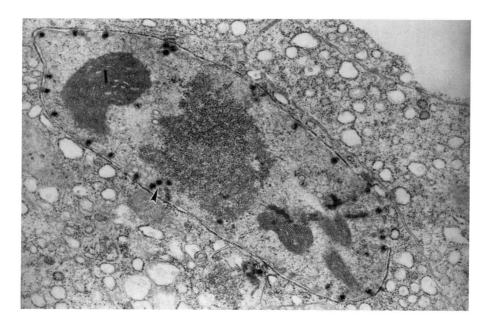

Figure 4   Channel catfish virus (arrow) in the nucleus of a spleen cell from experimentally infected channel catfish. Note the typical hyalin-like inclusion bodies (I). (From Plumb, J. A. et al., *J. Fish Biol.*, 6, 661, 1974. With permission.)

6. SIGNIFICANCE

The potential effects of CCVD on the channel catfish industry are great due to the exceptionally high mortality among young-of-the-year channel catfish. However, the incidence of CCVD comprises only 1 to 2% of the total cases at fish disease diagnostic laboratories. Thus, when and where CCVD occurs, it is highly significant, but the overall effect on channel catfish culture is not great.

7. MANAGEMENT

There are no available chemotherapeutic treatments for viral diseases of fishes. The only practical control measures are management, avoidance, quarantine of infected stocks, and sanitation. The mortality rate of CCV-infected fish is closely correlated to water temperature in laboratory experiments. Mortality decreased significantly when the water temperature was reduced from 28 to 19°C or below, 24 h after infection. Therefore, in certain circumstances a reduction of water temperature to 19°C will be beneficial. During periods when CCVD is most likely to occur (June through September), every effort should be made to eliminate adverse environmental conditions among fish of susceptible size or age. High stocking rates of fingerlings should be avoided, temperatures should be kept at moderate levels if possible, and transportation of susceptible fish during the summer should be minimized. When bacterial infections develop, especially "columnaris", they should be treated immediately.

An additional step that might be taken to avoid CCV could include surveying brood stocks for CCV-neutralizing antibodies. Also, detection of carrier fish may become more reliable as CCV-specific nucleic acid probes are perfected.

Ponds from which diseased fish are removed should be drained or disinfected with 40 mg/l of chlorine. Survivors of CCV epizootics may be grown to a marketable size, providing the fish are segregated from ponds with healthy, susceptible channel catfish. Under no circumstances should survivors of CCV epizootics be stocked in noninfected waters, nor should they be used as brood stock.

Vaccination is being considered to combat CCVD. CCV becomes attenuated when passed frequently and rapidly in K1K cells from walking catfish kidney.[12] When channel catfish are injected with, or bathed in, the attenuated virus, they become immune to wild type (virulent) CCV. This suggests that vaccination against CCV is possible, and Walczak et al.[32] reported the immunological properties of the attenuated vaccine. The potential for developing a CCV subunit vaccine was explored by Awad et al.[33] and Hanson[34] utilized genetic engineering in CCV vaccine development.

REFERENCES

1. **Fijan, N. N.,** Progress report on acute mortality of channel catfish fingerlings caused by a virus, *Bull. Off. Int. Epizool.,* 69(7–8), 1167, 1968.
2. **Fijan, N. N., Wellborn, T. L., Jr., and Naftel, J. P.,** An Acute Viral Disease of Channel Catfish, U.S. Fish Wildl. Ser. Tech. Pap. No. 43, Washington, D.C., 1970.
3. **Wolf, K. and Darlington, R. W.,** Channel catfish virus: a new herpesvirus of ictalurid fish, *J. Virol.,* 8, 525, 1971.
4. **Plumb, J. A. and Chappel, J.,** Susceptibility of blue catfish to channel catfish virus, *Proc. Annu. Conf. South. Assoc. Fish Wildl. Agen.,* 32, 680, 1978.
5. Southeastern Cooperative Fish Disease Laboratory, Auburn University, Alabama, case records, unpublished.
6. **Plumb, J. A., Hilge, V., and Quinlan, E. E.,** Resistance of the European catfish *(Silurus glanis)* to channel catfish virus, *J. Appl. Ichthyol.,* 1, 87, 1985.
7. **Plumb, J. A., Green, O. L., Smitherman, R. O., and Pardue, G. B.,** Channel catfish virus experiments with different strains of channel catfish, *Trans. Am. Fish. Soc.,* 104, 140, 1975.
8. **Bowser, P. R. and Plumb, J. A.,** Channel catfish virus: comparative replication and sensitivity of cell lines from channel catfish ovary and the brown bullhead, *J. Wildl. Dis.,* 16, 451, 1980.
9. **Davison, A. J.,** Channel catfish virus: a new type of herpesvirus, *Virology,* 186, 9, 1992.
10. **Robin, J. and Rodrique, A.,** Resistance of herpes channel catfish virus (HCCV) to temperature, pH, salinity and ultraviolet irradiation, *Rev. Can. Biol.,* 39, 153, 1980.
11. **Plumb, J. A., Wright, L. D., and Jones, V. L.,** Survival of channel catfish virus in chilled, frozen and decomposing channel catfish, *Prog. Fish Cult.,* 35, 170, 1973.
12. **Noga, E. J. and Hartmann, J. X.,** Establishment of walking catfish *(Clarias batrachus)* cell lines and development of a channel catfish *(Ictalurus punctatus)* virus vaccine, *Can. J. Fish. Aquat. Sci.,* 38, 925, 1981.
13. **Lewis, D. H.,** College of Veterinary Medicine, Texas A & M University, College Station, personal communication, 1988.
14. **Fernandez, R. D., Yoshimizu, M., Kimura, T., Ezura, Y., Inouye, K., and Takami, I.,** Characterization of three continuous cell lines from marine fish, *J. Aqua. Anim. Health,* 5, 127, 1993.
15. **McGlamery, M. H., Jr. and Gratzek, J. B.,** Stunting syndrome associated with young channel catfish that survived exposure to channel catfish virus, *Prog. Fish Cult.,* 36, 38, 1974.
16. **Plumb, J. A.,** Effects of temperature on mortality of fingerling channel catfish *(Ictalurus punctatus)* experimentally infected with channel catfish virus, *J. Fish. Res. Board Can.,* 30, 568, 1972.
17. **Plumb, J. A. and Gaines, J. L., Jr.,** Channel catfish virus disease, in *The Pathology of Fishes,* Ribelin, R. E. and Migaki, G., Eds., University of Wisconsin Press, Madison, 1975, 287.
18. **Plumb, J. A.,** Some Biological Aspects of Channel Catfish Virus, Ph.D. dissertation, Auburn University, Alabama, 1972.
19. **Bowser, P. R., Munson, A. D., Jarboe, H. H., Francis-Floyd, R., and Waterstrat, R. P.,** Isolation of channel catfish virus from channel catfish *Ictalurus punctatus* (Rafinesque) broodstock, *J. Fish Dis.,* 8, 557, 1985.
20. **Plumb, J. A.,** Neutralization of channel catfish virus by serum of channel catfish, *J. Wildl. Dis.,* 9, 324, 1973.
21. **Plumb, J. A. and Jezek, D. A.,** Channel catfish virus disease, in *Antigens of Fish Pathogens: Development and Production for Vaccines and Serodiagnostics,* Anderson, D. P., Dorson M., and Dubourget, Ph., Eds., Collection Marcel Merieux, Lyon, France, 1983, 33.
22. **Amend, D. F. and McDowell, T.,** Comparison of various procedures to detect neutralizing antibody to the channel catfish virus in California brood channel catfish, *Prog. Fish. Cult.,* 46, 6, 1984.
23. **Hedrick, R. P., Groff, J. M., and McDowell, T.,** Response of adult channel catfish to waterborne exposures of channel catfish virus, *Prog. Fish. Cult.,* 49, 181, 1987.
24. **Plumb, J. A., Thune, R. L., and Klesius, P. H.,** Detection of channel catfish virus in adult fish, *Dev. Biol. Stand.,* 49, 29, 1981.
25. **Wise, J. A. and Boyle, J. A.,** Detection of channel catfish virus in channel catfish, *Ictalurus punctatus* (Rafinesque): use of a nucleic acid probe, *J. Fish Dis.,* 8, 417, 1985.

26. **Wise, J. A., Bowser, P. R., and Boyle, J. A.,** Detection of channel catfish virus in asymptomatic adult catfish, *Ictalurus punctatus* (Rafinesque), *J. Fish Dis.,* 8, 495, 1985.

27. **Wise, J. A., Harrell, S. F., Busch, R. L., and Boyle, J. A.,** Vertical transmission of channel catfish virus, *Am. J. Vet. Res.,* 49, 1506, 1988.

28. **Bird, R. C., Nusbaum, K. E., Screws, E. A., Young-White, R. R., Grizzle, J. M., and Toivio-Kinnucan, M.,** Molecular cloning of fragments of the channel catfish virus (Herpesviridae) genome and expression of the encoded mRNA during infection, *Am. J. Vet. Res.,* 49, 1850, 1988.

29. **Wolf, K., Herman, R. L., and Carlson, C.,** Fish viruses: histopathologic changes associated with experimental channel catfish virus disease, *J. Fish. Res. Board Can.,* 29, 145, 1972.

30. **Plumb, J. A., Gaines, J. L., Mora, E. C., and Bradley, G. G.,** Histopathology and electron microscopy of channel catfish virus in infected channel catfish, *Ictalurus punctatus* (Rafinesque), *J. Fish Biol.,* 6, 661, 1974.

31. **Major, R. D., McCraren, J. R., and Smith, C. E.,** Histopathological changes in channel catfish *(Ictalurus punctatus)* experimentally and naturally infected with channel catfish virus disease, *J. Fish. Res. Board Can.,* 32, 563, 1975.

32. **Walczak, E. M., Noga, E. J., and Hartmann, J. X.,** Properties of a vaccine for channel catfish virus disease and a method of administration, *Dev. Biol. Stand.,* 9, 419, 1981.

33. **Awad, M. A., Nusbaum, K. E., and Brady, Y. J.,** Preliminary studies of a newly developed subunit vaccine for channel catfish virus disease, *J. Aquat. Anim. Health,* 1, 233, 1989.

34. **Hanson, L.,** Biochemical Characterization and Gene Mapping of the Channel Catfish Herpesvirus (CCV) Encoded Thymidine Kinase, a Selectable Site for Homologous Recombination, Ph.D. dissertation, Louisiana State University, Baton Rouge, 1990.

Chapter V

Carp and Minnows (Cyprinidae)

A. SPRING VIREMIA OF CARP

Spring viremia of carp (SVC) is a subacute to chronic disease of subadult, cultured common carp *(Cyprinus carpio)*. The virus was first isolated in tissue culture, in the former Yugoslavia, by Fijan et al.,[1] who successfully recovered a virus from diseased carp which had clinical signs of infectious dropsy (ID). Consequently, it was proposed to split the ID syndrome into two etiologically and clinically independent diseases: SVC, the causative agent of which was named *Rhabdovirus carpio,* and the carp erythrodermatitis (CE), a skin infection with necrotic lesions surrounded by hemorrhagic areas.[2] Subsequently, Fijan and Petrinec[3] presented additional evidence to differentiate between SVC and CE, while Bootsma et al.[4] and Pol et al.[5] demonstrated that CE was associated with the bacterium *Aeromonas salmonicida achromogenes* (atypical *A. salmonicida*), later shown to be also the causative agent of goldfish ulcer disease. It should be mentioned that *Aeromonas hydrophila (A. punctata),* historically thought to be the causative agent of ID, can complicate both SVC and the CE as a secondary infection.

1. GEOGRAPHICAL RANGE AND SPECIES SUSCEPTIBILITY

SVC occurs primarily in Europe, having been confirmed in Yugoslavia, Hungary, Czechoslovakia, Austria, Bulgaria, France, Germany, Romania, Spain, Great Britain, and the former U.S.S.R.[1,6-7] However, the virus is likely to occur in other countries of Europe where carp are cultured. Curiously, SVC has not been reported outside of this region in spite of the movement of carp to many parts of the world and of carp being a primary culture species in China.

Species of fish naturally infected with SVC are the common carp, bighead carp *(Aristichthys nobilis),* silver carp *(Hypophthalmichthys molitrix),* and grass carp *(Ctenopharyngodon idella).*[8] A population of sheatfish *(Silurus glanis),* i.e., European catfish, was also naturally infected,[9] while guppies *(Lebistes reticulata)* and pike *(Esox lucius)* fry have been experimentally infected.[10]

2. CLINICAL SIGNS AND FINDINGS

At the onset of SVC, fish are attracted to the water inlet. Seriously affected fish become moribund, respire slowly, and lie on their side.[2] Fish experimentally infected by cranial injection express similar behavior. External clinical signs include darker pigmentation, a pronounced enlargement of the abdominal area, exophthalmia, and a prolapsed and inflamed anus. The gills are very pale with distinct petechiae (Figure 1). Internally, a generalized hyperemia is apparent with peritontis, enteritis, and hemorrhages in the kidney, liver, and air bladder. Also, the liver is edematous with adhesions, and hemorrhages occur in the muscle tissue. In fry and small carp, the swim bladder is inflamed and shows focal hemorrhaging or petechiae.

3. DIAGNOSIS AND VIRUS CHARACTERISTICS

Diagnosis of SVC is made by isolation of *R. carpio* from diseased fish in cell cultures where replication of the virus occurs between 10 and 30°C with 21°C being the optimum. The fathead minnow (FHM) and epithelioma papillosum of carp (EPC) cell lines, or primary ovary cells of carp, are susceptible to SVCV and yields of 10^8 pfu/ml can be obtained. Lower yields of virus occur in BB cells. CPE consists of degeneration and rounding of cells which then detach from the vessel surface.[2] Focal CPE includes granulation of chromatin in the nucleus, with thickening or lysis of the nuclear membrane, and cytoplasmic degeneration (Figure 2). Rainbow trout gonad (RTG-2) and bluegill fry (BF-2) cells are also susceptible, but CPE develops more slowly.[1,11]

SVCV can be positively identified by neutralization, using specific SVCV antiserum, but cross-neutralization reactions may occur with other rhabdoviruses from fish (Table 1). The antigen can be detected in frozen liver, kidney, and spleen tissues by either the immunoperoxidase (ELISA) or fluorescent antibody procedures.[12] These serological techniques can also be used to identify SVCV in FHM cell cultures 12 h after inoculation as compared to 48 h required for detection of CPE in inoculated cell cultures. More recently, Way[13] used an ELISA system to detect SVCV antigen in inoculated cell

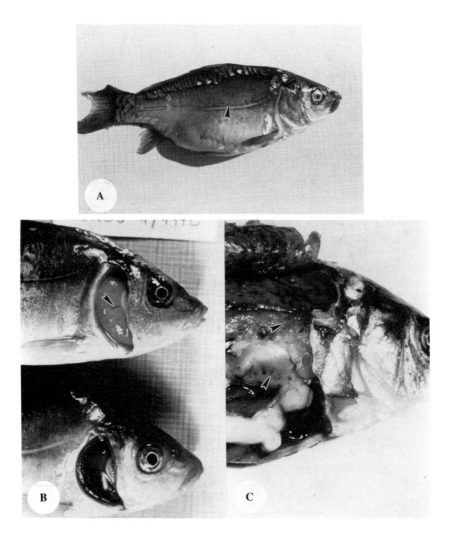

Figure 1 Common carp infected with spring viremia of carp virus. (A) Petechiae in the skin (arrow); (B) upper fish with pale gills and lower fish with normal gills [note the hemorrhage in the pale gill (arrow)]; (C) petechiae in the muscle and swim bladder (arrow) and pale liver. (Photos A and B by N. Fijan; photo C by P. Ghitino.)

cultures and extracts from clinically infected fish within 1 h. The same system detected the virus in subclinical fish, but it was not as accurate and some false positives were noted. The antibody in the SVCV ELISA reagents also reacted with the pike fry rhabdovirus, but at a very low level.

R. carpio is a bullet-shaped RNA virus which measures approximately 70×180 nm.[14] Activity of the virus is destroyed with exposure to 45°C, ether, or storage at pH 3;[11] however, in tissue culture medium containing 5% serum, the virus survives at a low level for 28 d at 23°C and for over 6 months at 4, -20, or -74°C.[13] Infectivity is reduced if serum is not included in the medium. Most disinfectants (e.g., formalin, sodium hydroxide, and chlorine) kill the virus in minutes, but in water and mud it remains infective for up to 42 d.

The virus isolate from carp with swim bladder inflammation, named swim bladder inflammation virus (SBIV), has the same characteristics as SVCV.[15-16] It was established later that SBI of carp has a protozoan etiology.[17] Hill et al.[18] compared SVC with the rhabdoviruses that infect other species of fish and showed that SVC and a virus isolated from carp with SBI were serologically identical, but both of these viruses were distinctly different from pike fry rhabdovirus, infectious hematopoietic necrosis virus, and viral hemorrhagic septicemia virus (Table 1).

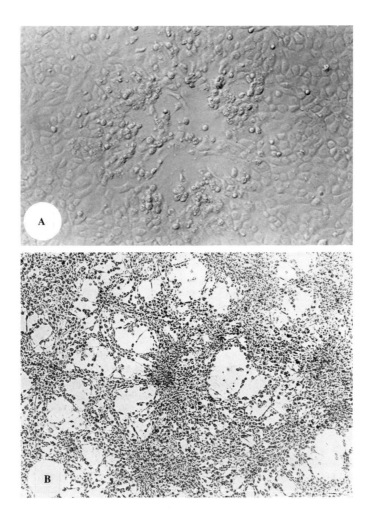

Figure 2 Focal cytopathic effect of *Rhabdovirus carpio* (spring viremia of carp). (A) Focal CPE in epithelioma papillossum cells. Infected cells are rounded and detached (original magnification × 230); (B) massive *R. carpio* CPE in FHM cells 72 h after infection. (Photo A by N. Fijan; photo B by W. Ahne.)

4. EPIZOOTIOLOGY

SVC occurs primarily in pond populations of carp that are 1-year-old or older. Epizootics due to SVC were initially confirmed during the springs of 1969 and 1970 in Yugoslavia[1] when water temperature ranged from 12 to 22°C, however, the optimum water temperature for this disease is 16 to 17°C. Using 50-g carp, Baudouy et al.[19] clarified the role of water temperature in the development of SVC following waterborne exposure. They found that infection occurred at temperatures above 18°C, but a protective immunity developed. Between 11 and 18°C clinical signs occurred with some fish becoming immune and surviving the disease while others died. Below 10°C, no immunity developed and the disease was always fatal. They suggested that the elevation of water temperature which naturally occurs in spring is not necessarily a factor in the disease process.

The severity of SVC varies from farm to farm, pond to pond on the same farm, and from year to year on a given farm.[2] After consecutive years with no SVC outbreaks, heavy losses may be experienced for several years. The cause of these variations in severity of SVC outbreaks is unknown, but probably results from the interactions of susceptible fish, the virulence of the virus, and environmental conditions. One important observation was that carp which survived a SVCV infection in the laboratory were resistant to repeated challenges with virulent virus, but remained susceptible to carp erythrodermatitis.[2] This would further support the idea that the etiologies of the two diseases are different.

Table 1　Cross-neutralization test of five rhabdoviruses from fish against homologous and heterologous antiserum

Antiserum	50% Plaque reduction titer of virus				
	SVC	SBI	PFV	IHN	VHS
SVC	6500	6200	20	20	<10
SBI	980	980	68	50	<10
PFV	33	28	200	27	<10
IHN	25	25	27	1700	<10
VHS	<10	<10	<10	<10	2300

Note: Acronym of viral antigen: SVC — spring viremia of carp, SBI — swim bladder infection virus of carp, PFV — pike fry rhabdovirus, IHN — infectious hematopoietic necrosis virus, VHS — viral hemorrhagic septicemia virus. Reciprocal of antiserum dilution giving 50% plaque reduction against the specified virus.

From Hill, B. et al., *J. Gen. Virol.*, 17, 376, 1975. With permission of Cambridge University Press.

SVC can be transmitted by cohabitation of infected and healthy fish, contaminated food organisms, and by intraperitoneal, intramuscular, or intracranial injection. Fish parasites (e.g., fish louse, *Argulus* sp., and *Pisciocola geometra*) can also transmit SVC to uninfected fish.[20] The incubation period for SVC varies from 6 to 60 d depending on water temperature and method of infection. Survival of carp experimentally infected with SVCV by injection in the swim bladder was 8.9 d, compared to 12.8 d for intraperitoneal injection, and 12.9 d for intracranial injection.[21]

Shedding of *R. carpio* by adult carp has been reported by Bekesi and Csontos.[22] They isolated *R. carpio* from 3 of 491 ovarian fluid samples assayed, but virus was not isolated from 211 seminal fluid samples. Whether or not the virus was actually inside the eggs was not demonstrated, but it does indicate that some adult survivors of SVCV infections can be carriers.

5. PATHOLOGICAL MANIFESTATIONS

Histopathological descriptions of SVC are limited, but Negele[23] reported on the histopathology of experimentally infected carp fingerlings. The liver has edema, necrotic blood vessels, and focal necrosis in the parenchyma. Vessels in the peritoneum are hyperemic and the intestine inflamed. The spleen is hyperemic and hyperplastic. Renal tubules of the kidney become clogged with casts and tubule cells develop cytoplasmic inclusions. The swim bladder has extensive focal areas of hemorrhage.

During waterborne exposure, SVC enters the fish through the gills where the virus replicates. The virus then spreads via the bloodstream with primary viremia becoming evident at 6 d postinfection.[19,24] Target organs are the kidney, liver, spleen, heart, and alimentary canal. Clinical disease is evident 7 d after exposure; SVCV is then shed by intestinal mucus casts and feces.

6. SIGNIFICANCE

Spring viremia of carp is an extremely important disease of cultured carp in Europe.[6] It appears that the disease is most prevalent in yearling fish with up to 70% mortality occurring in infected populations. Adult fish are also susceptible to the disease, but to a lesser degree. The overall impact of SVC has been somewhat offset by the fact that in years following epizootics in a given year class of carp, mortalities are greatly reduced, probably due to an increased immunity to the disease among survivors.

7. MANAGEMENT

SVC is best managed and controlled by quarantining infected fish from healthy populations and by vaccination when practical. Chemotherapeutics have no known benefit. Fish that have been exposed to SVCV produce antibodies to the virus, and they are resistant to the disease in differing degrees during subsequent exposures. Carp were first experimentally vaccinated for SVCV by Fijan et al.,[25] however, the immune response was only measured by serum neutralization. Fijan et al.[26] further demonstrated that a strong protective immunity was obtained in carp for 9 months after vaccination by a intraperitoneal injection which showed promise as a vaccination procedure. Protective immunity against SVC is temperature-dependent in that fish vaccinated by intraperitoneal injection develop protective immunity at 20 to 22°C, but not at 10°C. Successful mass vaccinations of carp by injection have been reported, but vaccination must be done when water temperature is 18°C or above and the practicality of such a procedure remains problematic.

B. FISH POX (HERPESVIRUS OF CARP)

Fish pox, or carp pox, is one of the oldest known diseases of fish, recorded as early as 1563.[27] It takes the form of a benign hyperplastic, epidermal papilloma which occurs on common carp *(Cyprinus carpio).* Based on electron microscopy, Schubert[28] proposed that the disease was caused by a virus, and Sonstegard and Sonstegard[29] further proposed that the disease, also known as epithelioma papillosum, was caused by a herpesvirus. Sano et al.[30-31] confirmed the previous findings when they isolated the virus from epithelioma growths of Japanese ornamental (Agasi) carp. Because the disease is not a true "pox", they proposed that the disease be called papillosum cyprini which is caused by *Herpesvirus cyprini,* also referred to as carp herpesvirus (CHV).

1. GEOGRAPHICAL RANGE AND SPECIES SUSCEPTIBILITY
Fish pox has been reported from most European countries, Japan, Russia, Israel, Korea, Malaysia, and the U.S. The common carp and Agasi carp (i.e., koi, fancy carp) are primary hosts. Other cyprinids, such as the crucian carp *(Carassius auratus),* willow shiner *(Gnathopogon elongatus),* and grass carp were not susceptible.[31-32] The disease was also reported for the first time in North America in the golden ide *(Leuiscus idus)* which had been imported from Europe 1 year prior to the appearance of the disease.[33] Hedrick et al.[34] also discovered carp pox in koi carp in California.

2. CLINICAL SIGNS AND FINDINGS
H. cyprini produces benign, hyperplastic, papillomatous growths in the epithelium of carp.[31] The tumors are milky-white to gray and are raised about 1 to 3 mm above the skin on the head, fins, or anywhere on the body surface (Figure 3). The lesions are generally smooth and begin as small raised growths which may increase in size and cover large areas. The lesions eventually regress and disappear. Infected fish show no distinct behavioral signs and there is no apparent morbidity among infected adult fish.

It has become apparent that juvenile carp are adversely affected by *H. cyprini* with high mortality.[32] Two-week-old common carp fry exhibited a loss of appetite and swimming in rigid lines with intermittent immobility. Infected fish had a distended abdomen, exophthalmia, darkened pigmentation, and hemorrhages on the operculum and abdomen.

3. DIAGNOSIS AND VIRUS CHARACTERISTICS
Prior to isolation of the virus, the only method of identifying fish pox was by histopathology and electron microscopy.[3] However, in the presence of typical clinical signs indicative of viremia in juvenile carp and tumors on older fish, isolation of virus should be attempted in FHM or EPC cells.[31] Virus titers in fish tissues range from $10^{5.8}$TCID$_{50}$/ml in FHM cells to $10^{2.5}$TCID$_{50}$/ml in the EPC cells. Optimum incubation temperature is 20°C; replication also occurs at 15°C, but does not occur at 10 or 25°C. CHV-infected FHM cells developed vacuolation, with rounding and slight pyknosis in focal areas 5 d after inoculation. Cellular disquamation occurs slowly, but expands to 60 to 70% of the cell sheet with occasional cellular recovery. CPE is less remarkable in the EPC cells. Syncytia formation, characteristics of cell lines infected with most herpesviruses, does not occur. However, intranuclear, Cowdry type A, inclusion bodies occur in infected FHM cells (Figure 4). Presently, identification of *H. cyprini* is made by serum neutralization, however, Sano et al.[35] were able to detect CHV in infected carp fry by *in situ* hybridization with biotinylated probes. Sano et al.[31] detected *H. cyprini* antigen in neoplastic tissues at all stages of tumor development, especially in older and recurrent tumors.

Herpesvirus cyprini is an icosahedral DNA virus with a 110- to 113-nm diameter capsid (Figure 4).[31] With its envelope, it measures 190 nm. *H. cyprini* is sensitive to ether, pH 3, 50°C for 1 h, and IUdR. Basically only one strain of CHV is recognized, but Sano et al.[36] noted slight differences in the viral genome of two different isolates by restriction endonuclease cleavage profiles.

4. EPIZOOTIOLOGY
The epizootiology of *H. cyprini* virus has been poorly understood until recently. Although enzootic in eastern and western Europe, indications are that international movements of carp have led to greater

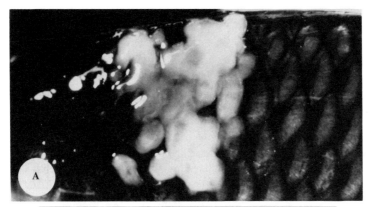

Figure 3 Herpesvirus of carp (fish pox). (A) Massive papilloma on naturally infected common carp; (B) papilloma on fingerling carp that survived experimental infection by immersion. (From Sano, T., Morita, N., Shima, N., and Akimoto, M., *J. Fish Dis.*, 14, 533, 1991. With permission of Blackwell Scientific Publications LTD.)

geographical range of the disease (e.g., the occurrence of the disease in golden ide that had been imported into the U.S. from Germany and its detection in koi carp in California).

Most data concerning carp pox have been generated from captive fish populations. The incidence of carp pox in wild populations is not known, however, it is assumed to be low. It has been suggested that young fish are infected from their parents, but whether this occurs vertically and/or horizontally is unknown. However, release of virus from ruptured cells that have sloughed from the skin of infected fish is probably instrumental in transmission. Evidence indicates that the disease can be transmitted by cohabitation or by injection of media from infected cell cultures. At 15°C, the incubation period between exposure and tumor development is from 5 months to a year.[32-33] Sano et al.[30] found that the virus can be transmitted to 30-day-old carp in which mortality occurred, while survivors of the infection developed papillomas 6 months later.

The foregoing epizootiological confusion of carp pox was clarified by Sano et al.[32] when they infected carp fry as young as 2 weeks of age by immersion at 20°C and produced a clinical, viremic disease. From 86 to 97% of these fish died because of the virus as compared to 3% mortality in the controls. *H. cyprini* was isolated from these fish for 3 weeks following exposure. In 4- and 8-week-old carp, mortality was 20 and 0%, respectively. Neoplasia developed in 55% of the surviving common carp 6 months after exposure to *H. cyprini*, and virus was reisolated from these tumors. Reoccurrence of a second set of neoplasms developed in 83% of the fish 7.5 months after the first tumors had disquamated, and again CHV was isolated from all necropsied moribund fish and all survivors. The virus induced papillomas in 13% of adult mirror carp and 10% of adult fancy carp 5 months after intraperitoneal injection.

CHV-induced papillomas occur at an optimum temperature of 15°C, however, regression of these lesions occurs when water temperature is raised to 20°C with the process accelerating as the temperature rises to 30°C.[37] The mechanism of regression was not known until Morito and Sano[38] showed that peripheral blood lymphocytes (PBL) of carp contribute to this regression. When carp with CHV papillomas were injected with anti-carp PBL serum, the regression was delayed.

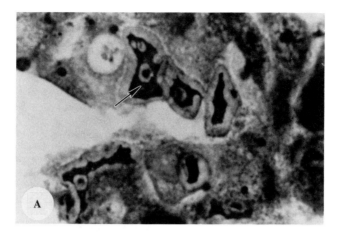

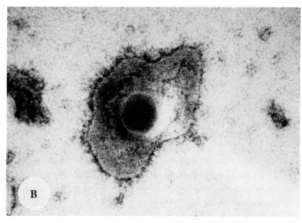

Figure 4   (A) Fathead minnow cells infected with herpesvirus of carp showing Cowdry Type A inclusion bodies (arrow); (B) *Herpesvirus cyprini* with electron dense nucleocapsid and envelope. (Photos by T. Sano.)

5. PATHOLOGICAL MANIFESTATIONS

Sano et al.[30-31] and McAllister et al.[33] described the histopathology of tumors in fish pox. Scaled areas of skin show hyperplastic epidermal cells. Similar hyperplastic epidermal lesions develop on the fins, but there are no mucus or alarm substance cells in the epidermis. The compact layer of the dermis is thickened, and eosinophils and lymphocytes in the loose connective tissue or the hyperdermis suggest an inflammatory response. The oncogenic nature of *H. cyprini* in carp was demonstrated by Sano et al.[30] when they experimentally produced tumors on the skin and fins of fish. Experimentally infected juvenile common carp that developed a viremia showed necrosis in liver parenchyma, kidneys, and the lamina propria of the intestinal mucosa. Moribund fish show severely necrotic hepatitis, epithelial disquamation and edema in the gills, and mild renal necrosis.

6. SIGNIFICANCE

In the tumor state, fish pox is regarded as benign and has not been associated with mortality. The lesions slough from the skin with little apparent harm to the host. However, from an aesthetic point of view, infected fish are unsightly and undesirable for food or display, particularly "fancy carp" in Japan. When very young carp are infected with the virus, high mortalities may occur and survivors of such outbreaks may develop tumors within a year of infection, although the growths usually regress to an inapparent condition.

Table 2 **Aquareoviruses and species of fish or shellfish from which they were isolated[40,48]**

Virus	Host species	Location
Golden shiner virus	Golden shiner (*Notemigonus chrysoleucas*)	U.S.
13$_p$2 Reovirus	American oyster (*Crassostrea virginica*)	U.S.
Chum salmon virus	Chum salmon (*Oncorhynchus keta*)	Japan
Catfish reovirus	Channel catfish (*Ictalurus punctatus*)	U.S.
Grass carp reovirus	Grass carp (*Ctenopharyngdon idella*)	China
Tench reovirus	Tench (*Tinca tinca*)	Europe
Chub reovirus	Chub (*Leusciscus cephalus*)	Europe
Turbot reovirus	Turbot (*Pleuronectus maxima*)	Spain
Coho salmon reovirus	Coho salmon (*Oncorhynchus kisutch*)	U.S.

7. MANAGEMENT

There is no control for *H. cyprini* and the only management practice is avoidance. Carp shipped from infected areas into regions where the disease has not been reported should be carefully selected from stocks historically free of the disease, because it can probably be transferred from one geographical area to another in covertly infected fish. Since tumor development follows a long incubation, it is advisable to quarantine newly transported carp for 1 year before releasing them into communal waters with other fish.

C. GOLDEN SHINER VIRUS

An increasing number of viruses that are members of the family Reoviridae have been isolated from fish (Table 2). Recently, the *Aquareovirus* genus has been formed to include those reoviruses isolated from fish and shellfish.[39] The most notable of these viruses that infect cultured fish is the golden shiner virus (GSV), the type species of the new genus. Generally aquareoviruses exhibit low or no known pathogenicity for fish, with the exception of GSV which has the capacity to cause significant mortality under certain circumstances. GSV was isolated during the mid 1970s from cultured golden shiners (*Notemegonus chrysoleucas*) in which it produces a mild viremia.[40]

1. GEOGRAPHICAL RANGE AND SPECIES SUSCEPTIBILITY

GSV appears to be confined to the U.S. where golden shiners are grown commercially in the southeastern region[40] and in California.[41] The virus has also been isolated from grass carp, but other species are apparently refractive.

2. CLINICAL SIGNS AND FINDINGS

Shiners infected with GSV may swim lethargically at the surface, diving when disturbed, or they may hold their fins close to the body and lie on the bottom of aquaria or tanks.[40] The back and head of infected fish are predominantly red (erythemia) as a result of hemorrhage in the skin and dorsal musculature. Internally, petechiae are present in visceral fat and intestinal mucosa. Occasionally, open ulcerative lesions will develop in the skin, however, these lesions may be due to secondary bacterial infections.

3. DIAGNOSIS AND VIRUS CHARACTERISTICS

GSV can most readily be isolated from the kidney, liver, and spleen by inoculation of filtrates from these organs into FHM cells. Virus replication in FHM cells will occur at temperatures from 20 to 30°C, with an optimum being 28 to 30°C. No replication occurs at 15 or 35°C.[42] Some virus replication with CPE will also occur in Chinook salmon embryo cells (CHSE-214) and BB cells, but the virus yield is less than in FHM.

Within approximately 24 h postinoculation, focal CPE is apparent in FHM monolayers that are incubated at 30°C. By 48 h, CPE will expand and envelop the entire cell sheet (Figure 5). Infected cells become large, rounded, and vacuolated structures which detach from the surface and float in the

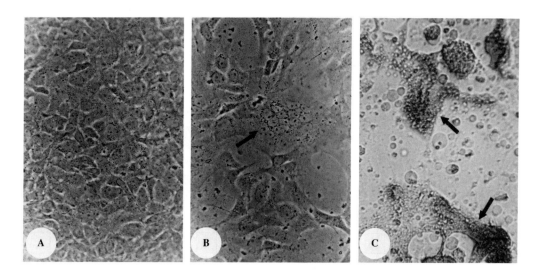

Figure 5   Golden shiner virus CPE in fathead minnow cells. (A) Noninfected control cells; (B) focal infection with multinucleated cells; (C) massive CPE with rounded, vaculated cells that have released from the substrate. (From Plumb, J. A., Bowser, P. R., Grizzle, J. M., and Mitchell, A. J., *J. Fish. Res. Board Can.*, 36, 1390, 1979. With permission.)

media as spherical, highly vacuolated debris. If a very low level of virus (less than 10^1 $TCID_{50}$/ml) is inoculated onto a cell sheet, the initial foci may be overgrown by apparently healthy cells leaving no evidence of infection other than a few large spherical cells floating in the media. This phenomenon could be induced by interferon production by the FHM cells.

When GSV was first isolated, it was suspected that the new virus was a strain of infectious pancreatic necrosis virus (IPNV), but using serology this was shown not to be the case.[43] The aquareoviruses, of which GSV is the type virus, are icosahedral particles with a double capsid and double-stranded RNA genome. The virus measures 70 to 75 nm in diameter.[40] Several reports have serologically compared GSV, catfish reovirus (CRV), chum salmon reovirus (CSV), and $13p_2$ reovirus.[44-45] These indicate a serological relationship between them, but additional research is required before this can be accurately determined.

4. EPIZOOTIOLOGY

Although not a major disease of fish, under certain circumstances GSV can cause significant losses. Normally the virus causes a chronic mortality of less than 5%, however, if the fish are stressed by overcrowding in holding tanks, mortality may increase to 75%. Crowding will increase the number of GSV-positive fish from very low to as high as 50% (Figure 6).[46] Intermediate size (7- to 10-cm) fish appear to be more severely affected than small fish.

All documented cases of GSV disease have occurred during summer when water temperatures exceeded 25°C. Limited GSV transmission studies indicate that the virus is not easily transmitted by cohabitation, but can be transmitted by injection. Possible vertical transmission has not been investigated.

5. PATHOLOGICAL MANIFESTATIONS

In addition to the petechiae in the skin and occasionally in internal organs, the most notable histopathology of GSV disease is invasion of the liver by lymphocytes which contain pyknotic nuclei.[47]

6. SIGNIFICANCE

The disease has little impact on the aquaculture industry based on its modest pathogenicity and low frequency of occurrence. It usually causes low mortality in culture ponds, but may contribute to high losses when golden shiners are held in tanks. This is especially true if a secondary infection of columnaris is present.

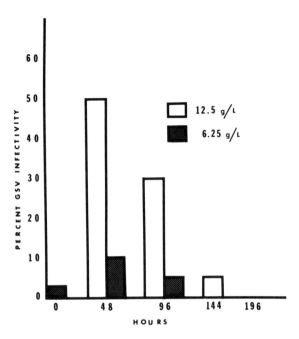

Figure 6   Effect of crowding on golden shiners infected with golden shiner virus and stocked in troughs at two different densities. (From Schwedler and Plumb.[46])

7. MANAGEMENT

Proper management of GSV centers around the elimination of overcrowding, in conjunction with treatment of external bacterial infections.

REFERENCES

1. **Fijan, N., Petrinec, Z., Sulimanović, D., and Zwillenberg, L. O.,** Isolation of the viral causative agent from the acute form of infectious dropsy of carp, *Vet. Arh.,* 41, 125, 1971.
2. **Fijan, N.,** Infectious dropsy in carp — a disease complex, *Symp. Zool. Soc. London,* 30, 39, 1972.
3. **Fijan, N. and Petrinec, Z.,** Mortality in a pond caused by carp erythrodermatitis, *Riv. Ital. Pisci. Ittiopath.,* 8(2), 45, 1973.
4. **Bootsma, R., Fijan, N., and Blommaert, J.,** Isolation and preliminary identification of the causative agent of carp erythrodermatitis, *Vet. Arh.,* 47, 291, 1977.
5. **Pol, J. M. A., Bootsma, R., and Berg-Blommaert, J. M.,** Pathogenesis of carp erythrodermatitis (CE): role of bacterial endo- and exotoxin, in *Fish Diseases,* Ahne, W., Ed., Springer-Verlag, Berlin, 1980, 120.
6. **Bucke, D. and Finlay, J.,** Identification of spring viraemia of carp (*Cyprinus carpio* L.) in Great Britain, *Vet. Rec.,* 104, 69, 1979.
7. **Wolf, K.,** *Fish Viruses and Fish Viral Diseases,* Cornell University Press, Ithaca, NY, 1988, 476.
8. **Shchelkunov, I. S. and Shchelkunova, T. I.,** *Rhabdovirus carpio* in herbivorous fishes: isolation, pathology, and comparative susceptibility of fishes, in *Viruses of Lower Vertebrates,* Ahne, W. and Kurstake, E., Eds., Springer-Verlag, Berlin, 1989, 333.
9. **Fijan, N., Matasin, Z., Jeney, Z., Olah, J., and Zwillenberg, L. O.,** Isolation of *Rhabdiovirus carpio* from sheatfish *(Silurus glanis), Symp. Biol. Hung.,* 23, 17, 1984.
10. **Ahne, W.,** Viral infection cycles in pike (*Esox lucius* L.), *J. Appl. Ichthyol.,* 1, 90, 1985.

11. **de Kinkelin, P. and Le Berre, M.,** Rhabdovirus des poissons. II. Proprietes in vitro du virus de la viremie printaniere de la carpe, *Ann. Microbiol. (Paris),* 125A, 113, 1974.

12. **Faisal, M. and Ahne, W.,** Spring viraemia of carp virus (SVCV): comparison of immunoperoxidase, fluorescent antibody and cell culture isolation techniques for detection of antigen, *J. Fish Dis.,* 7, 57, 1984.

13. **Way, K.,** Rapid detection of SVC virus antigen in infected cell cultures and clinically diseased carp by enzyme-linked immunosorbent assay (ELISA), *J. Appl. Ichthyol.,* 7, 95, 1991.

14. **Ahne, W.,** Untersuchungen uber die stabilitat des karpfen pathogenen virusstammes, *Fish. Umwelt.,* 2, 121, 1976.

15. **Bachmann, P. A. and Ahne, W.,** Isolation and characterization of agent causing swim bladder inflammation in carp, *Nature,* 244, 235, 1973.

16. **Bachmann, P. A. and Ahne, W.,** Biological properties and identification of the agent causing swim bladder inflammation in carp, *Arch. Gesamte Virusforsch.,* 44, 261, 1974.

17. **Casaba, G., Kovacs-Gayer, E., Békési, L., Bucsek, M., Szakolzai, J., and Molnar, K.,** Studies into possible protozoan aetiology of swim bladder inflammation in carp fry, *J. Fish Dis.,* 7, 39, 1984.

18. **Hill, B. J., Underwood, B. O., Smale, C. J., and Brown, F.,** Physioco-chemical and serological characterization of five rhabdoviruses infecting fish, *J. Gen. Virol.,* 27, 369, 1975.

19. **Baudouy, A. M., Danton, M., and Merle, G.,** Viremie printaniere de la carpe: etude experimentale de l'infection evoluant a differentes temperatures, *Ann. Virol. (Institute Pasteur),* 131E, 479, 1980.

20. **Ahne, W.,** *Argulus foliaceus* L. and *Philometra geometra* L. as mechanical vectors of spring viremia of carp virus (SVCV), *J. Fish Dis.,* 8, 241, 1985.

21. **Varovic, K. and Fijan, N.,** Osjetljivost sarana prema rhabdovirus carpio pri raznim nacinima inokulacije, *Vet. Arh.,* 43, 271, 1973 (English summary).

22. **Bekesi, L. and Csontos, L.,** Isolation of spring viraemia of carp virus from asymptomatic broodstock carp, *Cyprinus carpio* L., *J. Fish Dis.,* 8, 471, 1985.

23. **Negele, R. D.,** Histopathological changes in some organs of experimentally infected carp fingerlings with *Rhabdovirus carpio, Bull. Off. Int. Epizool.,* 87, 449, 1977.

24. **Ahne, W.,** Uptake and multiplication of spring viremia of carp virus in carp, *Cyprinus carpio* L., *J. Fish Dis.,* 1, 265, 1978.

25. **Fijan, N., Petrinec, Z., Stancl, Z., Dorson, M., and Le Berre, M.,** Hyperimmunization of carp with *Rhabdovirus carpio, Bull. Off. Int. Epizool.,* 87, 439, 1977.

26. **Fijan, N., Petrinec, Z., Stancl, Z., Kezic, N., and Teskeredzie, E.,** Vaccination of carp against spring viremia: comparison of intraperitoneal and peroral application of live virus to fish kept in ponds, *Bull. Off. Int. Epizool.,* 87, 441, 1977.

27. **Hedrick, R. P. and Sano, T.,** Herpesviruses of fishes, in *Viruses of Lower Vertebrates,* Ahne, W. and Kurstak, E., Eds., Springer-Verlag, Berlin, 1989, 161.

28. **Schubert, G.,** The infective agent in carp pox, *Bull. Off. Int. Epizool.,* 65, 1011, 1966.

29. **Sonstegard, R. A. and Sonstegard, K. S.,** Herpesvirus-associated epidermal hyperplasia in fish (carp), in *Proc. Int. Symp. Oncol. Herpesvirus,* de The, G., Henle, W., and Rapp, F., Eds., Int. Agency Res. Cancer Sci., Lyon, France, Publ. No. 24, 1978, 863.

30. **Sano, T., Fukuda, H., and Furukawa, M.,** *Herpesvirus cyprini:* biological and oncogenic properties, *Fish Pathol.,* 20, 381, 1985.

31. **Sano, T., Fukuda, H., Furukawa, M., Hosoya, H., and Moriya, Y.,** A herpesvirus isolated from carp papilloma in Japan, in *Fish and Shellfish Pathology,* Ellis, A. E., Ed., Academic Press, Orlando, 1985, 307.

32. **Sano, T., Morita, N., Shima, N., and Akimoto, M.,** *Herpesvirus cyprini:* lethality and oncogenicity, *J. Fish Dis.,* 14, 533, 1991.

33. **McAllister, P. E., Lidgerding, B. C., Herman, R. L., Hoyer, L. C., and Hankins, J.,** Viral disease of fish: first report of carp pox in golden ide *(Leuciscus idus)* in North America, *J. Wildl. Dis.,* 21, 199, 1985.

34. **Hedrick, R. P., Groff, J. M., Okihiro, M. S., and McDowell, T. S.,** Herpesviruses detected in papillomatous skin growths of koi carp *(Cyprinus carpio), J. Wildl. Dis.,* 26, 578, 1990.

35. **Sano, N., Sano, M., Sano, T., and Hondo, R.,** *Herpesvirus cyprini:* detection of the viral genome *in situ* hybridization, *J. Fish Dis.,* 15, 153, 1992.

36. **Sano, N., Honda, R., Fukuda, H., and Sano, T.,** *Herpesvirus cyprini:* restriction endonuclease cleavage profiles of the viral DNA, *Fish Pathol.,* 26, 207, 1991.

37. **Sano, T., Moriwake, M., Sano., N., Hasobe, M., and Fukuda, H.,** Thermal effect on pathogenicity and oncogenicity of cyprinid herpesvirus (CHV), in *Diseases of Fish and Shellfish,* European Association of Fish Pathologists, 1991, 91.

38. **Morita, N. and Sano, T.,** Regression effect of carp, *Cyprinus carpio* L., peripheral blood lymphocytes on CHV-induced carp papilloma, *J. Fish Dis.,* 13, 505, 1990.

39. **Holmes, I. H.,** Family Reoviridae, in Classification and Nomenclature of Viruses, Fraucki, R. I. B., Fauquet, C. M., Knudson, D. L., and Brown, F., Eds., *Arch. Virol.,* Suppl. 2, 186, 1991.

40. **Plumb, J. A., Bowser, P. R., Grizzle, J. M., and Mitchell, A. J.,** Fish viruses: a double-stranded RNA icosahedral virus from a North American cyprinid, *J. Fish. Res. Board Can.,* 36, 1390, 1979.

41. **Hedrick, R., Groff, J. M., McDowell, T., and Wingfield, W. H.,** Characteristics of reoviruses isolated from cyprinid fishes in California, USA, in *Viruses of Lower Vertebrates,* Ahne, W. and Kurstak, E., Eds., Springer-Verlag, Berlin, 1988, 241.

42. **Brady, Y. J. and Plumb, J. A.,** Replication of four aquatic reoviruses in experimentally infected golden shiners, *(Notemigonus crysoleucas), J. Wildl. Dis.,* 27, 463, 1991.

43. **Schwedler, T. E. and Plumb, J. A.,** Fish viruses: Serologic comparison of the golden shiner and infectious pancreatic necrosis viruses, *J. Wildl. Dis.,* 16, 597, 1980.

44. **Winton, J. R., Lannan, C. N., Fryer, J. L., Hedrick, R. P., Meyers, T. R., Plumb, J. A., and Yamamoto, T.,** Morphological and biochemical properties of four members of a novel group of reoviruses isolated from aquatic animals, *J. Gen. Virol.,* 68, 353, 1987.

45. **Brady, Y. L. and Plumb, J. A.,** Serological comparison of golden shiner virus, chum salmon virus, reovirus 13_p2 and catfish reovirus, *J. Fish Dis.,* 11, 441, 1988.

46. **Schwedler, T. E. and Plumb, J. A.,** Golden shiner virus: effects of stocking density on incidence of viral infection, *Prog. Fish Cult.,* 44, 151, 1982.

47. **Brady, Y. J.,** Comparative Serological Responses and Histopathology of Golden Shiner Virus, Chum Salmon Virus, Reovirus 13_p2, and Channel Catfish Reovirus Infection in Golden Shiners (*Notemigonus crysoleucas* Mitchell), PhD. dissertation, Auburn University, AL, 1985.

48. **Ahne, W. and Kolbl, O.,** Occurrence of reoviruses in European cyprinid fishes, *J. Appl. Ichthyol.,* 3, 139, 1991.

Chapter VI

Eels (Anguillidae)

Several viruses have been reported in European eel *(Anguilla anguilla)*, American eel *(A. rostrata)*, and Japanese eel *(A. japonica)*. Viruses isolated from eels have come mostly from Europe and Japan with sporadic occurrences in North America and elsewhere. The impact of these viruses on cultured eel populations is difficult to assess because many have been isolated from apparently normal fish. The numerous viruses reported from eels,[1] remain poorly classified, therefore, only four of these viruses will be discussed: eel virus European (EVE), eel iridovirus (EV-102), eel herpesvirus, and eel rhabdovirus.

A. EEL VIRUS EUROPEAN

Eel virus European (EVE) received its name because it was isolated from European eels being cultured in Japan. EVE has been isolated from eels and tilapia *(Tilapia mossambica)* in Taiwan and from eels in Japan,[2-3] and a virus similar to EVE, but characterized as IPNV strain Ab, was isolated from young normal eels in England.[4] In 1988, IPNV (Ab serotype) was isolated by the author from moribund cultured American eels in Florida (unpublished data). The serotype was confirmed by Hill[5] by plaque reduction, but whether or not the virus was adversely affecting these eels was not clear because they also had a concomitant *Aeromonas hydrophila* infection. Cultured young Japanese eels appear to be most susceptible to the EVE agent, but it is generally regarded as having low virulence for young rainbow trout.

EVE may be isolated from internal organs of eels. Isolation is best accomplished using BF-2, CHSE-214, or RTG-2 cells incubated at 10 to 20°C, but cell lines from eels should be equally satisfactory.[6]

The EVE agent has biophysical characteristics of IPNV-Ab and is closely related serologically.[7] Okamoto et al.[8] compared ten strains of IPNV to EVE and found the eel virus to be more closely related to the IPNV-Ab serotype than to any other viral strain. The similarity of EVE to other aquatic birnaviruses was demonstrated by Chi et al.[9] using monoclonal antibodies (MAB), that had been isolated from Japanese eels with branchionephritis, against EVE. Three of six MABs identified epitopes that were among members of the IPN-Ab virus only; one recognized an epitope on Ab and Sp serotypes; and two MABs exhibited epitopes on Ab, Sp, and VR-299 serotypes of IPNV.

Experimentally, EVE-infected eels become rigid with muscle spasms and the anal fins are congested.[10-11] The underside of the fish reveal slight petechiation. Ascitic fluid may be present in the abdominal cavity and, at times, the gills become hyperplastic with fusion of the lamellae giving them a swollen appearance.

EVE has a propensity for producing disease at temperatures from 8 to 14°C, especially in small eels.[3] Mortalities of EVE in 4- to 26-g eels may be as high as 60%, while Sano et al.[10] reported 55 and 75% mortality of experimentally infected 9- to 12-g Japanese eels. Ueno et al.[12] determined that an EVE isolate from Japanese eels was not infectious to common carp at 20 to 25°C. However, tilapia were most severely affected at 10 to 16°C which is not surprising because these low temperatures approach the stressful level for most tilapia, and they become susceptible to a variety of pathogens.

EVE can be transmitted by waterborne exposure and fish to fish transmission is very likely. Wolf[13] postulated that the virus was transmitted from Europe to Japan in elvers, and then to Taiwan in eyed trout eggs.

Japanese eels are affected by a gill disease commonly called brancionephritis, from which EVE has been isolated, however, the virus apparently has not caused the disease in experimentally infected fish.[12] The relationship of EVE to branchionephritis disease of Japanese eels is still inconclusive, although there is a very strong causal relationship.

The significance of EVE for cultured eels is not clear, but large mortalities of eels in Japan and Taiwan have been associated with the presence of the virus. Control of EVE can only be accomplished through avoidance.

Table 1   **Rhabdoviruses isolated from eels (*Anguilla*)**

Rhabdovirus group[14]	Isolate designation	Species of eel	Geographical origin
Lysovirus	B_{12}	*A. anguilla*	Europe[18]
	C_{26}	*A. anguilla*	Europe[14]
Vesiculovirus	Eel virus America (EVA)	*A. rostrata*[a]	Japan[3]
	Eel virus European X (EVEX)	*A. anguilla*[b]	Japan[11]
	C_{30}	*A. anguilla*	Europe[18]
	B_{44}	*A. anguilla*	Europe[14]
	D_{13}	*A. anguilla*	Europe[14]

[a]From elvers from Cuba. [b]From elvers from Europe.

B. EEL RHABDOVIRUSES

At least seven different isolates of rhabdoviruses of eels (*Anguilla* spp.) are designated: EVA, EVEX, B_{12}, B_{44}, C_{26}, C_{30}, and D_{13} (Table 1).[13] These viruses, several of which came from apparently healthy eels, were analyzed by Castric et al.[14] who reported that two of the viruses belonged to the *Lyssavirus* genus (B_{12} and C_{26}) and five were members of the *Vesiculovirus* genus (EVA, EVEX, C_{30}, B_{44}, and D_{13}).

For isolation of the eel rhabdoviruses, filtrates from tissue homogenates of brain, kidney, liver, and spleen should be inoculated on BF-2, eel kidney (EK-1), eel ovary (EO-2), FHM, or RTG-2 cells and incubated at 15 and 20°C depending upon the cell line used.[13] CPE consists of pyknosis, cytoplasmic granulation, and then cell lysis.

Since most of the eel rhabdoviruses have been isolated from normal, apparently healthy elvers, there are few clinical signs of infection. However, Sano[3] reported that naturally EVA-infected American eel elvers imported from Cuba, displayed hemorrhages and congestion on the underside of the body and on the anal and pectoral fins. The fish also turned their heads downward, and the gills showed hemorrhage and degradation of bone or cartilage.

Of all the viruses from eels, the rhabdoviruses may be of greatest interest to fish virologists. These viruses seem to be very prevalent in some established wild populations of European eels, but do not appear to cause any significant problem among cultured populations. As culture systems intensify, the role of these viruses in disease production could change.

C. EEL HERPESVIRUS

A previously unidentified herpesvirus isolated from Japanese eel and American eel in Japan was reported by Sano et al.[15] They designated the virus Anguillid herpesvirus 1 with *Herpesvirus anguillidae* (HVA) as the latinized name, but the disease is commonly called eel herpesvirus. HVA clearly conforms to the herpesvirus criteria while being serologically distinct from CCV and NeVTA. Although HVA produced CPE in several noneel cell lines from fish, only cells derived from eels (i.e., EK-1 and EO-2) replicated the agent. Optimum incubation for HVA in cell culture was 20 to 25°C, but replication occurred at 10 to 37°C.

HVA was initially isolated from Japanese eels with a 1% mortality rate at 1 facility and European eels with a 6.8% mortality rate at a 2nd facility in Japan.[15] Affected eels had varying degrees of erythema on the skin and gills. Irregular sized eosinophilic granulations appeared in epidermal tissue which also became necrotic with partial desquamation. In the dermis, blood capillaries increased but melanophores contracted. Branchial lamellae of the gills became fused, followed by mild necrosis. Necrotic hepatic cells were scattered throughout the liver.

The significance of eel herpesvirus is unknown. To date, it has been found in limited eel populations in Japan, but by electron microscopy, Békési et al.[16] identified a probable herpesvirus in skin lesions of European eels. The relationship, if any, of the Japanese isolate to the agent found in Europe is unknown.

D. EEL IRIDOVIRUS

Japanese eel iridovirus (EV-102) was isolated from cultured eels in Japan during a routine survey of eel farms.[3] After experimentally infecting small eels (up to 120 g) with EV-102 by immersion and injection at 15 to 23°C, clinical signs consisting of depigmentation of the skin, congestion of the fins, and increased mucus production appeared in 3 to 5 d.[18] Mortalities began shortly after the onset of clinical signs and continued for 2 weeks. Deaths stopped and clinical signs abated when the water temperature reached 23°C, but up to 70% mortality had occurred. EV-102 is pathogenic to Japanese eel, but not to European eel. EV-102 replicates in EPC and RTG-2 cells as well as EK-1 and EO-2 cells incubated at 15 to 20°C.

E. SIGNIFICANCE OF EEL VIRUSES

The frequency with which viruses have been isolated from symptomatic or asymptomatic cultured eels enhances their importance to aquaculture. The full impact of these viruses on eel culture is difficult to assess because some isolates are nonpathogenic, while others have been associated with extremely high mortalities but not necessarily proven to be the etiological agent. However, the potential for at least sporadic, significant eel die-offs in culture facilities due to viruses exists.

REFERENCES

1. **Ahne, W., Schwanz-Pfitzner, I., and Thomsen, I.,** Serological identification of 9 viral isolates from European eels *(Anguilla anguilla)* with stomatopapilloma by means of neutralization tests, *J. Appl. Ichthyol.,* 3, 30, 1987.
2. **Chen, S. N., Kou, G. H., Hedrick, R. P., and Fryer, J. L.,** The occurrence of viral infections of fish in Taiwan, in *Fish and Shellfish Pathology,* Ellis, T., Ed., Academic Press, London, 1985, 313.
3. **Sano, T.,** Viral diseases of cultured fishes in Japan, *Fish Pathol.,* 10, 221, 1976.
4. **Hudson, E. B., Bucke, D., and Forrest, A.,** Isolation of infectious pancreatic necrosis virus from eels, *Anguilla anguilla* in the United Kingdom, *J. Fish Dis.,* 4, 429, 1981.
5. **Hill, B. J.,** Fish Disease Laboratory, The Nothe Weymouth, England, personal communication, 1988.
6. **Chen, S. N. and Kou, G. H.,** A cell line derived from Japanese eel *(Anguilla japonica)* ovary, *Fish Pathol.,* 16, 129, 1981.
7. **Hedrick, R. P., Fryer, J. L., Chen, S. N., and Kou, G. H.,** Characteristics of four birnaviruses isolated from fish in Taiwan, *Fish Pathol.,* 18, 91, 1983.
8. **Okomoto, N., Sano, T., Hedrick R. P., and Fryer, J. L.,** Antigenic relationships of selected strains of infectious pancreatic necrosis and European eel virus, *J. Fish Dis.,* 6, 19, 1983.
9. **Chi, S. C., Chen, S. N., and Kou, G. H.,** Establishment, characterization and application of monoclonal antibodies against eel virus European (EVE), *Fish Pathol.,* 26, 1, 1991.
10. **Sano, T., Okamoto, N., and Nishimura, T.,** A new viral epizootic of *Anguilla japonica* Temminck and Schlegel, *J. Fish Dis.,* 4, 127, 1981.
11. **Sano, T., Nishimura, T., Okamoto, N., and Fukuda, H.,** Studies on viral diseases of Japanese fishes. VII. A rhabdovirus isolated from European eel *(Anguilla anguilla),* Bull. Jpn. Soc. Sci. Fish., 43, 491, 1977.
12. **Ueno, Y., Chen, S., Kou, G., Hedrick, R. P., and Fryer, J. L.,** Characterization of a virus isolated from Japanese eels *(Anguilla japonica)* with nephroblastoma, *Bull. Inst. Zool. Acad. Sin. (Taipei, Taiwan),* 23, 47, 1983.
13. **Wolf, K.,** *Fish Viruses and Fish Viral Diseases,* Cornell University Press, Ithaca, NY, 1988, 476.
14. **Castric, J., Rasschaert D., and Bernard, J.,** Evidence of Lyssaviruses among rhabdovirus isolates from the European eel *Anguilla anguilla,* Ann. Virol., 135E, 35, 1984.
15. **Sano, M., Fukuda, H., and Sano, T.,** Isolation and characterization of a new herpesvirus from eel, in *Pathology in Marine Science,* Perkins, F. O. and Cheng, C. T., Eds., Academic Press, Tokyo, 1990, 15.

16. **Békési, L. I., Horvath, I., Kovacs-Gayer, E., and Csaba, G.,** Demonstration of herpesvirus-like particles in skin lesions of the European eel *(Anguilla anguilla), J. Appl, Ichthyol.,* 4, 190, 1986.
17. **Sorimachi, M. and Egusa, S.,** Characteristics and distribution of viruses isolated from pond-cultured eels, *Bull. Nat. Res. Inst. Agua.,* 3, 97, 1982.
18. **Castric, J. and Chastel, C.,** Isolation and characterization attempts of three viruses from European eel, *Anguilla anguilla,* preliminary results, *Ann. Virol.,* 13E, 435, 1980.

Chapter VII

Pike (Esocidae)

A. PIKE FRY RHABDOVIRUS DISEASE

Pike fry rhabdovirus disease (PFRD) is an acute, highly contagious disease, commonly called "head disease" in cultured fry and "red disease" in cultured fingerling pike *(Esox lucius)* as described by Bootsma[1] in 1971 and de Kinkelin et al.[2] in 1973. The etiological agent, pike fry rhabdovirus (PFRV), was characterized by de Kinkelin et al.[2] and Bootsma.[3] It was noted that in The Netherlands, head disease had been previously reported in 1959 and red disease in 1956. Due to management practices initiated in The Netherlands, outbreaks of PFRD have diminished during the last two decades.[4]

1. GEOGRAPHICAL RANGE AND SPECIES SUSCEPTIBILITY

Pike fry rhabdovirus has been isolated from pike in hatcheries in The Netherlands. In Germany, a serologically identical rhabdovirus was isolated from grass carp *(Ctenopharyngodon idella)*, tench *(Tinca tinca)*, white bream *(Blicca bjoerna)*, and the gudgeon *(Pseudorasbora parva)*, an ornamental fish that had been accidentally introduced into Europe with grass carp.[5-7] These isolates have subsequently been shown to be PFRV. PFRD has also been reported to occur in Hungary.

2. CLINICAL SIGNS AND FINDINGS

Two forms of PFRD are known to occur, one of which is commonly known as head disease and affects swimming fry.[2] Infected fry show a raised cranium (Figure 1) that results from hydrocephalus, are exophthalmic, and exhibit poor growth. The second form of the disease occurs in larger fish where infected pike show severe hemorrhages along the lateral trunk musculature, thus the name red disease (Figure 1). The hemorrhagic areas are swollen, gills are very pale, and infected fish exhibit bilateral exophthalmia, abdominal distension, and ascites. In both forms of PFRD, the fish lose their schooling behavior and individuals either swim lethargically at the surface or lie listlessly on the bottom.

3. DIAGNOSIS AND VIRUS CHARACTERISTICS

PFR is detected by isolation of the virus in FHM or EPC cells.[2] The kidney, spleen, and intestine provide the best sources of tissue for virus isolation. Infected cells incubated at 20°C showed focal areas of rounded cells in 2 to 3 d. Identification can be obtained by serum neutralization, but Clerx et al.[8] demonstrated that PFRV neutralization by homologous rabbit antiserum is complement dependent. Also, other viruses, (e.g., spring viremia of carp — SVC) can be at least partially neutralized by PFRV-specific antiserum.[9] The partial cross-neutralization capability of the antiserum can be removed by reacting the serum with other rhabdoviruses of fish. PFRV is serologically distinct from viral hemorrhagic septicemia virus (VHSV) and infectious hematopoietic necrosis virus (IHNV), but some cross-reactivity occurs with SVCV (Chapter V, Table 1).[10-11]

The causative agent of PFRD is a rhabdovirus that measures 125 to 155 nm in length and 80 nm in diameter in negatively stained ultrathin sections.[3,12] It is enveloped and has a single-strand nonsegmented RNA genome.[10] PFRV is highly infectious to FHM cells where titers reach greater than 10^9 pfu/ml.[13] EPC, BB, and RTG-2 cells are also susceptible, but virus yield is lower than in FHM cells. PFRV replicates in FHM cells at 10 to 28°C. Infectivity persists in tissue culture medium at 4°C, but 70% of infectivity is lost in 3 days in water at 14°C.

4. EPIZOOTIOLOGY

Fry and small fingerling pike (up to 5 cm) are susceptible to PFRV with up to 100% mortality resulting.[1] The mechanism by which PFRV is transmitted naturally is not known. Bootsma et al.[14] were unable to isolate the virus from frozen visceral tissue (kidney, liver, spleen, intestine, and gonads) of brood pike, but progeny of these fish were infected. These investigators also experimentally infected pike eggs with PFRV that resulted in 100% mortality of the resulting fry within 3 weeks of hatching. However, definitive proof of vertical transmission is lacking. Horizontal transmission of PFRV has been experimentally demonstrated following waterborne exposure and injection of juvenile pike.[15] Fry (0.2 g) bathed in PFRV then fed to larger pike (60 g), did transmit the virus to recipient fish, but deaths

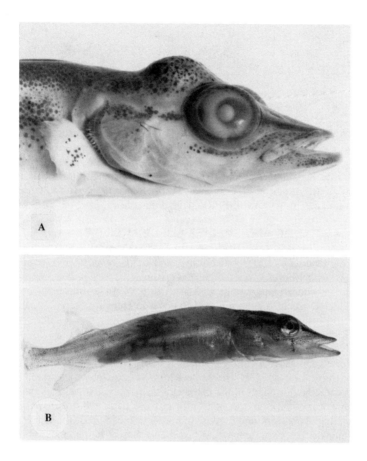

Figure 1 Pike fry rhabdovirus disease. (A) Pike with head disease showing hydrocephalus; (B) pike with red disease with hemorrhage in the lateral musculature. (Photos by R. Bootsma.)

or overt disease did not result. The disease occurs in the early spring when water temperatures are in the 10°C range.

PFR disease may be confused with other viruses of pike. Ahne[15] demonstrated that pike fry (0.2 g) could be infected by bathing them in infectious pancreatic necrosis virus (IPNV), SVCV, and VHSV that resulted in 42 to 90% mortality. Dorson et al.[16] further stated that pike fry from 10 d to 2 months old are susceptible to IPNV, IHNV, VHSV, and perch rhabdovirus which cause hemorrhaging, exophthalmia, and up to 74% mortality. Therefore, when a virus is isolated from pike fry, which appear susceptible to a variety of fish viruses, definitive identification should be completed. It was also postulated by Ahne[9] that pike could play the role of vector for other pathogenic viruses normally not found in pike.

5. PATHOLOGICAL MANIFESTATIONS

In head disease, there is an accumulation of cerebral fluid in the ventricle of the mesencephalon.[1] Petechia can be found in the brain, spinal cord, and spleen. The kidney tubules show degeneration with necrosis of the epithelial cells. Red disease is characterized by hemorrhages in connective tissue of the muscles and in the muscle fibers. Petechial hemorrhages occur throughout the spinal cord, pancreas, and hematopoietic tissue in the kidney. Severe hemorrhage in the muscle is the most dramatic histopathological lesion because erythrocytes and inflammatory cells congregate between bundles of necrotic muscle. The excretory tubules of the kidney are severely necrotic and the cerebral region becomes edematous. The liver, heart, and gastrointestinal tract appear normal in both forms of the disease.

6. SIGNIFICANCE

When PFRD occurs in small pike, it can be devastating. In 1972, only 0.6% of 1.85 million cultured young pike reached a length of 4 to 5 cm in The Netherlands. It was estimated that about 86% of the mortalities were due to PFRD, and in crowded conditions mortality may be even higher. PFRD has had a significant effect on pike fry culture in the past, but due to egg disinfection in iodine at spawning, the incidence of PFRV disease has been reduced.[4]

7. MANAGEMENT

Bootsma et al.[14] showed that 25 mg/l of iodine in the form of Wescodyne deactivated 99.99% of PFRV in 30 s. In view of these data, pike eggs in The Netherlands are routinely bathed in iodine when spawned, a practice that has successfully interrupted the infectious cycle of the virus.[4]

REFERENCES

1. **Bootsma, R.,** Hydrocephalus and red-disease in pike fry *Esox lucius* L., *J. Fish Biol.,* 3, 417, 1971.
2. **de Kinkelin, P., Galimard, B., and Bootsma, R.,** Isolation and identification of the causative agent of "red disease" of pike (*Esox lucius* L. 1766), *Nature,* 241, 465, 1973.
3. **Bootsma, R. and van Vorstenbosch, C. J. A. H. V.,** Detection of a bullet-shaped virus in kidney sections of pike fry (*Esox lucius* L.) with red-disease, *Neth. J. Vet. Sci.,* 98, 86, 1973.
4. **de Kinkelin, P.,** Laboratoire d' Ichtyopathologie, Thiverval-Grignon, France, personal communication, 1990.
5. **Ahne, W.,** A rhabdovirus isolated from grass carp (*Ctenopharyngodon idella* Val.), *Arch. Virol.,* 48, 181, 1975.
6. **Ahne, W., Mahnel, H., and Steinhagen, P.,** Isolation of pike fry rhabdovirus from tench, *Tinca tinca* L., and white bream, *Blicca bjoerkna* (L.), *J. Fish Dis.,* 5, 535, 1982.
7. **Ahne, W. and Thomsen, I.,** Isolation of pike fry rhabdovirus from *Pseudorasbora parva* (Temminck & Schlegel), *J. Fish Dis.,* 9, 555, 1986.
8. **Clerx, J. P. M., Horzinek, M. C., and Osterhaus, A. D. M. E.,** Neutralization and enhancement of infectivity of non-salmonid fish rhabdoviruses by rabbit and pike immune sera, *J. Gen. Virol.,* 40, 297, 1978.
9. **Ahne, W.,** Unterschiedliche biologische eigenschaften 4 cyprinidenpathogener rhabdovirusisolate, *J. Vet. Med.,* 33, 253, 1986.
10. **Ahne, W.,** Biological properties of a virus isolated from grass carp (*Ctenopharyngoden idella* Val.), in *Wildlife Diseases,* Page, L. A., Ed., Plenum, NY, 1975, 135.
11. **Clerx, J. P. M., van der Zijst, B. A. M., and Horzinek, M. C.,** Some physicochemical properties of pike fry rhabdovirus RNA, *J. Gen. Virol.,* 29, 133, 1975.
12. **Cohen, J. and Lenoir, G.,** Ultrastructure et morphologie de quatre rhabdovirus de poissons, *Ann. Rech. Veter.,* 5, 443, 1974.
13. **de Kinkelin, P., Le Berre, M., and Lenoir, G.,** Rhabdovirus des poissons, *Ann. Microbiol., (Inst. Pasteur),* 125 A, 93, 1974.
14. **Bootsma, R., de Kinkelin, P., and Le Berre, M.,** Transmission experiments with pike fry (*Esox lucius* L.) rhabdovirus, *J. Fish Biol.,* 7, 269, 1975.
15. **Ahne, W.,** Viral infection cycles in pike (*Esox lucius* L), *J. Appl. Ichthyol.,* 1, 90, 1985.
16. **Dorson, M., de Kinkelin, P., Torchy, C., and Monge, D.,** Sensibilite du brochet *(Esox lucius)* a differents virus de salmonides (NPI, SHV, NHI) et au rhabdovirus de la perche, *Bull. Fr. Piscic.,* 307, 91, 1987.

Chapter VIII

Trout and Salmon (Salmonidae)

A. INFECTIOUS HEMATOPOIETIC NECROSIS VIRUS

Infectious hematopoietic necrosis (IHN) is a highly contagious viral disease primarily of rainbow trout and Pacific salmon. The etiological agent of IHN is a rhabdovirus, infectious hematopoietic necrosis virus (IHNV).[1] IHN dates back to the 1940s and 1950s when unexplained mortalities of juvenile salmon were reported at hatcheries from California to Washington state. Rucker et al.[2] suggested a viral etiology for these deaths, and Watson et al.[3] presented nearly conclusive evidence that this was the case. Disease outbreaks in the Pacific Northwest were known at one time or another as infectious hematopoietic necrosis, Columbia River sockeye salmon disease, Oregon sockeye virus disease, sockeye salmon virus disease, and Sacramento River chinook disease.[1] Comparative studies of histopathology and viral biochemical, biophysical, and serological properties determined that the viruses causing these diseases were in essence the same virus — IHNV.[1,4-6]

1. GEOGRAPHICAL RANGE AND SPECIES SUSCEPTIBILITY

Originally, IHNV was confined to anadromous salmon populations in tributaries that flowed directly into the Pacific Ocean from northern California to Alaska. By the late 1960s, epizootics of IHN were confirmed in other regions of North America including Minnesota,[7] South Dakota, Montana, and West Virginia, and later the virus was reported in rainbow trout in a closed system in New York. These IHN outbreaks were believed to be due to movement of virus-infected eggs from hatcheries in which the virus was enzootic,[8] but apparently it failed to become established in these areas. In addition to being enzootic on the west coast of North America, IHNV has become established in the Hagerman Valley, Idaho, and has been confirmed in Japan, Taiwan, Germany, France, and Italy.[9-12]

Species most susceptible to IHNV are rainbow trout *(Oncorhynchus mykiss)* (including steelhead), chinook *(O. tshawytscha),* and sockeye (kokanee) salmon *(O. nerka).*[1-2] Wingfield and Chan[13] listed the coho salmon *(O. kisutch)* as being refractive to IHNV; however, LaPatra et al.[14] isolated IHNV from adult coho but alevins of the species were resistant to the virus. Sano[9] reported that amago *(O. rhodurus)* and masou (yamame) *(O. masou)* were susceptible. IHNV will infect brook trout *(Salvelinus fontinalis)* and brown trout *(Salmo truta),* but to a lesser degree than rainbow trout.[14] Mulcahy[15] reported Atlantic salmon *(Salmo salar)* mortalities from a natural infection of the virus.

2. CLINICAL SIGNS AND FINDINGS

Clinical signs have been described by several authors.[1,15] The signs may vary somewhat from species to species. The larger fish in an infected population of the same age usually exhibit clinical signs before smaller fish. IHNV-infected fry and fingerlings are lethargic, avoid currents, and move to the edge of ponds or raceways, or lie on the bottom respiring weakly. In the final stages, they swim in circles and hang vertically or flash in a frenzy followed quickly by death.

Affected fish are dark, exophthalmic, have swollen abdomens, and pale gills. The base of fins, mouth, and body surface may be hemorrhaged, and chevron-like hemorrhages occur along the lateral musculature (Figure 1). An opaque, mucoid fecal cast may trail from the vent. An abnormally dark (red) area can develop behind the head and/or in the abdominal region.[16] Amend et al.[1] reported spinal deformities that included misshapen heads, scoliosis, or lordosis in 5 to 60% of survivors.

Internally, the organs appear anemic with petechiae in the mesenteries, peritoneum, air sac, liver, and kidney. The digestive tract is void of food, but contains mucoid fluid. The body cavity may contain pale, yellowish clear fluid.

3. DIAGNOSIS AND VIRUS CHARACTERISTICS

Clinical IHN is diagnosed by isolation of the virus in cell culture using conventional methods. Whole fry, or viscera of fingerlings, are suitable for virus isolation. In rainbow trout infected by immersion, IHNV titers can reach 10^9 pfu/g of body tissue within 5 d.[17] Significant viral infectivity can survive in tissue samples for up to 2 weeks at 4°C with some ovarian fluid, eggs, serum, and brain preparations yielding virus for 5 weeks.[18] Frozen samples stored at $-20°C$ will retain infectivity for 3 to 5 months.

Figure 1   Coho salmon infected with infectious hematopoietic necrosis virus. Note the chevron-shaped hemorrhages in the musculature (arrow), the hemorrhage behind the head and on the abdomen. (Photograph by J. Rohovec.)

The CHSE-214 and EPC cell lines are the systems of choice for isolation, but IHN can also be isolated in RTG-2, FHM, and other cells.[19] Yoshimizu et al.[19] indicated that IHNV would replicate in all 17 cell lines derived from salmonids and in 12 of 15 nonsalmonid fish cell lines tested. Cultures should be incubated at 15°C (never above 20°C) for up to 14 d. If clinical IHN is present in necropsied fish, CPE may appear in 2 to 3 d at 15°C, but usually typical CPE appears in 4 to 5 d. CPE is characterized initially by the presence of pyknotic cells followed by rounding and sloughing of these cells from the surface (Figure 2).[20] Treatment of monolayers of IHNV susceptible cells with 7% polyethylene glycol, prior to inoculation with virus-suspect preparations, will enhance plaque forming titers by 4- to 17-fold.[21] Blind passages are recommended in negative primary inoculations. Once the IHNV is isolated, positive identification can be made by conventional serum neutralization, enzyme-linked immunosorbent assay (ELISA), and other serological methods.

Winton[22] reviewed recent advances in the identification of IHNV and these include monoclonal antibodies, enzyme-linked immunosorbent assay, development of DNA probes, and polymerase chain reaction amplification. LaPatra et al.[23] found that polyclonal and monoclonal antibodies can be used to detect IHNV in blood and organ smears of juveniles, and in ovarian fluids of infected adult salmonids. All isolates of the virus were equally sensitive and were detectable in cell cultures 48 h after inoculation. The FAT method is as sensitive as the plaque assay, but requires less time to obtain a confirmed diagnosis.[24-25] By using monoclonal antibody to IHNV nucleoprotein and glycoprotein, Arnzen et al.[25] detected the early stages of virus replication of cell cultures as soon as 6 to 8 h after inoculation. Dixon and Hill[26] developed an ELISA technique for rapid identification of IHNV in infected tissues and results were obtained in 2 h. The immunodot blot method of detecting IHNV protein was used by Medina et al.[27] and McAllister and Schill.[28]

Molecular technology has recently been used extensively in detection of IHNV in fish tissues and in cell cultures. Arakawa et al.[29] utilized the nucleic acid probe in combination with polymerase chain reaction (PCR), which amplifies the nucleoprotein gene of the virus to develop a rapid, highly specific and sensitive detection method for IHNV. Deering et al.[30] developed a nonradioactive DNA probe that detects and identifies IHNV using a dot blot format with virus infected tissue cultures. This technique proved to be sensitive in detecting very small concentrations of target mRNA. Any of these detection and identification procedures may be used with IHNV, but they are technically intensive and require specific equipment. However, all are more rapid than tissue culture methods and some are just as sensitive.

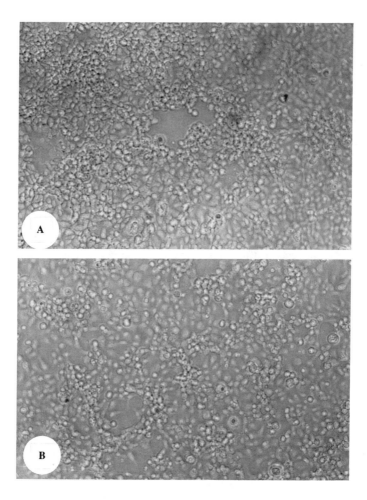

Figure 2 Cytopathic effect (CPE) of infectious hematopoietic necrosis virus in CHSE-214 cells incubated at 15°C for 48 h. (A) Focal CPE; (B) total CPE 60 h after inoculation. (Original magnification × 500.)

IHNV is a rhabdovirus which has a mean diameter of about 70 nm and length of 170 nm.[5,31] The enveloped RNA virus is heat, acid, and ether labile. Studies testing the stability of IHNV are contradictory. Pietsch et al.[32] reported that the virus survived for 5 d or less in sea water and HBSS, but with the addition of 10% fetal calf serum viability was extended to 12 d. Barja et al.[33] demonstrated that IHNV survived for more than 30 d in fresh water at 15°C and for about 17 d at 20°C and for more than 8 d in distilled water. In sea water, the survival was 17 d at 20°C and 22 d at 15°C. Infectivity of IHNV is reduced in distilled water at pH 5 and 9, but at pH 6 to 8 it remains infective for over 10 d. IHNV does not survive well under dry conditions at any temperature and freezing has little adverse effect on the virus.

McCain et al.[6] demonstrated by plaque reduction that there was only one IHNV serotype, and Engelking et al.[34] confirmed these results. Based on molecular weights of nucleocapsid proteins and glycoproteins, Hsu et al.[35] classified the virus into 5 different types: Type 1, 2, 3, 4, and 5. Using monoclonal antibodies Winton et al.[36] was able to separate IHNV isolates from Alaska, California, Idaho, Oregon, Washington, and Japan into four separate groups. Although only one serological strain of IHNV has been identified, Arkush et al.[37] showed that the isolate from the U.S. had different polypeptide profiles than the isolates from Europe and were antigenically different. In view of these differing results, it appears that the issue of one or more strains of IHNV is unsettled.

Table 1 **Effect of water temperature on mortality of rainbow trout (0.2 to 0.3 g) infected with infectious hematopoietic necrosis virus by waterborne exposure**[44]

Treatment	Percent mortality at water temperature						
	21°C	18°C	15°C	12°C	9°C	6°C	3°C
Control	67	30	6	1	0	0	0
Infected	63	80	66	68	76	81	83

4. EPIZOOTIOLOGY

In the aquaculture environment, IHNV is generally considered to cause high mortality among young salmonids. The percentage of deaths may vary from low to nearly 100% depending upon species, stock, age and size of the fish, and environmental conditions, particularly water temperature.

Amend and Nelson[38] demonstrated that sockeye salmon families, which had undergone selective breeding, suffered from 52 to 98% mortality after artificial infection with IHNV. They concluded that these variances were due to genetic differences of the host. A 30% heritability to IHNV resulted from selective breeding suggesting that more IHNV-resistant strains of fish could be developed through selective breeding.[39]

Mortality from IHNV is related to the strain of virus involved.[39] Using rainbow trout as the host, Idaho, Oregon, and California strains of IHNV were compared. At 10°C, a 62% mortality occurred among groups infected with the Idaho strain, 4% among groups infected with the Oregon strain, and 67% among groups infected with the California strain.

Most hatchery epizootics involve fry or small fingerlings with high mortality.[1,7,40] Pilcher and Fryer[41] reported that fry up to 2 months of age suffered mortality exceeding 90% at 10°C. In 2- to 6-month-old fish, the mortality is often more than 50%, but may be as low as 10% in yearlings. LaPatra et al.[42] reported that kokanee salmon were susceptible to IHNV for up to 210 d (7.2 g) and rainbow trout were susceptible up to 170 d (13.1 g). However, outbreaks have occurred in kokanee salmon that were 2 years old but usually the loss of fish is comparatively low.

IHN disease usually does not occur in normal aquaculture conditions at temperatures above 15°C.[12,43] However, Hetrick et al.[44] showed that IHNV could cause 72 to 88% mortality in 0.2- to 0.3-g rainbow trout at 18°C and 72 to 94% at 3°C (Table 1). The mean day to death ratio was 15 to 18 d at 3°C and 7 to 10 d at 18°C. IHNV was isolated from infected, dead fish at 21°C, but mortalities among groups of infected fish and noninfected control fish at the higher temperature were similar.

The mortality pattern of IHNV is typical of highly virulent viruses of young fish. Under optimum conditions, the death rate accelerates rapidly and often 70% or greater mortality will accrue in 10 d.

Survivors of IHNV infections are thought to become carriers of the virus, but it is not clear if these fish continue to possess IHNV after surviving an epizootic as juveniles or become reinfected at a later time. Amos et al.[45] found that adult sockeye salmon, captured as they left salt water during their spawning migration, were free of IHNV. If these fish were allowed to become sexually mature in IHNV-free fresh water, they remained uninfected until spawning. Adults that were allowed to migrate naturally to spawning grounds developed an IHNV positive prevalence of 90 to 100%. The data strongly suggest that returning adults become infected horizontally rather than by reappearance of lifelong latent virus. In a study by Wingfield and Chan,[13] 34% of females and 5% of males assayed were IHNV carriers. Mulcahy et al.[46] surveyed females from 7 adult salmon populations and found that active infections in fish ranged from 39 to 100% (average of 85%). The samples were taken from ovarian fluid or postspawning tissue and many virus titers exceeded $10^5 TCID_{50}$/ml. Grischkowsky and Amend[47] reported that prevalence of IHNV-positive adult female Alaskan sockeye salmon varied from 7 to 94% (average of 44%) and positive adult males varied from 0 to 48% (average of 13%). Mulcahy et al.[48] studied the IHNV carrier rate of female sockeye salmon from August (prespawning) through October (peak spawning). In August, no assayed fish were positive for IHNV, but in October 100% of assayed fish were positive (Table 2). As females progress from a prespawning to a spawning condition, they shed virus more readily and the potential for vertical transmission increases. Mulcahy and Pascho[49] reported two instances of vertical transmission of IHNV. Sperm could be a vehicle for virus to enter the egg at fertilization because experimentally, IHNV was found to absorb onto sperm.[50]

Table 2 **Incidence of IHNV in prespawning and
spawning female sockeye salmon in 1980[48]**

Date	Number of fish assayed	% Positive IHNV	Condition
August 17–30	18	0	Prespawning
September 11–25	31	0	Prespawning
October 6–7	87	21.5	Spawning
October 15	69	96.5	Spawning
October 20–29	65	100	Spawning

The foregoing studies present convincing, albeit circumstantial, evidence that IHNV can be transmitted vertically. In an attempt to clarify the role of eggs in this process, Yoshimnizu et al.[51] showed that IHNV does not survive in the yolk of fertilized eggs prior to the eyed stage and no virus was present on the surface of these eggs. However, when eyed eggs were injected with IHNV, virus concentration increased and 90% of the eggs died. Also, 90% of hatched masou salmon died and 20 to 30% of chum salmon died. These authors doubted that vertical transmission of IHNV occurs because of the inability of the virus to survive in eggs before they become eyed. IHNV can be experimentally transmitted horizontally via waterborne exposure, feeding, or injection.[52]

Detection of specific antibody in the serum of fish may be used to determine prior exposure. Jorgensen et al.[53] collected serum from rainbow trout and screened them for IHNV antibody by serum (plaque) neutralization test (PNT), immunofluorescence, and enzyme-linked immunosorbent assay (ELISA) and showed that positive antibody were detected in from 9 to 18 of 20 serum samples by the 3 methods. ELISA proved the most sensitive and PNT the least sensitive. These assays could be useful in epizootiological studies of IHNV but because some of the sera cross-reacted with VHSV, the procedure probably could not be used when specific pathogen-free determinations are required.

The reservoir of IHNV in the environment is uncertain, but Mulcahy et al.[54] isolated high concentrations of the virus from a leech *(Piscicola salmositica)* and a copepod *(Salminocola* sp.) on sockeye salmon. High concentrations of IHNV were also found in detached leeches taken from the bottom gravel of a spawning area. Though the role of these parasites in the transmission of IHNV is not known, the presence of IHNV in leeches possibly facilitates the infection rate during spawning.

IHNV can also cause significant losses in natural spawning runs. Williams and Amend[55] confirmed an IHNV epizootic in a naturally spawned sockeye salmon fry population in British Columbia. They speculated that an abnormally low survival of eggs in Chilko Lake was due to IHNV. More recently, IHNV caused significant mortality in the Weaver Creek spawning channel, where an estimated 50% of 16.8 million migrating sockeye salmon fry succumbed to IHNV.[56]

5. PATHOLOGICAL MANIFESTATIONS

The pathology of IHNV was summarized by Yasutake.[57] The hematopoietic tissue of the kidney and spleen is the most severely involved tissue with extensive degeneration and necrosis, especially in the anterior kidney. Pleomorphic intracytoplasmic inclusions frequently occur in pancreatic cells and focal necrosis occurs in the liver. Necrosis also occurs in the granular cells of the lamina propria, the stratum compactum, and stratum granulosum of the intestine. A sloughing of mucus membrane (possibly catarrhal inflammation) in the intestine may be the origin of fecal cast observed in moribund fish.[1] LaPatra et al.[40] compared the pathogenesis of three different strains of IHNV and found that the most virulent strain produced severe tissue injury in 3 to 4 d, while the other strains required 8 to 10 d to develop similar pathological changes.

Pathogenesis studies by Yasutake and Amend[58] confirmed that the target tissue of IHNV was the hematopoietic portion of the kidney, where they found a close correlation between histopathological progression and virus concentration. About 3 d postinfection, focal concentrations of macrophages were observed in the kidney followed by an involvement of the pancreas, liver, and the presence of granular cells in the intestinal wall. Amend and Smith,[59] in hematological studies of IHNV, concluded that death was probably the result of severe electrolyte and fluid imbalance caused by renal failure.

LaPatra et al.[60] suggested that infection from waterborne exposure resulted from virus invading the integument and gills because detectable virus increased in these tissues for 3 d after exposure. They further postulated that detectable virus in the mucus resulted from replication in the integument. Yamamoto and Clermont[61] described the sequential spread of IHNV to tissues of rainbow trout infected by immersion. Some tissues, gills for example, had virus titers 16 to 20 h after exposure, and the virus spread rapidly to the kidney and spleen which are target organs. Using whole-body assay and immunohistochemistry, Yamamoto et al.[17] determined that the earliest lesions following infection were in the epidermis of the pectoral fins, opercula, and ventral surface of the body. LaPatra et al.[60] also found that virus is present in the mucus of juvenile and adult fish infected with IHNV and postulated that the epithelium is a portal of entry and site of virus replication. Lesions did not become apparent in the kidneys until the third day. These experiments indicated that the epidermis and gills of fish constitute important sites of early IHNV replication.

6. SIGNIFICANCE

IHN is the most serious viral disease of trout and salmon in the Pacific Northwest. It also poses serious problems in Japan, Taiwan, and Europe. Sano et al.[62] reported that in Japan in 1975, 70 million young salmon succumbed to IHNV infection. Clearly the virus has been responsible for the death of millions of hatchery juvenile salmon and rainbow trout, and large numbers of juvenile salmon spawned in the wild have likely succumbed to the virus.

7. MANAGEMENT

Because IHNV is an untreatable viral disease, the approach to control is management and avoidance of the virus.[8] This management strategy requires the use of eggs from brood stock that are free of IHNV and that fish hatched from these eggs are raised in virus-free (fish-free) water. For dependable "IHNV-free eggs", brood stock that are repeatedly spawned should be shown to be free of the virus for 3 consecutive years. If potential IHNV-positive brood stock are unavoidable, they should be tested for virus at the time of spawning. The eggs of each mating should be identified and incubated independently, so that if a particular spawn is known to have come from an IHNV-positive fish, those eggs are destroyed.[46] Mortality of hatched fry can be reduced from as high as 97% to as low as 4% by this type of brood stock management.

Because IHN responds to management, Alaska has devised a program that allows reasonable culture success of sockeye salmon.[63] (1) IHNV-free water is used for egg incubation and rearing fry. If fish-free water is not available, depuration is accomplished by UV treatment. (2) A strict disinfection policy is used for utensils, facilities, field clothing, personnel, and external surfaces of brood fish during egg and milt collection. (3) Eggs from each female are fertilized separately with milt from one or two males. Each spawn of eggs is water-hardened separately in 100 mg/l solution of iodophor for 60 min. (4) Eggs from 80 to 100 females are pooled into stacked incubators with upwelling water. (5) Eggs in each hatching unit are kept physically isolated from eggs in other hatching units. (6) Alevins are transported and released when yolk is depleted or after rearing fry for 3 to 6 weeks.

Amend and Pietsch[64] recommended that eggs be surface disinfected with 100 mg/l iodine at pH 6.0 for 15 min to kill any virus that might be present on the surface, but the efficacy of this practice has been poor. Amend[43] also suggested that increasing water temperature to 18°C will help arrest IHNV epizootics. However, Hetrick et al.[44] pointed out that increasing water temperature does not stop mortality in fish that have already become infected, and that fish so treated may still become virus carriers.

Several antiviral agents have been used experimentally to attempt to retard the development of the virus in fish and in tissue cultures, but none have proven practical.[65] The addition of 0.14 mg/l of free iodine to hatchery waters will help prevent infection of rainbow trout with waterborne IHNV.[66]

Considerable effort has been expended in the development of a vaccine for IHNV utilizing several different preparations. Attenuated and killed virus, and subunit preparations have been developed and tested experimentally. When exposed to an attenuated IHNV and then challenged, significant protection was detectable.[67] A formalin-killed preparation was shown to be protective when administered by intraperitoneal injection or by hyperosmotic infiltration.[68] Engelking and Leong[69] protected kokanee salmon from five strains of IHNV by injecting them with a glycoprotein from a single strain of the virus. Gillmore et al.[70] reported successful preliminary results of a genetically engineered IHNV subunit vaccine, composed of viral glycoprotein produced through recombinant DNA introduced in *Escherichia coli*. Oberg et al.,[71] expressed the ribonucleoprotein gene of IHNV in *E. coli* through plasmid induction.

Table 3 **Country and year in which IPN virus was confirmed**[76-78,81,129]

Year	Country	Year	Country
1955	U.S.	1975	Germany, Greece
1965	France	1976	Norway
1969	Denmark	1982	Canada
1971	Japan, Scotland	1983	Taiwan
1972	Italy	1984	Chile, Korea
1973	Sweden	1988	Spain
1974	Yugoslavia	1989	China

Although this viral protein alone does not induce immunity or antibody, resistance to IHNV is induced in immunized fish when administered with the bacterial glycoprotein lysate.

In spite of encouraging early vaccination results, it may be difficult to successfully vaccinate against IHNV for several reasons: (1) very young fish are most susceptible, (2) it takes time to develop a protective immunity, and (3) the temperature at which salmonids hatch and grow is not very conducive to the development of a rapid or strong immune response.

B. INFECTIOUS PANCREATIC NECROSIS

Infectious pancreatic necrosis (IPN) is generally considered to be an acute disease of fry and fingerling salmonids, but IPN or IPN-like viruses have been isolated from a variety of marine and freshwater fish and invertebrates. IPN virus is, in all probability, the etiological agent of "acute catarrhal enteritis" described by M'Gonigle[72] in 1941 in Canada. Wood et al.[73] were the first to describe the disease as infectious pancreatic necrosis and to propose that IPN was of possible viral etiology, and Snieszko et al.[74] supported their observations and conclusions. Wolf et al.[75] proved the viral etiology of IPN by isolating and propagating IPN virus in cell culture. IPN was the first proven viral disease of fish.

1. GEOGRAPHICAL RANGE AND SPECIES SUSCEPTIBILITY

IPNV is one of the most widely distributed aquatic viruses, essentially occurring worldwide. Wolf[76] lists 16 countries in which IPNV has been found including those in North and South America, Europe, Asia, the British Isles, and most recently, in China[77] and Spain[78] (Table 3).

There are 16 different species of Salmonidae from which IPNV has been isolated.[76] Brook and rainbow trout have the highest susceptibility to IPNV, but virtually all the salmonids are susceptible, including Pacific and Atlantic salmon.[79-81] Wolf[76] lists 19 families of nonsalmonid fish ranging from the primitive Petromyzonidae (lampreys) to phylogenetically advanced Cichlidae (cichlids), from which IPNV or IPN-like virus has been isolated. Estuarine fishes are especially susceptible to the virus.[80]

IPNV has been isolated from marine mollusks,[82] but it has not been determined if the health of these animals is affected by the virus or if they are simply biological carriers of the agent. An IPN-like virus was isolated from dead Asian clams (*Corbicula* sp.) taken from the Tallapoosa River in central Alabama at the Southeastern Cooperative Fish Disease Laboratory, Auburn University in 1986 (unpublished data).

Clearly, IPNV is not host specific, but it does appear to affect salmonids more severely than nonsalmonids. In a survey of fish farms in England and Wales, Buck et al.[83] found that of 29 salmonid farms surveyed, 17% were rearing fish infected with IPNV. They found no evidence that nonsalmonid fishes in the effluent from these farms were infected with the virus.

2. CLINICAL SIGNS AND FINDINGS

The larger, more robust fry and small fingerlings are usually the first to develop clinical signs of the disease. Externally the fish have overall darker pigmentation, abdominal distention, exopthalmia, hemorrhages on the ventral surface and fins, and pale gills (Figure 3).[74-75] Rapid whirling on the long axis followed by quiescence is a typical behavior of IPNV-infected trout. Internally, a general hemorrhagic or erythemic appearance is obvious with petechiae throughout the viscera, especially in the pyloric cecae and adipose tissue. Internal organs (spleen, heart, liver, and kidneys) are unusually pale

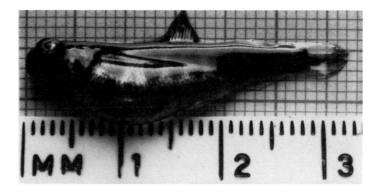

Figure 3   Rainbow trout infected with infectious pancreatic necrosis virus. Note the enlarged abdomen, exophthalmia, and very dark pigmentation.

in advanced cases. The body cavity is filled with a clear yellow fluid. The digestive tract is void of food, but the posterior stomach contains a gelatinous, mucoid (clear or milky) plug which is pathognomonic for IPN. The gelatinous plug remains intact in 10% formalin, therefore, it is of diagnostic value in examining preserved specimens.

The clinical signs described above are evident primarily in IPNV-infected salmonids and are not always seen in other IPNV-infected fish. Sano et al.[84] reported that Japanese eels infected with IPNV had muscle spasms, retracted abdomen, and congestion of the anal fin, abdomen, and gills. These fish also had ascites and mild hypertrophy of the kidneys. IPN-infected larval sea bass displayed spiral swimming, swim bladder distention, fecal casts, exophthalmia, and sloughing of gut epithelium.[85] Striped bass *(Morone saxatilis)* infected with IPN showed no overt clinical signs of disease other than darker pigmentation.[86] Atlantic menhaden *(Brevortia tyrannus)* from the Chesapeake Bay demonstrated a spinning behavior.[87] Gizzard shad *(Dorosoma cepedianum),* from which an IPN-like virus was isolated, showed only dark pigmentation.

3. DIAGNOSIS AND VIRUS CHARACTERISTICS

IPNV infects a wide range of fish species. In trout, presumptive clinical diagnosis of IPN can be made based on mortality patterns and clinical signs. In nonsalmonid fish, IPN infection may not result in clinical disease or mortality. Diagnosis of IPN is confirmed by virus isolation and serological identification. Histological examination can be used to support the clinical diagnosis by detecting typical IPNV associated histopathology.[88-89]

Virus isolation from clinical specimens is best obtained by homogenizing tissue and organs (kidney, pyloric cecae, liver, etc.) from suspect fish. Homogenates are decontaminated by filtraton or by antibiotic treatment. The treated homogenates are diluted at least to 1:100 and inoculated onto susceptible cells, e.g., RTG-2, CHSE-214, BF-2, etc. IPN virus will infect a variety of cell lines which have been derived from salmonids and nonsalmonids. Yoshimizu et al.[19] tested 32 salmonid and nonsalmonid cell lines and found 26 to be susceptible to IPN virus. Inoculated cells incubated at about 20°C develop CPE in 18 to 36 h.[75] Typical CPE consists of cellular pyknosis, elongation and lysis of infected cells (Figure 4). Temperature range for IPNV replication is 4 to 27.5°C. Incubation temperature should be appropriate to the cell line used, e.g., neither CHSE-214 nor RTG-2 should be incubated at above 20°C. A blind passage at dilutions greater than 1:10 should be made in the absence of CPE. Clinically IPNV-infected fish will have titers of 10^6 to 10^9 $TCID_{50}$/ml, but in the carrier state, titers can be as low as 10^1 $TCID_{50}$/ml.

Serum neutralization is usually used to positively identify IPNV. Other types of identification techniques have been developed, e.g., fluorescent antibody, ELISA, and immunoperoxidase systems can be used to identify IPNV in cell cultures, cell culture fluids, and infected tissues.[90-91] Nucleic acid probes and monoclonal antibodies have advanced the rapid detection and identification of IPNV.[92-93] Babin et al.[94] described a capture immunodot employing monoclonal antibodies to detect IPNV that can be use for field identification.

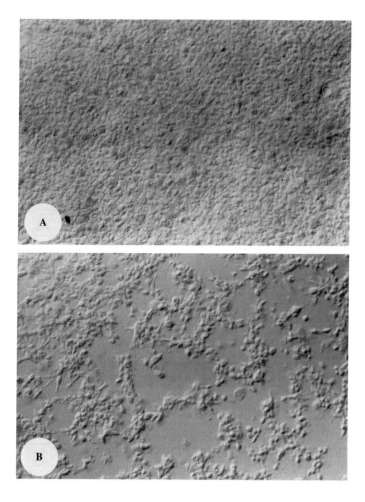

Figure 4   CHSE-214 cells infected with infectious pancreatic necrosis virus and incubated at 15°C. (A) Noninfected control; (B) massive IPN CPE 48 h after inoculation. (Photographs by J. Grizzle.)

Procedures for nondestructive virus detection in carrier fish involve the collecting of ovarian fluids, seminal fluids, or fecal material.[95] McAllister et al.[96] found that ovarian fluids yield greater amounts of virus when the fluids were centrifuged into a pellet, and the pellet sonicated before inoculating cell cultures. When detecting IPN carrier fish, a more accurate and reliable method is to process tissues from internal organs.

IPNV is a member of the Birnaviridae.[97] IPNV is a nonenveloped icosahedral particle containing a single capsid layer and a genome containing two strands of double-stranded RNA. The virus replicates in the cytoplasm and has an average diameter of 60 nm (ranges from 55 to 75 nm).[98-99] The virion is stable in acid, ether, and glycerol and is relatively heat stable. The virus is stable at 4°C for 4 months in cell culture medium containing serum, but a temperature of -20°C or lower is recommended for long-term storage. At -70°C, about 1 $\log_{10}$ of infectivity is lost per year. Most IPN infectivity is lost in fresh water in 10 weeks at 4°C, with residual infectivity remaining for 24 weeks. IPN virus remains infective for up to 20 d in fresh water at 15°C and 15 d at 20°C, compared to a 20-d survival time at both temperatures in sea water.[100] The virus will survive drying for more than 8 weeks.[101] IPNV may be lyophilized in lactoalbumin hydrolysate, lactose, powdered milk, or saline with 10% fetal calf serum for up to 1 year.[102]

Isolates of IPNV have been placed into distinct serotypes (e.g., VR-299, Ab, and Sp) (Table 4).[103-105] Most isolates in the U.S. are of the VR-299 serotype, while most salmonid isolates from Europe are the Ab and Sp serotypes. Many isolates from nonsalmonid species tend to be of lower virulence for trout and are of the Ab serotype.

Table 4 **Major serological groups of IPN virus**[76,104-106]

IPN serological designation	Species of origin	Geographical origin
VR-299 (ATCC)	Brook trout	U.S.
Ab (Abild)	Rainbow trout	Denmark
Sp (Spjarup)	Rainbow trout	Denmark

Using polyvalent rabbit antisera and the 50% plaque neutralization technique, Hill and Way[106] proposed that 200 isolates of IPNV and other aquatic birnaviruses be divided into Serogroups I and II. Furthermore, Serogroup I could be divided into nine different serotypes, all of which are related. Isolates that showed no relatedness to Serogroup I, but did cross-react with each other, constituted Serogroup II.

4. EPIZOOTIOLOGY

IPNV is one of the more fascinating fish viruses because of its broad adaptibility, geographical range, and diverse host susceptibility. The severity of an IPN infection is closely related to age, physiological condition, strain and species of fish, strain of virus, and environmental conditions. In a hatchery population, manifestation of IPN infection can vary from subclinical to epizootic with nearly 100% mortality.

Optimum water temperature for development of IPN is 10 to 15°C. When trout fry are naturally exposed to IPNV, clinical signs of disease can occur from 1 week to 6 months of age. Under experimental conditions, the incubation period at 15°C can be as little as 3 to 5 d. Brook trout reach peak susceptibility between 30 and 50 d of age, and mortality due to IPN infection becomes less severe as fish become older. At 10°C 1-month-old brook trout can suffer 85% cumulative mortality, but by 6 months of age, mortality is less than 5%.[107]

The mortality of nonsalmonid species infected with IPN can also vary from none to high. For example, Bonami et al.[85] reported up to 95% mortality in cultured sea bass fry, and Schutz et al.[86] reported that 2 million 4-week-old striped bass died in association with IPN infection. Both of these reports carefully avoided claiming that the IPNV killed the fish. An attempt to experimentally infect 4 strains of striped bass of 1 to 120 d of age with IPNV, did not produce clinical disease.[108] No clinical signs of disease were noted in any fish, and virus was subsequently isolated only from those fish that had been infected by injection. Although it is questionable if striped bass are killed by IPNV, McAllister and McAllister[109] successfully transmitted the virus from asymptomatic IPNV-carrier striped bass to juvenile brook trout, and 8% of the surviving brook trout became long-term virus carriers. Because striped bass are susceptible to IPNV infection under certain conditions, all striped bass (or hybrids) which are stocked into IPNV-free waters should be free of the virus.

Numerous reports describe isolation of IPN or IPN-like virus from asymptomatic fish during routine virus assays. Wolf[76] states that subclinically infected fishes of any species can be a source of IPNV for more susceptible species. McAllister and McAllister[109] demonstrated that IPN virus could be transmitted from carrier striped bass to brook trout, and the virus can also be transported by fish-eating birds and mammals and released in the excreta.[109-112]

Although adverse environmental conditions are not necessarily associated with IPN epizootics, Roberts and McKnight[89] linked temperature and transport stress to recurrent outbreaks of IPN in a population of yearling rainbow trout. Frantsi and Savan[113] described a similar pattern in yearling IPN carrier fish following exposure to low oxygen concentrations. These observations suggest that stressful environmental conditions may cause a reocurrence of the disease, especially in older fish populations.

Multiple infections involving two or more pathogens in the same fish is not uncommon. Yamamoto[114] demonstrated that IPNV and *R. salmoninarum* (bacterial kidney disease — BKD) could occur in the same population of rainbow and brook trout and electron microscopy indicated that both disease agents could occupy the same cell. It has also been shown that IHNV and IPNV could occur in the same rainbow trout fry, but the concentration of IHNV (2.8×10^5 pfu/g) was greater than IPNV (1.2×10^4 pfu/g).[115]

Transmission of IPNV occurs horizontally from infected fish through the water to susceptible fish, where it is absorbed by the gills and the gut. During an epizootic, virus titers in water can be as high as 10^5 TCID$_{50}$/l.[100] Carrier trout shed IPNV in their reproductive products and feces and vertical transmission of IPN was strongly suggested by early IPN work.[116] Some evidence suggests the potential for IPNV to be transmitted vertically from adults to progeny via the gametes.[117-119] Circumstantial evidence and some experimental data suggests that vertical transmission is a possible but unpredictable event. Seemingly, when IPNV is present in ovarian fluid and absorbed onto sperm, the likelihood of vertical transmission could be enhanced.[120] An unusual means of IPN dissemination could be by injection of pituitary extracts from infected trout to stimulate spawning of other species.[121]

When an IPN outbreak occurs in trout, some survivors become virus carriers and are reservoirs of the virus probably for life. The apparent prevalence of IPNV-carrier fish in a population can vary with the season. Mangunwiryo and Agius[122] showed that the percentage of fish shedding IPNV increased between March and June, and declined during the latter part of the year. The heightened carrier state of these fish was inversely related to their IPN antibody levels implying that high antibody serum titers reduced the active virus production. Billi and Wolf[123] found that when an IPNV-carrier population was sampled over time, more than 90% of the individuals were IPNV carriers, but not all fish shed virus at each sampling period. A model for the IPNV-carrier state in trout and cell cultures was described by Hedrick and Fryer.[124] They found that virus was produced by less than 1% of the cell population of either persistently infected cell lines or kidney cells of carrier trout. Further, in cell culture neither antibody nor interferon interfered with viral replication, but defective virus may control or interfere with virus replication. A similar mechanism may function in IPNV-carrier fish.

The importance of IPNV as a pathogen of cultured trout, salmon, and occasionally other species is unquestioned, but the significance of IPNV or IPNV-carrier fish in the wild is uncertain. Survivors of IPN epizootics are often stocked into streams or lakes for sport fishing, and at times these populations become reproductively self-sustaining. Yamamoto[125] found that a hatchery population of brook trout had an IPNV-carrier prevalence of 90%, but at 2.5 years after stocking in a lake, the carrier prevalence was 69%. These results suggest that the carrier state may last for years in IPNV-infected fish and that these fish could be a reservoir for viral transmission to noninfected fish.

5. PATHOLOGICAL MANIFESTATIONS

Pathological changes resulting from IPNV are typical of a viremia.[57] The most profound histopathological change is the obliteration of most of the acinar tissue of the pancreas, and its replacement by necrotic detritus containing fragmented acinar cells, pyknotic nuclei, and zymogen granules. In advanced cases, monocytes, and polymorphonuclear lymphocytes increase indicating the presence of some inflammation. Initially, foci of necrosis are scattered throughout the acinar tissue, and necrosis rapidly spreads to surrounding tissue. At the periphery of the necrotic areas, many acinar cells will contain intracytoplasmic inclusions. Fatty tissues surrounding the pancreatic tissue often become necrotic, and necrosis frequently occurs in the islet of Langerhans.

Slight to severe necrosis occurs in the voluntary muscle and is accompanied by mild inflammation. Excretory and hematopoietic tissues of the kidney become necrotic to various degrees. Congestion, hemorrhage, and edema occur in the epithelium of renal tubules. Sano[126] found congestion and necrosis in liver tissue of rainbow trout.

Using rainbow trout as an experimental model, McKnight and Roberts[88] described the sequence of histopathological changes occurring in the pancreas, renal tissue, and intestinal mucosa. Changes in the pancreatic tissue were characterized by destruction of acinar tissue which degenerated into an amorphous eosinophilic mass. Contrary to earlier descriptions, islet of Langerhans tissue of the spleen and fatty tissue appeared normal and no inflammation was noted. The kidney showed focal degenerative changes in the hematopoietic tissue. The intestinal mucosa was more severely affected than previously reported and exhibited complete necrosis, sloughing of the epithelium, and accumulation of a pinkish or whitish catarrhal exudate in the lumen of the intestine. The lesions persisted to a diminishing degree for over a year after the infection. These authors postulated that the intestinal lesions contributed more to the death of IPNV-infected fish than did the pancreatic lesions.

6. SIGNIFICANCE

Because mortalities in susceptible fish populations are moderate to very high, IPNV is one of the more significant diseases of cultured fishes. Despite extensive fish health inspection efforts, IPNV has continued to spread to most regions of the world where trout are cultured, and the known geographic and host range of IPNV continues to expand with the aggressive search for viruses associated with aquatic animals.

7. MANAGEMENT

Avoidance is the most effective and economical way of controlling IPN.[76] Eggs should come from sources known to be IPNV-free, and progeny must be hatched and reared through the susceptible fingerling stage in a closed, fish-free water supply. Because nonsalmonid fish can carry IPNV, fish that are to come in contact with cultured salmonids should be free of IPNV.

Where IPNV is indigenous, and therefore unavoidable, more resistant strains or species should be considered. Reducing the water temperature to 4 or 5°C during the period of peak susceptibility can moderate mortality due to IPN virus.[113] Immunization as a method for disease prevention and control is an area of active research. Bootland et al.[127] showed that immersion of 2-, 3-, and 6-week-old posthatch brook trout in a formalin-inactivated IPNV preparation enhanced the relative percent survival, but the protection decreased as fish increased in age and size. Vaccination did not prevent IPNV infection in any age group of fry. In an effort to develop a viral subunit vaccine for IPNV, Manning and Leong[128] inserted the cDNA of IPNV into *Escherichia coli* and the resultant bacterial lysate was effective in inducing protective immunity to IPNV in rainbow trout. The fish were vaccinated by immersion in the bacterial lysate for 20 min and challenged 20 d later. IPNV is an excellent antigen, and at some future time a practical vaccine may become available.

C. VIRAL HEMORRHAGIC SEPTICEMIA

Viral hemorrhagic septicemia (VHS) is an acute, highly infectious viral disease of cultured trout in Europe. It has a long history with references being made to the condition dating back to the mid 1930s.[76] The disease was originally named Egtved disease after the village in Denmark where the virus was first found.[130] There have been many common names associated with the disease, most of which reflect its viral nature or the kidney and liver tropism of the disease. The etiological agent of viral hemorrhagic septicemia is viral hemorrhagic septicemia virus (VHSV).

1. GEOGRAPHICAL RANGE AND SPECIES SUSCEPTIBILITY

Since the establishment of its viral etiology in the early 1960s,[131] clinical VHS is confined to the European continent; however, in 1988 VHSV was isolated from adult salmon returning to hatcheries in Washington state.[132-133] Wolf[76] lists 13 European countries where VHS has been found and now others have been added, including Spain.[78]

Although rainbow trout is the species most susceptible to VHSV, brown trout are also considered a natural host.[134-135] Dorson et al.[136] demonstrated that hybrids of rainbow and brook trout and Arctic char *(Salvelinus alpinus)* were resistant to VHSV, but that lake trout *(S. namaycush)* were susceptible to the virus resulting in clinical disease and significant mortality. It was reported by Meier[137] that juvenile pike *(Esox lucius)* were more susceptible to VHSV than rainbow trout. At least ten additional species of fish, including a hybrid salmonid, are experimentally or naturally susceptible to the virus.[76] Of these ten, eight are salmonids and two are marine species: sea bass *(Dicentarchus labrax)* and turbot *(Scophthalmus maximus)*.[138] In 1990 and 1991, VHSV was isolated from Pacific cod *(Gadus macrocephalus)* with ulcerative lesions in Prince William Sound, Alaska.[139] The virus had been previously isolated from Atlantic cod *(Gadus morhua)* in Denmark, which would lead to speculation that a reservoir of VHSV may exist among nonsalmonid fish in the marine environment. Wolf[76] also listed nine species of fish that are refractive to overt VHSV infection. Ironically, this list included chinook salmon[140-141] and coho salmon,[142] the two species of Pacific salmon in Washington from which VHSV was isolated.

2. CLINICAL SIGNS

Viral hemorrhagic septicemia of salmonids is categorized into acute, chronic, and nervous forms (Table 5). The description of each reflects the behavior, overt pathological changes, and mortality pattern

Table 5   Characteristics of the different forms of viral hemorrhagic septicemia of salmonids[141,181]

Form	Behavior	External signs	Internal lesions
Acute	Poor feeding, erratic swimming, high mortality	Dark pigmentation, anemia, pale gills, exophthalmia	Multiple hemorrhages in skeletal muscle, eye and viscera; liver pale, gray; kidney hyperemic; swollen intestine and no food
Chronic	Lower mortality, reduced activity	Dark pigmentation, exophthalmia	Liver pale to gray, kidney rough
Nervous	No or negligible mortality, erratic swimming	Occasional anemia and sunken abdomen	None

typifying each particular form of the disease. Typically, infected salmonids do not feed, and their swimming behavior ranges from lethargic to hyperactive.

The acute form of VHS is the most serious because of the high mortality associated with it. In this form, affected fish exhibit erratic and spiral swimming, the skin is dark, they are exophthalmic, have pale gills indicative of anemia, and the gills may be flecked with petechiae (Figure 5). Internally there are multiple hemorrhages in the skeletal muscle and swim bladder; the kidney is swollen, hyperemic, and more or less necrotic and liquefied; and the liver is either hyperemic, gray, or yellowish. Chronic VHS is characterized by a lower mortality. Affected fish have dark pigmentation (melanosis), exophthalmia, and swollen abdomens. Internally, the liver is pale. In the nervous stage (i.e., neurovegetative N-form), there is very low to no mortality, but the fish do exhibit poor balance, swim in a circle, and are somewhat anemic. Environmental or handling stress can cause a less severe form of VHS to advance into a more severe form.

Nonsalmonid fishes will also show clinical signs when infected with VHSV. Naturally infected whitefish (*Coregonus* sp.) showed hemorrhages in the skeletal muscle and in the swim bladder.[143] Naturally infected pike had exophthalmia; extensive hemorrhage in the skin, muscle, and kidney;[144] and the body cavity contained ascitic fluid. Meier and Wahli[145] found that VHSV-infected grayling (*Thymallus thymallus*) in Switzerland exhibited anemia, enteritis, and subcutaneous and intermuscular hemorrhage.

3. DETECTION AND VIRUS CHARACTERISTICS

During epizootics, VHSV can be readily isolated from clinically sick fish, but in later stages of infection it becomes more difficult. Often after clinical disease disappears, the virus can no longer be isolated from survivors. The kidney or spleen are the organs of choice when attempting VHSV isolation.[144-145] The BF-2 cells are most susceptible to VHSV, but CHSE-214, EPC, RTG-2, or STE-137 cells may also be used. It is important to maintain medium at pH 7.6 to 7.7 and incubation temperature at approximately 15°C or less, because these factors will help presumptively distinguish VHSV from other fish viruses.[130,146] CPE of VHSV develops as focal areas of rounded cells followed by lysis and involvement of the entire cell sheet (Figure 6).

Positive identification can be obtained by serological differentiation from other fish viruses, including IHNV, SVCV, and PFR and other fish rhabdoviruses. Conventional serum neutralization, or plaque reduction, using virus specific reagents may be used to identify isolated virus. Fluorescent antibody[146] may be used for tissue culture isolates or on sections of frozen tissues from infected animals. However, Ahne et al.[147] reported a VHSV strain that could not be neutralized by VHSV antiserum, but was positive with immunofluorescence.

It has been noted that in a given fish population VHSV and IPNV may co-exist in the same fish[148] as may VHSV and IHNV;[10] therefore, detection and identification for all three viruses should be used when assaying trout populations. Monoclonal antibody that identifies a conserved epitope on VHSV or IHNV, respectively, can be used to easily distinguish IHNV and VHSV from each other regardless of their origin (North America or Europe) when employed in an immunodot assay.[149] Neither of these monoclonal antibody bond to any other fish rhabdovirus.

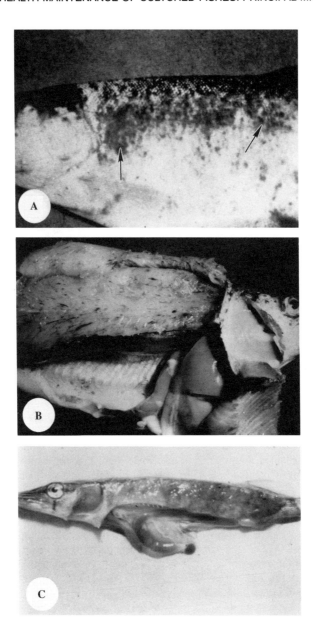

Figure 5 Fish infected with viral hemorrhagic septicemia infection. (A) Hemorrhage (arrow) in skin of rainbow trout; (B) petechia in muscle, peritoneum, and pale liver and gills; (C) petechiae in the muscle and pale gills of pike infected with VHSV. (Photographs A and B by P. Ghiteino; photograph B reprinted with permission from CRC Press; photograph C by W. Ahne.)

Detection of VHSV antibody has become a method of determining possible virus carrier populations of trout in Europe. Virus neutralization, immunofluorescence antibody (IFA), and ELISA have been used for this purpose, and Olesen et al.[150] showed that the ELISA techniques were more accurate than serum neutralization or IFAT, but less specific than serum neutralization. Like IHNV, VHSV neutralization is compliment dependent. Following the detection of IHNV in France, Hattenberger-Baudouy et al.[10] serologically surveyed an IHNV-infected rainbow trout population known to be carrying VHSV, and found that some of these fish had IHNV and IHNV neutralizing antibody. Jorgensen et al.[151] cautioned against using antibody detection in fish sera to indicate the presence of VHSV or IHNV in populations where both may co-exist because of the presence of cross-reactive antibody in the same

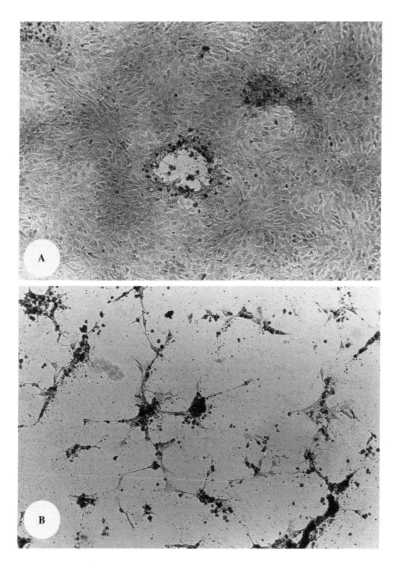

Figure 6 Viral hemorrhagic septicemia cytopathic effect in FHM cells. (A) Focal CPE consisting of rounding of cells and becoming pyknotic; (B) advanced CPE involving entire cell sheet (iodine stain). (Photographs by P. de Kinkelin.)

fish. Olesen and Jørgensen[152] emphasized that the ELISA tests for VHSV detection is a valuable addition to traditional virus isolation in cell culture when a rapid diagnosis is essential, but should be used in conjunction with isolation procedures for confirmation. In their tests, 24 of 40 tested trout were VHSV-positive by culture, but only 20 (80%) of the 24 culture-positive fish were also positive by the ELISA method.

Viral hemorrhagic septicemia virus is a member of the family Rhabdoviridae. The virus is an enveloped, bullet-shaped particle which measures about 65 nm in diameter and 180 nm in length.[153] The virus has a single-stranded RNA genome. VHSV is glycerin, heat, ether, and acid labile, but very stable at pH 5 to 10.[154] There are at least two and possibly three serological strains of VHSV;[135,155] however, Olesen et al.[156] identified four strains of VHSV using monoclonal antibody and neutralization.

The virus is completely inactivated by 3% formalin, 2% NaOH, and 0.01% I_2 (Actomar) in 5 min, and in 2 min by 500 mg/l of chlorine. VHSV is 90% inactivated after 14 d in stream or tap water at 10°C, 99% inactivated in mud at 4°C after 10 d, and drying at 15°C for 14 d. Gamma and UV irradiation will completely inactivate the virus.[154,157]

4. EPIZOOTIOLOGY

Originally VHS was thought to be a nutritional disease.[158] However, with the isolation of VHSV agent by Jensen[130] and more detailed clarification of the etiology by Zwillenberg,[153] the cause was clearly shown.

Winton et al.[159] showed that the VHSV isolates from chinook and coho salmon in Washington were similar serologically and biochemically to the European isolates. It was later stated that the Washington strain was significantly less virulent than the European strain. The Washington VHSV did not kill significant numbers of eight different species of salmonids (including rainbow trout) that were exposed to virus either by waterborne exposure or injection. This fact would appear to indicate the North American strain of VHSV was different from typical European strains. The difference was confirmed by Bernard et al.[160] who sequenced the nucleoprotein gene of a North American strain and compared it with a sequence from a European strain. The sequence differences suggested the North American strains are not of European origin. The stability of the Washington virus was greater in sea water than in fresh water.[161] Attempts were made to eliminate VHSV from the continental U.S. in 1988 by the destruction of infected populations and disinfection of facilities; however, the virus was again isolated from coho salmon at different facilities in subsequent years.

There is no conclusive evidence that VHSV is transmitted vertically from adult to offspring by virus in the egg. However, VHSV has been isolated from egg surfaces, probably as a result of virus present in the ovarian fluid.[183] Attempts to experimentally infect eggs with VHSV were unsuccessful.

Degree of mortality in infected fish will vary with species, size, method of infection, and water temperature. The virus can be transmitted horizontally by injection, by brushing the gills with infected material, cohabitation of naive fish with infected fish, and by simple waterborne exposure.[162] Oral infection is difficult, although pike fry were successfully infected when fed virus-contaminated trout flesh.[163]

The younger the fish, the more susceptible they are to VHSV, and the weight of the most susceptible trout range from 0.5 to 4 g each.[162] As the fish become larger, they gain greater resistance, but fish up to 200 g have been reported to develop VHS.[164] Ironically, the healthier appearing fish in a population are more susceptible.

Incubation time between virus exposure and overt appearance of VHS is temperature dependent and may vary from 5 d to 4 weeks. VHS disease occurred 5 d postinjection at 9°C[165] and 8 to 10 d following exposure by immersion at 12°C. de Kinkelin[166] stated that incubation time was 3 to 10 d at 7 to 14°C and that an inapparent infection occurred at 2°C. Rasmussen[167] found that VHS outbreaks in rainbow trout were associated with low water temperatures of 5 to 10°C, and that the water temperature directly affected the duration and severity of the disease. Optimum water temperature is between 8 and 12°C, but temperature fluctuations of a few degrees seem to trigger deaths. Clinical disease in hatchery populations normally does not occur at temperatures above 14°C.

Castric and de Kinkelin[164] reported an 80% mortality of rainbow trout 1 month after they were moved from fresh water to full strength sea water. Also, in Denmark mariculture facilities, mortalities of 15 to 50% occurred in naturally infected rainbow trout 18 months after stocking.[168] In a study in Switzerland, Meier and Wahli[145] noted that mortality in rainbow trout and grayling was 56 to 100% in both species and the degree of mortality was age-dependent. VHS may cause mortalities of over 85% in rainbow trout depending on the strain of virus and fish.[144] When sea bass and turbot were injected with the virus, over 90% mortality resulted.[138] Meier et al.[169] compared mortality in whitefish (28 to 50%) and rainbow trout (40 to 80%) at a temperature of 11 to 14°C. Mortalities from VHSV in nonsalmonid fish vary, but Ahne[163] induced 90% mortality in 0.2-g pike infected by waterborne exposure to VHSV. When these virus infected pike were fed to 1-year-old, 60-g pike, clinical disease and 30% mortality ensued in the older fish. Although the smaller pike were susceptible to IPNV, PFRV, and SVCV, the older fish were refractive to them. Giorgetti[170] postulated that up to 30% of cultured trout in Italy were killed by VHSV.

Virus is shed from infected individuals in the feces, urine, and reproductive fluids, but the kidney, spleen, brain, and digestive tract are sites in which the virus is the most abundant. Virus sources, during the absence of clinical disease, are most likely infected fish which have survived an epizootic or a species of fish that can be infected, but does not develop clinical disease.[171] Wolf[76] points out that the more VHSV-resistant salmonids, such as brown trout, may be reservoirs for the virus. Jørgensen[171] reported that VHSV-infected rainbow trout (45% of those tested) were found in a stream below the outfall of a hatchery in Denmark, while fish above the hatchery were not infected. It could not be determined if the infected fish had escaped from the hatchery or had become infected via contaminated water. In 1966, Wolf[172] theorized that uncooked marine fish which was fed to rainbow trout may have been the original reservoir of the virus. The isolation of VHSV from marine fish in Europe and North America supports Wolf's earlier theory on the source of the original VHSV. Opinions differ as to whether birds are a source of VHSV. Eskildsen and Jorgensen[173] reported that VHSV was not present in the feces of 41 sea gulls, but other reports claim that herons could be a vector for the virus.

VHSV is normally found in the milt and ovarian fluid of salmonids, and Eaton et al.[174] isolated the virus from both milt and ovarian fluid of spawning salmonids in North America. They also isolated VHSV and IHNV from reproductive products and kidneys of coho salmon in the same watershed in Washington.

5. PATHOLOGICAL MANIFESTATIONS
VHSV produces hemorrhages, usually in the form of petechiae, throughout the musculature, visceral organs, and gills of affected fish.[57,144,166] These hemorrhages occur because the endothelial cells are major target cells for VHSV replication in fish.[162] Kidneys are the most severely affected organ, particularly the hematopoietic tissue, which is characterized by hyperplasia and necrosis. Melanophor centers are destroyed and leukopenia is present along with congestion. The liver exhibits focal necrosis and degeneration with vacuolated pyknotic and karyolytic hepatic cells. While small hemorrhages are present in the musculature, muscle fibers are essentially normal.

In natural infections of VHSV, gill epithelium is one of the possible routes of virus entry where viral replication may or may not take place.[175] VHSV also replicates near its portal of entry, in the wall of blood capillaries of skin and underlying muscle. The virus is then transported to the kidney and spleen where active viral replication also occurs. According to protocol presented by de Kinkelin et al.[162] VHSV takes no more than 24 h between waterborne exposure to the virus and recovery of the agent from internal organs.

6. SIGNIFICANCE
Historically, VHSV has been the most serious viral disease affecting European trout culture. With its detection along the northwest coast of North America in 1988, new emphasis has been placed on its potential impact on salmon culture throughout the world. It is now recognized that where VHSV occurs, there is the potential for a highly communicable disease outbreak.

7. MANAGEMENT
Avoidance and strict sanitation are the best prevention and control measures for VHSV.[166,176] Prevention of wide fluctuations of water temperature and control of other environmental stressors are essential. Wolf[76] suggested that fish-free water supplies be used and that eggs be obtained from certified virus-free adults. Eggs should be disinfected in 100 mg/l of iodophor at pH 6.5 for 10 min because the virus can adhere to the surface of the egg. VHS control programs have been initiated in European countries.[177] The general approach involves certification of fish stocks free of VHSV, controlled movement, destruction of infected populations, disinfection of facilities, and restocking farms with virus-free fish.

Resistance of fish to various diseases may also be part of a management program, and some strains of rainbow trout are more resistant to VHSV than others. Using male and female rainbow trout from a resistant stock, Kaastrup et al.[178] showed that the resistance to VHSV was carried by the male rather than the female. In view of this, a breeding program to enhance resistance is feasible, and a program to exploit this factor has begun in Danish rainbow trout farms. Also, Dorson et al.[136] showed that rainbow and brook trout hybrids, which were triploids, were resistant to VHSV and IHNV, but susceptible to IPNV.

Immunization is a possibility as noted by de Kinkelin and Le Berre[134] and Jørgensen[179] but to date, it has not evolved into an effective practice. de Kinkelin[176] reviewed experimental vaccination programs for VHSV in Europe that employed both killed and live preparations. Killed-VHS vaccines required injection, but they are effective in eliciting protection. Live, attenuated vaccines can be applied via immersion and also provide a moderate to high degree of protection in laboratory trials. The potential of subunit vaccines for VHS are being investigated employing monoclonal antibodies against the VHS viral G protein.[180] Two of these monoclonal antibodies showed strong protection in fish following passive immunization, but obviously much work needs to be done in this area.

D. ERYTHROCYTIC INCLUSION BODY SYNDROME

A viral infection that causes anemia among stocks of cultured coho and chinook salmon was first recognized in 1982 in the northwestern U.S.[182-183] The disease, previously referred to as Pacific salmon anemia virus, is most commonly known as erythrocytic inclusion body syndrome (EIBS),[182] and is caused by the erythrocytic inclusion body syndrome virus (EIBSV).

1. GEOGRAPHICAL RANGE AND SPECIES SUSCEPTIBILITY

EIBS occurs naturally in cultured populations of coho and chinook salmon along the Columbia River and its tributaries in Oregon and Washington.[183] The disease has also been reported in Canada, Chile, Ireland, Japan, and Norway.[184-186] Okamoto et al.[187] found that chum and masou salmon were as susceptible to EIBS virus as coho. Rainbow trout and cutthroat trout *(O. clarki)* can be experimentally infected, but the severity of the disease in trout is mild compared to that in salmon.[188]

2. CLINICAL SIGNS AND FINDINGS

Other than severe erythrocytic anemia, characterized by pale gills and internal organs, no behavioral or external clinical signs have been described.[183] Takahashi et al.[185] found that EIBSV-infected fish had hyperemia of the intestine, dilation of the spleen, and hyperemia and hemorrhage of the atrium, and that the stomach was filled with mucus. The livers appeared yellowish in color. Infected erythrocytes develop cytoplasmic inclusions that stain more lightly than the cell nucleus (Figure 7). Hematocrits of normal salmon are 40 to 45%, while those of EIBS-infected fish are often in the mid 30% range and can plummet to as low as 15%.

3. DIAGNOSIS AND VIRUS CHARACTERISTICS

EIBS is presumptively diagnosed by staining erythrocytes with pinacyanol chloride to detect basophilic cytoplasmic inclusions that measure 1 to 8 μm in diameter (Figure 7).[189] The etiological agent is, as yet, not isolated in tissue culture. With filtrates from EIBS-infected fish, 9 different piscine cell lines were inoculated and during a 30-d incubation period, no CPE was detected in any of the cell lines.[183,185]

Infected red blood cells contain spherical viral particles that are 75 to 100 nm in diameter and are scattered randomly in the cytoplasm[189] (Figure 7). These viruses are associated with viroplasm-like material or contained within membrane-bound organelles. The EIBS virus is distinctly different from the iridovirus of viral erythrocytic necrosis. Because of its size, possession of a lipid envelope, and its presence in organelles, the EIBS virus seems most like a member of the Togaviridae.[187]

4. EPIZOOTIOLOGY

EIBS infection, per se, is not particularly lethal. Piacentini et al.[188] described 5 (I through V) stages of EIBS when fish were held at 12°C (Table 6). Takahashi et al.[190] described similar stages of EIBS development in coho salmon in Japan. Each stage is characterized by specific changes in hematocrit value and in the presence of erythrocytic cytoplasmic inclusions. The entire cycle of infection from exposure to recovery requires about 45 d, after which surviving fish appear normal and are resistant to reinfection. Although Takahashi et al.[185] attributed a mortality of about 23% directly to EIBS, affected fish become severely infected with opportunistic pathogens such as external fungi, and suffer much higher than normal mortality due to the enzootic pathogens *Renibacterium salmoninarum, Yersinia ruckeri,* and *Cytophaga psychrophila.* The majority of deaths are due to complicating bacterial and fungal infections, and infected fish generally survive if they do not succumb to other pathogens. However, Takahashi et al.[190] claimed that EIBS is a major contributor to high mortality of coho salmon reared in saltwater net-pens in Japan and that the virus is responsible for great economic losses to aquaculturists.

Water temperature effects the progression and duration of EIBS. The disease cycle begins more quickly at 15 to 18°C (5 d) and is more prolonged at 6 to 9°C (32 d). Takahashi et al.[185] observed that the prevalence of EIBS increased at temperatures below 10°C, was highest at 8 to 10°C, and did not occur when the temperature exceeded 13°C.

Experimental transmission of EIBS can be achieved by injecting naive fish with filtrates (0.22 μm) of homogenized kidney, spleen, or blood from afflicted fish.[188] Typical inclusions appear in erythrocytes in about 14 d. The disease was also successfully transmitted to coho salmon by cohabitation with infected fish, in which case EIBS developed 14 to 28 d after exposure. A wide size-range of fish are

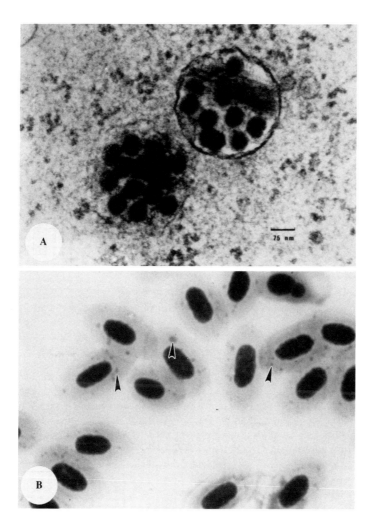

Figure 7   Erythrocytic inclusion body syndrome of salmon. (A) Electron micrograph of EIBS virus within membrane-bound organelle; (B) erythrocytes with multiple cytoplasmic inclusion bodies (arrows) typical of EIBS-infected fish. (Photograph by S. Leek.) (From Arakawa, C. K., Hursh, D. A., Lannan, C. N., Rohovec, J. S., and Winton, J. R., in *Viruses of Lower Vertebrates,* Ahne, W. and Kurstak, E., Eds., Springer-Verlag, Berlin, 1988, 442. With permission.)

susceptible to EIBS virus because Okamoto et al.[187] found that 1.2- to 220-g coho salmon are equally sensitive. Takahashi et al.[185] speculated that the EIBS virus could have been introduced into Japan via infected eggs.

5. PATHOLOGICAL MANIFESTATIONS

Pathology of EIBS has not been fully described. The major histopathological change is the lysis of infected erythrocytes, and about 35% of the erythrocytes of an infected fish will have the cytoplasmic inclusion bodies.[185] The number of erythrocytes in the anterior kidney is reduced and the parenchyma and hematopoietic areas show some necrosis. Phagocytes, many containing eosinophilic vacuoles, infiltrate the kidney, and epithelial cells of the renal tubules show hyaline droplet degeneration. Similar pathology occurs in the spleen with vacuolation and partial necrosis of hepatocytes in liver sinusoids. The pathogenesis of EIBS is poorly understood, but the virus infection, per se, can cause moderate mortality.

Table 6   **The progressive stages of erythrocytic inclusion body syndrome in coho salmon held at 12°C**[188]

Stage	Days post-infection	Hematocrit (%)	Number of cytoplasmic inclusions	Condition of fish
I	10	45	None	Normal
II	10–13	35–40	Low	Normal
III	18–25	25 ±	High	Lysis of RBC
IV	30–40	35–40	Low	Recovery
V	45	45	None	Normal

6. SIGNIFICANCE

By itself the EIBS virus may not be a significant pathogen, but virus infection presumably reduces natural resistance. Secondary infections become invasive, and the affected fish often succumb to these secondary infections. In some aquaculture situations, EIBS virus is thought to result in significant economic loss.[190]

7. MANAGEMENT

Currently, no management or control procedures, aside from avoidance, are used to prevent dissemination of the EIBS virus.

E. *HERPESVIRUS SALMONIS* INFECTION (TYPE 1)

Five different herpesviruses of trout and salmon have been described, and Hedrick et al.[191] determined that *Herpesvirus salmonis* from steelhead trout in the U.S. were different from herpesviruses isolates (i.e., *Oncorhynchus masou* virus) from salmonids in Japan. In view of this, the salmonid herpesviruses from the U.S. are discussed in this section, while those of Japan are discussed in the following section.

     H. salmonis infection is a mild viral disease of rainbow trout that is caused by *H. salmonis* virus (HPV), which was originally isolated by Wolf and Taylor.[192] A similar, or nearly identical virus, referred to as "steelhead herpesvirus" (SHV) was isolated from steelhead trout in California.[193-194]

1. GEOGRAPHICAL RANGE AND SPECIES SUSCEPTIBILITY

HPV has been found only in Washington state and California. The original HPV isolate was in Washington,[192] but between early 1985 and the middle of 1986 the virus, designated as SHV, was detected in adult steelhead and rainbow trout in seven California hatcheries.[194] Although steelhead and rainbow trout are the natural hosts of *H. salmonis,* chum salmon fry and chinook salmon are experimentally susceptible, while coho and Atlantic salmon and brook and brown trout are resistant.[192,194]

2. CLINICAL SIGNS AND FINDINGS

Adult fish from which the original isolate of HPV was obtained showed a distinctly darkened pigmentation, but no other clinical signs were noted. Approximately 2 weeks after exposure, experimentally infected juvenile rainbow trout stopped eating[195] and became lethargic. The fish lay on the bottom of the tanks, but responded to physical stimulation with erratic swimming. Some fish became dark and most developed exophthalmia accompanied by hemorrhage. Gills were pale and abdomens distended, hemorrhages developed on the fins of some fish and mucoid fecal casts were also evident. Internally, the peritoneal cavity had abundant ascites, which sometimes was bloody and gelatinous. The intestines were flaccid; the liver was hemorrhagic, mottled, or friable, and the kidneys were pale.

3. DIAGNOSIS AND VIRUS CHARACTERISTICS

H. salmonis disease virus is diagnosed by isolation of the infective agent in susceptible cell cultures incubated at 10°C. The virus replicates in CHSE-214, rainbow trout fin (RTF-1), and RTG-2 cells, but not in 3 nonsalmonid cell lines (BB, BF-2, or FHM).[192] Replication of the virus occurred at 5 to 10°C with 10°C being optimum. CPE, consisting principally of syncytia, was depressed or variable at 15°C with complete inhibition at higher temperatures. Identification of HPV is obtained by serum neutralization.[193] The herpesvirus isolate from steelhead trout in California by Hedrick et al.[193] proved to be similar, but distinquishable from the initial isolate. Hedrick et al.[191] serologically compared the North American isolates of *H. salmonis* (HPV and SHV) to herpesviruses of salmonids from Japan (OMV, YFV, NeVTA), and found that isolates from the U.S. were distinctly different from those of Japan. Because of serological differences, the two North American isolates were named salmonid herpesvirus Type 1, and the Japanese isolates salmonid herpesvirus Type 2.[196] The difference between the two North American herpesviruses (Type 1) is unresolved, therefore, for the purposes of the following discussion they will be considered as the same virus — HPV.

HPV is a typical herpesvirus with a capsid that measures 90 to 95 nm. The agent is heat, chloroform, ether, and acid labile. The virion is enveloped and contains a DNA genome.[193,197]

4. EPIZOOTIOLOGY

The first isolation of HPV was from ovarian fluid taken from postspawning rainbow trout that experienced 50% mortality at a federal hatchery in Washington state.[192] All of the brood stock infected with HPV were destroyed and the hatchery disinfected. The virus was not isolated again until more than 10 years later in California.[193] During the outbreaks in 1985 to 1986, virus was isolated in 5.8% of ovarian fluid samples and in 4.5% of pooled tissue samples from steelhead trout at 1 hatchery.[194]

HPV appears to have a very low pathogenicity because even intraperitoneal injection of steelhead alevins, juvenile steelhead, and rainbow trout did not precipitate death, but virus was reisolated from 20 to 40% of these fish over a 20-week postexposure period.[194] Eaton et al.[194] also transmitted HPV (SHV) by waterborne exposure to alevin chinook salmon, followed by isolation from 20% of these fish.

H. salmonis is thought to be transmitted vertically from adults to progeny because of its presence in ovarian fluids, but this means of conveyance is unproven. The disease occurs between January and April when water temperatures are 10°C or less. Incubation in experimental infections is 25 d or more. Wolf and Smith[195] reported mortality continuing for about 50 d after onset of the disease, but Eaton et al.[194] were unable to kill fish with the virus.

5. PATHOLOGICAL MANIFESTATIONS

Wolf and Smith[195] described the histopathologic changes in fish experimentally infected with HPV. Heart tissue was edematous and necrotic with loss of striations in muscle fibers. Kidneys, livers, and spleens were edematous, hyperplastic, congested, and necrotic. Syncytia were seen in hepatocytes of livers by Eaton et al.[194] and in pancreatic tissue by Wolf and Smith.[195] Edematous, hypertrophied, and hemorrhaged gill epithelium became separated from the connective tissue. Eaton et al.[194] also found that fused liver cells had enlarged nuclei containing Feulgen positive, eosinophilic Cowdry Type A inclusion bodies. Skeletal muscle of some infected fish showed hemorrhage and necrosis.

6. SIGNIFICANCE

Because HPV has very low pathogenicity and it is an infrequently encountered disease, its importance to trout culture is minimal.

7. MANAGEMENT

Wolf[76] suggested that hatchery sanitation practices be followed to control *H. salmonis*.

F. *ONCORHYNCHUS MASOU* VIRUS (TYPE 2)

Oncorhynchus masou virus (OMV) disease, originally discovered in adult masou salmon in 1978, also occurs in young members of several species of the genus *Oncorhynchus*.[198-199] The virus is lethal to young fish, but in older, surviving fish it may manifest itself as a neoplastic disease. Several other isolates from fish in Japan are similar to OMV, two of which are Nerka virus Towada Lake, Akita Prefecture (NeVTA) and yamame tumor virus (YTV).

1. GEOGRAPHICAL RANGE AND SPECIES SUSCEPTIBILITY

OMV is known only in Japan, where adult masou and coho salmon may carry the virus and develop tumors. Juvenile masou and coho salmon are susceptible to the virulent phase of OMV. Experimentally sockeye (kokanee), chum, and coho salmon and rainbow trout are susceptible.[200-201]

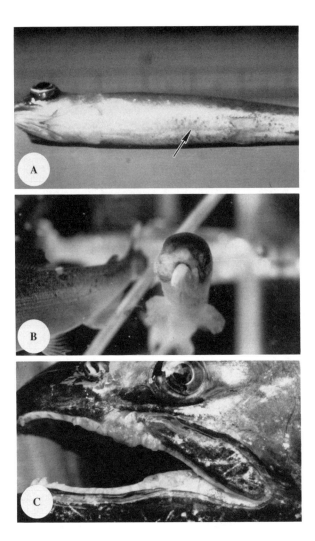

Figure 8 *Oncorhynchus masou* virus in chum salmon. (A) Experimentally infected young fish with petechiael hemorrhage on underside (arrow) and exophthalmia; (B) OMV tumor on mandible of young fish. (From Yoshimizu, M., Tanaka, M., and Kimura, T., *Fish Pathol.*, 22, 7, 1987. With permission of Springer-Verlag.) (C) Tumor on upper jaw of adult fish. (Photograph by M. Yoshimizu. With permission of CRC Press.)

2. CLINICAL SIGNS AND FINDINGS

Clinically OMV manifests itself two different ways; in juvenile fish OMV results in a lethal infection and in older fish as a neoplastic condition in some fish. The virus kills fish that are 30 to 150 d old. Infected juvenile fish lose their appetite and some show exophthalmia and petechiae on the body surface, especially under the jaw (Figure 8). Internally, the liver of infected salmon fry is mottled with white areas, and in advanced cases the liver is totally white. The spleen may be swollen, but the kidney appears normal to the naked eye. No food is present in the digestive tract.[198,200]

Beginning about 4 months postexposure to the virus and persisting for at least 1 year, neoplastic tumors will develop on surviving kokanee, chum, coho, and masou salmon and rainbow trout. These tumors occur mainly around the mouth in the maxillary and mandibular (Figure 8) regions, but may also be found on gill covers, body surface, cornea, and sometimes but rarely, in the kidney.[199-200,202]

3. DIAGNOSIS AND VIRUS CHARACTERISTICS

During active infections, OMV is diagnosed by isolation of the virus in cell cultures from clinically ill salmon less than 6 months old, normal adult salmon, or fish with tumors. The virus will replicate in all cell lines of salmonid origin including CHSE-214 and RTG-2, where the yield is about 10^6 $TCID_{50}$/ml.[198] Optimum cell culture incubation temperature for OMV is 10 to 15°C, but viral replication will occur at 5 to 16°C. CPE which develops in 5 to 7 d at 10 to 15°C is characterized by cells becoming rounded and then marked by syncytium formation and eventual lysis. Cowdry type A inclusion bodies are present in the syncytial cells. Inoculated cultures that do not develop CPE in 10 to 14 d should be blind passed.

Isolates of OMV can be distinguished from other herpesviruses of fish by serum neutralization.[191] Gou et al.[203] prepared recombinant plasmids as a DNA probe for OMV, that detected OMV DNA in tissues of salmonids 2 weeks before the virus could be isolated from asymptomatic or dead fish. Tumors may be confirmed by histological sectioning.

The causative agent is the herpesvirus *Oncorhynchus masou* virus.[198] Without the envelope, the icosahedral nucleocapsid measures 100 to 115 nm in diameter. The enveloped particle measures 220 to 240 nm.[204] The OMV agent, salmonid herpesvirus Type 2, is serologically and morphologically similar, and possibly identical to the YTV and NeVTA agents.[205] Using DNA homology, Eaton et al.[206] showed that OMV and YTV are nearly identical and similar, but distinct from NeVTA. Although OMV and the NeVTA agents are serologically related, the NeVTA virus has not been directly associated with tumor development. OMV is also distinctly different from *H. salmonis* of North America which is classified as salmonid herpesvirus Type 1.[191,193,197]

4. EPIZOOTIOLOGY

OMV was first isolated from a population of mature, normal masou salmon in Hakkaido, Japan. Most research on the virus has been done in Japan where several herpesviruses have been isolated from salmonid fishes.[207] Yoshimizu et al.[208] reported OMV was widespread in northern Japan between 1976 and 1987, but due to management practices the disease has disappeared from anadromous runs in some regions.

OMV is probably transmitted during spawning via contaminated ovarian fluid and horizontally via the water. Initial studies of OMV by Kimura et al.[198] reported that 35 to 60% of 3- to 5-month-old chum salmon died during the ensuing 60 d after infection by immersion in a solution containing 100 $TCID_{50}$/ml. Kimura et al.[200] compared the susceptibility of 0 to 7-month-old chum, masou, kokanee, and coho salmon and rainbow trout at 10 to 15°C. Cumulative mortalities in the chum salmon ranged from 3% for 0 age fish to as high as 98% for 3-month-old fish, but only 7 and 2% for 6- and 7-month-old chum salmon. Similar cumulative mortalities were recorded for the kokanee and masou salmon, but the 1-month-old kokanee suffered 100% mortality. Coho salmon and rainbow trout were less susceptible to OMV infection as their highest mortalities were 39 and 29%, respectively. Sano et al.[205] killed 65% of 5-month-old yammame (masou) when they were injected intraperitoneally with 10^3 $TCID_{50}$ of the YTV, but only 15% when injected with 10^2 $TCID_{50}$. Only 15 and 0% of 5-month-old chum salmon were killed when injected with identical doses, respectively. Tanaka et al.[209] noted an 87% mortality of masou salmon over a 4 month period following infection at 1 month of age. Masou salmon infected at 3 months of age suffered 63% mortality.

The oncogenic nature of OMV provides a model for studies of tumors induced by herpesviruses. Of the masou surviving OMV epizootics, 60% exhibited tumors and 45% of the tumors occurred around the mouth.[198-199] Sano et al.,[205] working with the YTV agent, found that tumors developed in masou and chum salmon in 10 to 13 months after exposure. These tumors developed primarily on the mandibles or premaxillae. A comparison study of tumor incidence in 4 different species of salmonids at different ages conducted by Yoshimizu et al.[202] showed that tumors developed in 100% of 3-month-old masou, 12% of rainbow trout, 35% of coho salmon, and 40 to 71% of chum salmon all of which were between 1 and 7 months of age at the time of exposure (Figure 9). The tumors took 120 to 360 d to develop. Isolation of virus from tumorous tissue is not consistent between OMV and the YTV. OMV was isolated from primary tumor cell cultures after spontaneous degeneration, and the YTV was isolated directly from homogenized tumors.

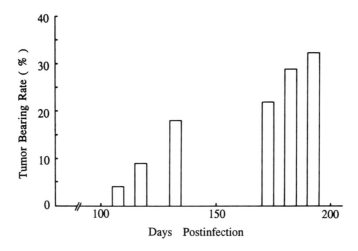

Figure 9   Increasing incidence of tumor development on chum salmon following experimental infection with OMV.[202]

5. PATHOLOGICAL MANIFESTATIONS

Histopathologically, OMV disease varies with species and age of infected fish.[201,209] In masou, chum, and coho salmon, the kidney is usually the principal target organ as judged by the severity of histopathological changes. The epithelium of the mouth, jaw, operculum, and head in 1-month-old masou was necrotic, with focal hepatic necrosis in the liver, spleen, and pancreas developing later. In 3-month-old masou, necrosis of the hematopoietic tissue was seen simultaneously with hyperplasia, and syncytial cells were also present in the kidney. As the infection progressed, the kidney returned to normal, but the liver, which was initially marked only with focal necrosis, became more severely necrotic. Also, liver hepatocytes displayed margination of chromatin. Cell degeneration occurred in the spleen, pancreas, cardiac muscle, and brain.

Tumors of differentiated epithelium developed in 12 to 100% of surviving chum, coho, and masou salmon and rainbow trout between 4 months and 1 year after infection. Generally, tumors developed in chum and coho salmon before they did in rainbow trout and masou salmon.[208] Most tumors were around the mouth and head with tumors also present in decreasing order of frequency: caudal fin, gill cover, body surface, corneas, and kidney. Tumors may include abundantly proliferative, well-differentiated squamous epithelial cells with the papilloma supported by a stroma of fine connective tissue.[210] Several layers of squamous epithelial cells are present in a papillomatous array.[202]

6. SIGNIFICANCE

OMV disease is a major problem, particularly in northern Japan, among land-locked salmon populations and pen-reared fish, especially coho salmon in the sea.[208] OMV was discovered in all but four locations sampled in northern Japan between 1976 and 1987. A significant number of 1- to 5-month-old salmon may be killed during an epizootic with the survivors becoming virus carriers, developing tumors, and the fish being reservoirs of OMV. Since OMV was isolated from ovarian fluids of adult fish, transmission during spawning is most likely, but treatment of eggs with iodophores significantly reduces this possibility.

7. MANAGEMENT

Screening of brood stock for OMV and iodophore disinfection of eggs are strongly suggested because avoidance is the most effective means of preventing disease.[211] Virus is inactivated on the egg surface by immersion in iodophores at 50 mg/l for 20 min. Eggs should be incubated in water from a fish-free source; or in hatcheries with open water supplies, eggs should be incubated in UV irradiated water as OMV is completely inactivated with 3.0×10^3 $\mu W \cdot sec/cm^2$.[212] Fingerlings should also be grown in UV treated water. Yoshimizu et al.[208] stated that incidence of OMV has declined in Japan since the recommendation of using UV irradiated water. Some antiviral agents have been noted to reduce mortality to OMV. Kimura et al.[213] showed that Acyclovir quanine reduced tumor development when fish were immersed for 30 min each d in 25 μg/ml for 2 months. Iododeoxyuridine gave even better results with an oral dosage of 15 μg per fish per d for 2 months, but neither drug is approved for use in food fish.

G. EPIZOOTIC EPITHELIOTROPIC DISEASE

Epizootic epitheliotropic disease (EED) is an acute viral disease of lake trout.[214-215] Since 1983, severe and acute mortalities have occurred in juvenile lake trout populations in several state and federal hatcheries in the Great Lakes basin. Bradley et al.[215] estimated that, in this area, 15 million cultured lake trout died from undetermined disease between 1985 and 1989. Much speculation was made as to the cause of these mortalities, but no definitive etiology was established until transmission studies showed that the disease was infectious in nature.

In 1989, McAllister and Herman[214] showed that a filterable agent was involved in affected lake trout, and presented electron micrographs showing putative herpesvirus particles in tissues from naturally and experimentally infected fish. Similarly, Bradley et al.[215] demonstrated that a filterable agent was involved and that putative herpesvirus capsids could be recovered from affected fish by ultra centrifugation of homogenates. It is now accepted that the etiological agent of EED is a virus and has been named "epizootic epitheliotropic disease virus" (EEDV).

1. GEOGRAPHICAL DISTRIBUTION AND SPECIES SUSCEPTIBILITY

Epizootic epitheliotropic disease is known to occur only in the Great Lakes region of North America, specifically in the states of Michigan and Wisconsin. The only known hosts of EEDV are lake trout and lake trout × brook trout hybrids. The virus has not been observed in wild or feral stocks, although they are considered a possible reservoir. Brown, brook, and rainbow trout as well as chinook and Atlantic salmon are refractive to EEDV.

2. CLINICAL SIGNS AND FINDINGS

Clinical signs of EED include lethargy, riding high in the water, and sporadic flashing and corkscrew diving.[214-215] Hemorrhages occur in the eyes and occasionally in the mouth. Gray to white mucoid blotches appear on the skin and fins of infected fish. Secondary fungal infections (presumably *Saprolegnia* sp.) develop on the eyes, fins, and body surface. The only visible internal lesion is a swollen spleen. Hematocrits are elevated, averaging about 45% compared to about 40% in normal fish.

3. DIAGNOSIS AND VIRUS CHARACTERISTICS

EED is presumptively diagnosed by recognition of clinical signs in lake trout that are experiencing acute mortality. Electron micrographs of epithelial tissues will reveal typical EEDV particles if the virus is present, and the virus can be recovered from skin tissue by ultracentrifugation and visualized by electron microscopy. Attempts to isolate the virus in CHSE-214, EPC, FHM, and RTG-2 cell cultures failed, therefore, no *in vitro* virological techniques are currently available for proper identification.

The etiological agent of EED is almost certainly a virus, having been transmitted by filtrates from infected fish. Although the virus has not been characterized due to a lack of isolation in cell culture, it has tentatively been placed in the family Herpesviridae. Electron microscopy of material harvested from infected tissues by ultracentrifugation revealed hexagonal, unenveloped viral particles (presumably viral capsids) that measured 100 to 105 nm and enveloped particles that measured 220 to 235 nm.[215] McAllister and Herman[214] found 2 types of ellipsoid to spherical viral particles; 1 type measured 130 to 175 nm in diameter with a diffused electron-dense nucleoid that was 75 to 100 nm in diameter. The other type of particle measured 150 to 200 nm in diameter with a diffused nucleoid measuring 75 to 100 nm in diameter.

4. EPIZOOTIOLOGY

Epizootics of EED occur in spring or fall when water temperatures are 6 to 15°C. However, the disease tends to occur primarily in the spring when the water temperature is 8 to 10°C. EED has been diagnosed only in cultured lake trout, and mortality can approach 100% in fish up to 14 months of age. The survivors of an epizootic shed the infectious agent. Experimentally, EEDV was transmitted to naive lake trout and lake trout × brook trout hybrids by cohabitation, and via waterborne exposure in a contaminated water bath containing filtered (0.22μm) material from homogenates of diseased fish.[214-215]

EED is stress mediated because almost any stressful condition will precipitate the disease. The process of anesthesia and fin clipping for identification prior to stocking appears to be a major predisposing factor in precipitating mortality in fingerling and yearling lake trout.[216]

5. PATHOLOGICAL MANIFESTATIONS

Histopathology is the same in EEDV-infected lake trout whether the disease occurs naturally or is experimentally induced.[215] The epithelium contains hypertrophied cells and areas of hyperplasia with lymphocytic infiltration, hydrophic degeneration, and necrosis. Macrophages laden with cellular debris are common in the kidneys and sinusoids of the liver. In advanced cases of EEDV, gills develop inconsistent edema and generally this tissue is eosinophilic. Intranuclear inclusion bodies are sometimes seen in epidermal cells.

6. SIGNIFICANCE

A filterable, infectious agent was implicated in at least some EED cases, and strong evidence points to a herpesvirus. The virus is capable of killing an enormous number of juvenile lake trout, but probably does not affect other species.

7. MANAGEMENT

Management and control of EEDV relies upon avoidance and prevention. Hatcheries where the disease has occurred were disinfected with chlorine and stringent regulations have been imposed on reintroduction of fish. Hatcheries which were so treated following EED epizootics in 1988 have had no reoccurrences of the disease as of 1991.[216] Experimentally, treating water with high doses of UV irradiation will prevent horizontal transmission of EEDV.

REFERENCES

1. **Amend, D. F., Yasutake, W. T., and Mead, R. W.,** A hematopoietic virus disease of rainbow trout and sockeye salmon, *Trans. Am. Fish. Soc.,* 98, 796, 1969.
2. **Rucker, R. R., Whipple, W. J., Parvin, J. R., and Evans, C. A.,** A contagious disease of salmon possibly of virus origin, *U.S. Fish Wildl. Serv. Fish. Bull.,* No. 76, 46, 1953.
3. **Watson, S. W., Guenther, R. W., and Rucker, R. R.,** A virus disease of sockeye salmon: interim report, *U.S. Fish Wildl. Serv. Spec. Sci. Rep.* 138, 1954, 36.
4. **Yasutake, W. T.,** Comparative histopathology of epizootic salmonid virus disease, in *Symposium on Diseases of Fishes and Shellfishes,* Snieszko, S. F., Ed., American Fisheries Society Special Publ. No. 5, Washington, D.C., 1970, 351.
5. **Amend, D. F. and Chambers, V. C.,** Morphology of certain viruses of salmonid fishes. I. *In vitro* studies of some viruses causing hematopoietic necrosis, *J. Fish. Res. Board Can.,* 27, 1285, 1970.
6. **McCain, B. B., Fryer, J. L., and Pilcher, K. S.,** Antigenic relationship in a group of three viruses of salmonid fish by cross neutralization, *Proc. Soc. Exp., Biol. Med.,* 137, 1042, 1971.
7. **Plumb, J. A.,** A virus-caused epizootic of rainbow trout *(Salmo gairdneri)* in Minnesota, *Trans. Am. Fish. Soc.,* 101, 121, 1972.
8. **Amend, D. F.,** Detection and transmission of infectious hematopoietic necrosis virus in rainbow trout, *J. Wildl. Dis.,* 11, 471, 1975.
9. **Sano, T.,** Viral diseases of cultured fishes in Japan, *Fish Pathol.,* 10, 221, 1976.
10. **Hattenberger-Baudouy, A. M., Danton, M., Merle, G., Torchy, C., and de Kinkelin, P.,** Serological evidence of infectious hematopoietic necrosis in rainbow trout from a French outbreak of disease, *J. Aquat. Anim. Health,* 1, 126, 1989.
11. **Bovo, G., Giorgetti, G., Jørgensen, P. E. V., and Olesen, N. J.,** Infectious hematopoietic necrosis: first detection in Italy, *Bull. Eur. Assoc. Fish Pathol.,* 7, 24, 1987.
12. **LaPatra, S. E., Fryer, J. L., Wingfield, W. H., and Hedrick, R. P.,** Infectious hematopoietic necrosis virus (IHNV) in coho salmon, *J. Aquat. Anim. Health,* 1, 277, 1989.
13. **Wingfield, W. H. and Chan, L. D.,** Studies on the Sacramento River chinook disease and its causative agent, in *Symposium on Diseases of Fishes and Shellfishes,* Snieszko, S. F., Ed., American Fisheries Society Special Publ. No. 5, Washington, D.C., 307, 1970.

14. **LaPatra, S. E. and Fryer, J. L.,** Susceptibility of brown trout *(Salmo trutta)* to infectious hematopoietic necrosis virus, *Bull. Eur. Assoc. Fish Pathol.,* 10, 125, 1990.

15. **Mulcahy, D.,** Atlantic Salmon *(Salmo salar)* are naturally susceptible to infectious hematopoietic necrosis (IHN) virus, U.S. Fish Wildl. Serv. Res. Inf. Bull., Washington, D.C., No. 84-84, 1984, 2.

16. **Burke, L. and Grischkowsky, R.,** An epizootic caused by infectious hematopoietic necrosis virus in an enhanced population of sockeye salmon, *Oncorhynchus nerka* (Walbaum), smolts at Hidden Creek, Alaska, *J. Fish Dis.,* 7, 421, 1984.

17. **Yamamoto, T., Batts, W. N., Arakawa, C. K., and Winton, J. R.,** Multiplication of infectious hematopoietic necrosis virus in rainbow trout following immersion infection: whole-body assay and immunohistochemistry, *J. Aquat. Anim. Health,* 2, 271, 1990.

18. **Burke, F. and Mulcahy, D.,** Retention of infectious hematopoietic necrosis virus infectivity in fish tissue homogenates and fluids stored at three temperatures, *J. Fish Dis.,* 6, 543, 1983.

19. **Yoshimizu, M., Kamei, M., Dirakbusarakom, S., and Kimura, T.,** Fish cell lines: susceptibility to salmonid viruses, in *Invertebrate and Fish Tissue Culture,* Kuroda, Y., Kurstak, E., and Maramorosch, K., Eds., Japan Scientific Societies Press, Tokyo, 1988, 207.

20. **Wingfield, W. H., Fryer, J. L., and Pilcher, K. S.,** Properties of the sockeye salmon virus (Oregon strain) (33719), *Proc. Soc. Exp. Biol. Med.,* 130, 1055, 1969.

21. **Batts, W. N. and Winton, J. R.,** Enhanced detection of infectious hematopoietic necrosis virus and other fish viruses by pretreatment of cell monolayers with polyethylene glycol, *J. Aquat. Anim. Health,* 1, 284, 1989.

22. **Winton, J. R.,** Recent advances in detection and control of infectious hematopoietic necrosis virus in aquaculture, *Annu. Rev. Fish Dis.,* 1991, 83.

23. **LaPatra, S. E., Robert, K. A., Rohovec, J. S., and Fryer, J. L.,** Fluorescent antibody test for the rapid diagnosis of infectious hematopoietic necrosis, *J. Aquat. Anim. Health,* 1, 29, 1989.

24. **Ristow, S. S. and Arnzen, J. M.,** Development of monoclonal antibodies that recognize a type-2 specific and a common epitope on the nucleoprotein of infectious hematopoietic necrosis virus, *J. Aquat. Anim. Health,* 1, 119, 1989.

25. **Arnzen, J. M., Ristow, S. S., Hesson, C. P., and Lientz, J.,** Rapid fluorescent antibody tests for infectious hematopoietic necrosis virus (IHNV) utilizing monoclonal antibodies to the nucleoprotein and glycoprotein, *J. Aquat. Anim. Health,* 3, 109, 1991.

26. **Dixon, P. F. and Hill, B. J.,** Rapid detection of fish rhabdoviruses by the enzyme-linked immunosorbent assay (ELISA), *Aquaculture,* 42, 1, 1984.

27. **Medina, D. J., Chang, P. W., Fradley, T. M., Yeh, M. T., and Sadasiv, E. C.,** Diagnosis of infectious hematopoietic necrosis virus in Atlantic salmon *Salmo salar* by enzyme-linked immunosorbent assay, *Dis. Aquat. Org.,* 13, 147, 1992.

28. **McAllister, P. E. and Schill, W. B.,** Immunoblot assay: A rapid and sensitive method for identification of salmonid fish viruses, *J. Wildl. Dis.,* 22, 468, 1986.

29. **Arakawa, C. K., Deering, R. E., Higman, K. H., Oshima, R. J., O'Hara, P. J., and Winton, J. R.,** Polymerase chain reaction (PCR) amplification of a nucleoprotein gene sequence of infectious hematopoietic necrosis virus, *Dis. Aquat. Org.,* 8, 165, 1990.

30. **Deering, R. E., Arakawa, C. K., Oshima, K. H., O'Hara, P. J., Landolt, M. L., and Winton, J. R.,** Development of a biotinylated DNA probe for detection and identification of infectious hematopoietic necrosis virus, *Dis. Aquat. Org.,* 11, 57, 1991.

31. **Darlington, R. W., Trafford, R., and Wolf, K.,** Fish rhabdoviruses: morphology and ultrastructure of North American salmonid isolates, *Arch. Gesamte. Virusforsch.,* 39, 257, 1972.

32. **Pietsch, J. P., Amend, D. F., and Miller, C. M.,** Survival of infectious hematopoietic necrosis virus held under various environmental conditions, *J. Fish. Res. Board Can.,* 34, 1360, 1977.

33. **Barja, J. L., Toranzo, A. E., Lemos, M. L., and Hetrick, F. M.,** Influence of water temperature and salinity on the survival of IPN and IHN viruses, *Bull. Eur. Assoc. Fish Pathol.,* 3(4), 47, 1983.

34. **Engelking, H. M., Harry, J. B., and Leong, J. C.,** Comparison of representative strains of infectious hematopoietic necrosis virus by serological neutralization and cross-protection assays, *Appl. Environ. Microbiol.,* 57, 1372, 1991.

35. **Hsu, Y. L., Engelking, H. M., and Leong, J. C.,** Occurrence of different types of infectious hematopoietic necrosis virus in fish, *Appl. Environ. Microbiol.,* 52, 1353, 1986.

36. **Winton, J. R., Arakawa, C. K., Lannan, C. N., and Fryer, J. L.,** Neutralizing monoclonal antibodies recognize antigenic variants among isolates of infectious hematopoietic necrosis virus, *Dis. Aquat. Org.,* 4, 199, 1988.

37. **Arkush, K. D., Bovo, G., de Kinkelin, P., Winton, J. R., Wingfield, W. H., and Hedrick, R. P.,** Biochemical and antigenic properties of the first isolates of infectious hematopoietic necrosis virus from salmonid fish in Europe, *J. Aquat. Anim. Health,* 1, 148, 1989.

38. **Amend, D. F. and Nelson, J. R.,** Variation in the susceptibility of sockeye salmon *Oncorhynchus nerka* to infectious hematopoietic necrosis virus, *J. Fish Biol.,* 22, 567, 1977.

39. **McIntyre, J. D. and Amend, D. J.,** Heritability of tolerance for infectious hematopoietic necrosis in sockeye salmon *(Oncorhynchuys nerka), Trans. Am. Fish. Soc.,* 107, 305, 1978.

40. **LaPatra, S. E., Groff, J. M., Fryer, J. L., and Hedrick, R. P.,** Comparative pathogenesis of three strains of infectious hematopoietic necrosis virus in rainbow trout *Oncorhynchus mykiss, Dis. Aquat. Org.,* 8, 105, 1990.

41. **Pilcher, K. S. and Fryer, J. L.,** The viral diseases of fish: a review through 1978. Part I: diseases of proven etiology, *Crit. Rev. Microbiol.,* 7, 289, 1980.

42. **LaPatra, S. E., Groberg, W. J., Rohovec, J. S., and Fryer, J. L.,** Size-related susceptibility of salmonids to two strains of infectious hematopoietic necrosis virus, *Trans. Am. Fish. Soc.,* 119, 25, 1990.

43. **Amend, D. F.,** Control of infectious hematopoietic necrosis virus disease by elevating the water temperature, *J. Fish. Res. Board Can.,* 27, 265, 1970.

44. **Hetrick, F. M., Fryer, J. L., and Knittel, M. D.,** Effect of water temperature on the infection of rainbow trout *Salmo gairdneri* Richardson with infectious hematopoietic necrosis virus, *J. Fish Dis.,* 2, 253, 1972.

45. **Amos, K. H., Hooper, K. A., and LeVander, L.,** Absence of infectious hematopoietic necrosis virus in adult sockeye salmon, *J. Aquat. Anim. Health,* 1, 281, 1989.

46. **Mulcahy, D., Pascho, R. J., and Jenes, C. K.,** Titer distribution patterns of infectious hematopoietic necrosis virus in ovarian fluids of hatchery and feral salmon populations, *J. Fish Dis.,* 6, 183, 1983.

47. **Grischkowsky, R. S. and Amend, D. F.,** Infectious hematopoietic necrosis virus: Prevalence in certain Alaskan sockeye salmon, *Oncorhynchus nerka, J. Fish. Res. Board Can.,* 33, 186, 1976.

48. **Mulcahy, D., Jenes, C. K., and Pascho, R.,** Appearance and quantification of infectious hematopoietic necrosis virus in female sockeye salmon *(Oncorhynchus nerka)* during their spawning migration, *Arch. Virol.,* 80, 171, 1984.

49. **Mulcahy, D. and Pascho, R. J.,** Vertical transmission of infectious hematopoietic necrosis virus in sockeye salmon, *Oncorhynchus nerka* (Walbaum): isolation of virus from dead eggs and fry, *J. Fish Dis.,* 8, 393, 1985.

50. **Mulcahy, D. and Pascho, R. J.,** Adsorption of fish sperm of vertically transmitted viruses, *Science,* 225, 333, 1984.

51. **Yoshimizu, M., Sami, M., and Kimura, T.,** Survivability of infectious hematopoietic necrosis virus in fertilized eggs of masou and chum salmon, *J. Aquat. Anim. Health,* 1, 13, 1989.

52. **Mulcahy, D., Pascho, R. J., and Jenes, C. K.,** Detection of infectious hematopoietic necrosis virus in river water and demonstration of waterborne transmission, *J. Fish Dis.,* 6, 321, 1983.

53. **Jørgensen, P. E. V., Olesen, N. J., Lorenzen, N., Winton, J. R., and Ristow, S. S.,** Infectious hematopoietic necrosis (IHN) and viral hemorrhagic septicemia (VHS): detection of trout antibodies to the causative viruses by means of plaque neutralization, immunofluorescence and enzyme-linked immunosorbent assay, *J. Aquat. Anim. Health,* 3, 100, 1991.

54. **Mulcahy, D., Klaybor, D., and Batts, W. N.,** Isolation of infectious hematopoietic necrosis virus from a leech *(Piscicola salmositica)* and a copepod *(Salmincola* sp.), ectoparasites of sockeye salmon *Oncorhynchus nerka, Dis. Aquat. Org.,* 8, 29, 1990.

55. **Williams, I. V. and Amend, D. F.,** A natural epizootic of infectious hematopoietic necrosis in fry of sockeye salmon *(Oncorhynchus nerka)* at Chilko Lake, British Columbia, *J. Fish. Res. Board Can.,* 33, 1564, 1976.

56. **Traxler, G. S. and Rankin, J. B.,** An infectious hematopoietic necrosis epizootic in sockeye salmon *Oncorhynchus nerka* in Weaver Creek Spawning channel, Fraser River system, B.C., Canada, *Dis. Aquat. Org.,* 6, 221, 1989.

57. **Yasutake, W. T.,** Fish viral diseases: Clinical, histopathological, and comparative aspects, in *The Pathology of Fishes,* Ribelin, W. E. and Migaki, G., Eds., University of Wisconsin Press, Madison, 1975, 247.

58. **Yasutake, W. T. and Amend, D. F.,** Some aspects of pathogenesis of infectious hematopoietic necrosis (IHN), *J. Fish Biol.,* 4, 261, 1972.

59. **Amend, D. F. and Smith, L.,** Pathophysiology of infectious hematopoietic necrosis virus disease in rainbow trout: Hematological and blood chemical changes in moribund fish, *Inflam. Immun.,* 11, 171, 1975.

60. **LaPatra, S. E., Rohovec, J. S., and Fryer, J. L.,** Detection of infectious hematopoietic necrosis virus in fish mucus, *Fish Pathol.,* 24, 197, 1989.

61. **Yamamoto, T. and Clermont, T. J.,** Multiplication of infectious hematopoietic necrosis virus in rainbow trout following immersion infection: organ assay and electron microscopy, *J. Aquat. Anim. Health,* 2, 261, 1990.

62. **Sano, T., Nishimura, T., Okamoto, N., Yamazaki, T., Hanada, J., and Watanabe, Y.,** Studies on viral diseases of Japanese fishes. VI. Infectious hematopoietic necrosis (IHN) of salmonids in the mainland Japan, *J. Tokyo Univ. Fish.,* 63, 63, 1977.

63. **Meyers, T. R., Thomas, J. B., Follett, J. E., and Saft, R. R.,** Infectious hematopoietic necrosis virus: trends in prevalence and the risk management approach in Alaskan sockeye salmon culture, *J. Aquat. Anim. Health,* 2, 85, 1990.

64. **Amend, D. F. and Pietsch, J. P.,** Virucidal activity of two iodophors to salmonid viruses, *J. Fish. Res. Board Can.,* 29, 61, 1972.

65. **Hasobe, M. and Saneyoshi, M.,** On the approach to the viral chemotherapy against infectious hematopoietic necrosis virus (IHN) *in vitro* and *in vivo* on salmonid fishes, *Fish Pathol.,* 20, 343, 1985.

66. **Batts, W. N., Landolt, M. L., and Winton, J. R.,** Inactivation of infectious hematopoietic necrosis virus by low levels of iodine, *Appl. Environ. Microbiol.,* 57, 1379, 1991.

67. **Fryer, J. L., Rohovec, J. S., Tebbit, G. L., McMichael, J. S., and Pilcher, K. S.,** Vaccination for control of infectious diseases in Pacific salmon, *Fish Pathol.,* 10, 155, 1976.

68. **Nishimura, T., Sasaki, H., Ushiyama, M., Inoue, K., Suzuki, Y., Ikeya, F., Tanaka, M., Suzuki, H., Kohara, M., Arai, M. Shima, N., and Sano, T.,** A trial vaccination against rainbow trout fry with formalin killed IHN virus, *Fish Pathol.,* 20, 435, 1985.

69. **Engelking, H. M. and Leong, J. C.,** Glycoprotein from infectious hematopoietic necrosis virus (IHNV) induces protective immunity against five IHNV types, *J. Aquat. Anim. Health,* 1, 291, 1989.

70. **Gillmore, R. D., Jr., Engelking, H. M., Manning, D. S., Leong, J. C.,** Expression in *Escherichia coli* of an epitope of the glycoprotein of infectious hematopoietic necrosis virus protects against viral challenge, *Bio/Tech.,* 6, 295, 1988.

71. **Oberg, L. A., Wirkkula, J., Mourich, D., and Leong, J. C.,** Bacterially expressed nucleoprotein of infectious hematopoietic necrosis virus augments protective immunity induced by the glycoprotein vaccine in fish, *J. Virol.,* 65, 4486, 1991.

72. **M'Gonigle, R. H.,** Acute catarrhal enteritis of salmonid fingerlings, *Trans. Am. Fish. Soc.,* 70, 297, 141.

73. **Wood, E. M., Snieszko, S. F., and Yasutake, W. T.,** Infectious pancreatic necrosis in brook trout, *Am. Med. Assoc. Arch. Pathol.,* 60, 26, 1955.

74. **Snieszko, S. F., Wolf, K., Camper, J. E., and Pettijohn, L. L.,** Infectious nature of pancreatic necrosis, *Trans. Am. Fish. Soc.,* 88, 289, 1959.

75. **Wolf, K., Snieszko, S. F., Dunbar, C. E., and Pyle, E.,** Virus nature of infectious pancreatic necrosis in trout, *Soc. Exp. Biol. Med.,* 104, 105, 1960.

76. **Wolf, K.,** *Fish Viruses and Fish Viral Diseases,* Cornell University Press, Ithaca, NY, 1988, 476.

77. **Jiang, Y. and Zhengqiu, L.,** Isolation of IPN virus from imported rainbow trout *(Salmo gairdneri)* in the P.R. of China, *J. Appl. Ichthyol.,* 3, 191, 1987.

78. **Jimenez, F. J., Marcotequi, M. A., San-Juan, M. L., Basurco, B.,** Diagnosis of viral diseases in salmonid farms in Spain, *Bull. Eur. Assoc. Fish. Pathol.,* 8, 1, 1988.

79. **Christie, K. E. and Haverstein, L. L.,** A new serotype of infectious pancreatic necrosis virus (IPN N1), in *Viruses of Lower Vertebrates,* Ahne, W. and Kurstak, E., Eds., Springer-Verlag, Berlin, 1989, 279.

80. **McAllister, P. E., Newman, M. W., Sauber, J. H., and Owens, W. J.,** Isolation of infectious pancreatic necrosis virus (serotype Ab) from diverse species of estuarine fish, *Helg. Wiss. Meeresunters.,* 37, 317, 1984.

81. **McAllister, P. E.,** Salmonid fish viruses, in *Fish Medicine,* Stoskopf, M. K., Ed., W.B. Saunders, Philadelphia, 1993, 380.

82. **Hill, B. J.,** Properties of a virus isolated from the bivalve mollusc *Tellina tenus* (da Costa), in *Wildlife Diseases,* Page, L. A., Ed., Plenum Press, NY, 1976, 445.

83. **Buck, D., Finlay, J., McGregor, D., and Seagrave, C.,** Infectious pancreatic necrosis (IPN) virus: its occurence in captive and wild fish in England and Wales, *J. Fish Dis.,* 2, 549, 1979.

84. **Sano, T., Okamoto, N., and Nishimura, T.,** A new viral epizootic of *Anguilla japonica* Temminck and Schlegel, *J. Fish Dis.,* 4, 127, 1981.

85. **Bonami, J. R., Cousserans, F., Weppe, M., and Hill B. J.,** Mortalities in hatchery-reared sea bass fry associated with Birnavirus, *Bull. Eur. Assoc. Fish Pathol.,* 3, 41, 1983.

86. **Schutz, M., May, E. B., Kraeuter, J. N., and Hetrick, F. M.,** Isolation of infectious pancreatic necrosis virus from an epizootic occurring in cultured striped bass, *Morone saxatilis* (Walbaum), *J. Fish Dis.,* 7, 505, 1984.

87. **Stephensen, E. B., Newman, M. W., Zachary, A. L., and Hetrick, F. M.,** A viral aetiology for the annual spring epizootics of Atlantic menhaden *Brevoortia tyrannus* (Latrobe) in Chesapeake Bay, *J. Fish Dis.,* 3, 387, 1980.

88. **McKnight, I. J. and Roberts, R. J.,** The pathology of infectious pancreatic necrosis. 1. The sequential histopathology of the naturally occurring condition, *Br. Vet. J.,* 132, 76, 1976.

89. **Roberts, R. J. and McKnight, I. J.,** The pathology of infectious pancreatic necrosis. 2. Stress-mediated recurrence, *Br. Vet. J.,* 132, 209, 1976.

90. **Hattore, M., Kodama, H., Ishiguro, S., Honda, A., Mikami T., and Izawa, H.,** *In vitro* and *in vivo* detection of infectious pancreatic necrosis virus in fish by enzyme-linked immunosorbent assay, *Am. J. Vet. Res.,* 45, 1876, 46.

91. **Rodak, L., Pospisil, Z., Tomanek, J., Vesely, T., Obr, T., and Valicek, L.,** Enzyme-linked immunosorbent assay (ELISA) detection of infectious pancreatic necrosis virus (IPNV) in culture fluids and tissue homogenates of the rainbow trout, *Salmo gairdneri* Richardson, *J. Fish Dis.,* 11, 225, 1988.

92. **Rimstad, E., Krona, R., Hornes, E., Olsvik, O., and Hyllseth, B.,** Detection of infectious pancreatic necrosis virus (IPNV) RNA by hybridization with an oligonucleotide DNA probe, *Vet. Micro.,* 23, 211, 1990.

93. **Dominguez, J., Babin, M., Sanchez, C., and Hedrick, R. P.,** Rapid serotyping of infectious pancreatic necrosis virus by one-step enzyme-linked immunosorbent assay using monoclonal antibodies, *J. Virol. Meth.,* 31, 93, 1991.

94. **Babin, M., Hernandez, C., Sanchez, C., and Dominquez, C.,** Deteccion rapida del virus de la necrosis pancreatica infecciosa por enzimo-imuno-adsorcion de captura, *Med. Vet.,* 7, 557, 1990.

95. **Wolf, K., Quimby, M. C., Carlson, C. P., and Bullock, G. L.,** Infectious pancreatic necrosis: selection of virus-free stock from a population of carrier trout, *J. Fish. Res. Board Can.,* 25, 383, 1967.

96. **McAllister, P. E., Owens, W. J., and Ruppenthal, T. M.,** Detection of infectious pancreatic necrosis virus in pelleted cell and particulate components from ovarian fluid of brook trout *Salvelinus fontinalis, Dis. Aquat. Org.,* 2, 235, 1987.

97. **Dobose, P., Hill, B. J., Hallett, R., Kells, D. T. C., Becht, H., and Teninges, D.,** Biophysical and biochemical characterization of five animal viruses with bisegmented double-stranded RNA genomes, *J. Virol.,* 32, 593, 1977.

98. **Malsberger, R. G. and Cerini, C. P.,** Characteristics of infectious pancreatic necrosis virus, *J. Bacteriol.,* 80, 1283, 1963.

99. **Cohen, J., Poinsard, A., and Scherrer, R.,** Physico-chemical and morphological features of infectious pancreatic necrosis virus, *J. Virol.,* 21, 485, 1973.

100. **Barja, J. L., Toranzo, A. E., Lemos, M. L., and Hetrick, F. M.,** Influence of water temperature and salinity on the survival of IPN and IHN viruses, *Bull. Eur. Assoc. Fish Pathol.,* 3(4), 47, 1983.

101. **Desautels, D. and Mackelvie, R. M.,** Practical aspects of survival and destruction of infectious pancreatic necrosis virus, *J. Fish. Res. Board Can.,* 32, 523, 1975.

102. **Wolf, K., Quimby, M. C., and Carlson, C. P.,** Infectious pancreatic necrosis virus: Lyophilization and subsequent stability in storage at 4°C, *Prog. Fish. Cult.,* 17, 623, 1969.

103. **Hill, B. J.,** Infectious pancreatic necrosis virus and its virulence, in *Microbial Diseases of Fish,* Roberts, R. J., Ed., Academic Press, London, 1982, 81.

104. **Nicholson, B. L. and Pochebit, S., 1981.** Antigenic analysis of infectious pancreatic necrosis viruses (IPNV) by neutralization kinetics, in *Developments in Biological Standardization – Fish Biologicals: Serodiagnostics and Vaccines,* Anderson, D. P. and Hennessen, W., Eds., S. Karger, Basel, 1981, 35.

105. **Ishiguro, S., Izawa, H., Kodama, H., Onuma, M., and Mikami, T.,** Serological relationships among five strains of infectious pancreatic necrosis virus, *J. Fish Dis.,* 7, 127, 1984.

106. **Hill, B. J. and Way, K.,** Proposed standardization of the serological classification of aquatic Birnaviruses, International Fish Health Conference, Vancouver, B.C., Canada, 1988, 151 (Abstr.).

107. **McAllister, P. E. and Owens, W. J.,** Infectious pancreatic necrosis virus: Protocol for a standard challenge in brook trout, *Trans. Am. Fish. Soc.,* 115, 466, 1986.

108. **Wechsler, S. J., Schults, C. L., McAllister, P. E., May, E. B., and Hetrick, F. M.,** Infectious pancreatic necrosis virus in striped bass *Morone saxatilis:* experimental infection of fry and fingerlings, *Dis. Aquat. Org.,* 1, 203, 1986.

109. **McAllister, K. W. and McAllister, P. E.,** Transmission of infectious pancreatic necrosis virus from carrier striped bass to brook trout, *Dis. Aquat. Org.,* 4, 101, 1988.

110. **Eskildsen, U. K. and Jørgensen, P. E. V.,** On the possible transfer of trout pathogenic viruses by gulls, *Rev. Ital. Pisci. Ittio.,* 8(4), 104, 1973.

111. **McAllister, P. E. and Owens, W. J.,** Recovery of infectious pancreatic necrosis virus from a faeces of wild piscivorous birds, *Aquaculture,* 106, 227, 1992.

112. **Peters, F. and Newkirch, M.,** Transmission of some fish pathogenic viruses by the heron, *Ardea cinerea, J. Fish Dis.,* 9, 539, 1986.

113. **Frantsi, C. and Savan, M.,** Infectious pancreatic necrosis virus — temperature and age factors in mortality, *J. Wildl. Dis.,* 7, 249, 1971.

114. **Yamamoto, T.,** Infectious pancreatic necrosis virus and bacterial kidney disease appearing concurrently in populations of *Salmo gairdneri* and *Salvelinus fontinalis, J. Fish. Res. Board Can.,* 32, 1, 1975.

115. **Mulcahy, D. M. and Fryer, J. L.,** Double infection of rainbow trout fry with IHN and IPN virus, *Fish Health News,* 5, 5, 1976.

116. **Wolf K., Quimby, M. C., and Bradford, A. D.,** Egg-associated transmission of IPN virus of trout, *Virology,* 21, 317, 1963.

117. **Bullock, G. L., Rucker, R. R., Amend, D., Wolf, K., and Stucky, H. M.,** Infectious pancreatic necrosis: transmission with iodine-treated and nontreated eggs of brook trout *(Salvelinus fontinalis), J. Fish. Res. Board Can.,* 33, 1197, 1976.

118. **Ahne, W. and Negele, R. D.,** Studies on the transmission of infectious pancreatic necrosis virus via eyed eggs and sexual products of salmonid fish, in *Fish and Shellfish Pathology,* Ellis, A. E., Ed., Academic Press, London, 1985.

119. **Bootland, L. M., Dobos, P., and Stevenson, R. M. W.,** The IPNV carrier state and demonstration of vertical transmission in experimentally infected brook trout, *Dis. Aquat. Org.,* 10, 13, 1991.

120. **Mulcahy, D. and Pascho, R. J.,** Absorption on fish sperm of vertically transmitted viruses, *Science,* 225, 33, 1984.

121. **Ahne, W.,** Presence of infectious pancreatic necrosis virus in the seminal fluid of rainbow trout, *Salmo gairdneri* Richardson, *J. Fish Dis.,* 6, 377, 1983.

122. **Mangunwiryo, H. and Agius, C.,** Studies on the carrier state of infectious pancreatic necrosis virus in rainbow trout, *Salmo gairdneri* Richardson, *J. Fish Dis.,* 11, 125, 1988.

123. **Billi, J. L. and Wolf, K.,** Quantitative comparison of peritoneal washes and feces for detecting infectious pancreatic necrosis (IPN) virus carrier brook trout, *J. Fish. Res. Board Can.,* 26, 1459, 1969.

124. **Hedrick, R. P. and Fryer, J. L.,** Persistent infections of salmonid cell lines with infectious pancreatic necrosis virus (IPNV): A model for the carrier state in trout, *Fish Pathol.,* 16, 163, 1982.

125. **Yamamoto, T.,** Frequency of detection and survival of infectious pancreatic necrosis virus in a carrier population of brook trout *(Salvelinus fontinalis)* in a lake, *J. Fish. Res. Board Can.,* 32, 568, 1975.

126. **Sano, T.,** Studies on viral diseases of Japanese fishes — II. Infectious pancreatic necrosis of rainbow trout: Pathogenicity of the isolants, *Bull. Jpn. Soc. Sci. Fish.,* 37, 499, 1971.

127. **Bootland, L. M., Dobos, P., and Stevenson, R. M. W.,** Fry age and size effects on immersion immunization of brook trout, *Salvelinus fontinalis* Mitchell, against infectious pancreatic necrosis virus, *J. Fish Dis.,* 13, 113, 1990.

128. **Manning, S. D. and Leong, J. C.,** Expression in *Escherichia coli* of the large genomic segment of infectious pancreatic necrosis virus, *Virology,* 179, 16, 1990.

129. **McAllister, P. E. and Reyes, X.,** Infectious pancreatic necrosis virus: isolation from rainbow trout, *Salmo gairdneri* Richardson, imported into Chile, *J. Fish Dis.,* 7, 319, 1984.

130. **Jensen, M. H.,** Research on the virus of Egtved disease, *Ann. N.Y. Acad. Sci.,* 126, 422, 1965.

131. **Jensen, M. H.,** Preparation of fish tissue cultures for virus research, *Bull. Off. Int. Epizool.,* 59(1–2), 131, 1963.

132. **Hooper, K.,** The isolation of VHSV from chinook salmon at Glenwood Springs, Orcas Island, Washington, *Fish Health Sect. Am. Fish. Soc. News Lett.,* 17(2), 1, 1989.

133. **Brunson, R., True, R., and Yancey, J.,** VHS virus isolated at Makah National Fish Hatchery, *Fish Health Sect. Am. Fish. Soc. News Lett.,* 17(2), 3, 1989.

134. **de Kinkelin, P. and Le Berre, M.,** Demonstration de la protection de la truite arc-enciel contre la SHV, par l'administraion d'un virus inactive, *Bull. Off. Int. Epizool.,* 87(5–6), 401, 1977.

135. **Jørgensen, P. E. V.,** Egtved virus: The susceptibility of brown trout and rainbow trout to eight virus isolates and the significance of the findings of VHS control, in *Fish Diseases,* Ahne, W., Ed., Springer-Verlag, Berlin, 1980, 3.

136. **Dorson, M., Chevassus, B., and Torhy, C.,** Comparative susceptibility of three species of char and of rainbow trout X char triploid hybrids to several pathogenic salmonid viruses, *Dis. Aquat. Org.,* 11, 217, 1991.

137. **Meier, Von W.,** Virale haemorrhagic septikaemie: empfanglichkeit un epizootiologische rolle des hechts *(Esox lucus* L.), *J. Appl. Ichthyol.,* 4, 171, 1985.

138. **Castric, J. and de Kinkelin, P.,** Experimental study of the susceptibility of two marine fish species, sea bass *(Dicentrarchus labrax)* and turbot *(Scophthalmus maximus),* to viral haemorrhagic septicemia, *Aquaculture,* 41, 203, 1984.

139. **Meyers, T., Sullivan, J., Emmenegger, E., Follett, J., Short, S., Batts, W. N., and Winton, J. R.,** Isolation of viral hemorrhagic septicemia virus from Pacific cod *Gadus macrocephalus* in Prince William Sound, Alaska, *Proceedings of the 2nd Int. Symp. of Viruses of Lower Vertebrates,* Oregon State University Press, Corvallis, OR, 1991, 83.

140. **Ord, W.,** Resistance of chinook salmon *(Oncorhynchus tschawytscha)* fingerlings experimentally infected with viral hemorrhagic septicemia virus, *Bull. Fr. Pisci.,* 257, 149, 1975.

141. **Ord, W., Le Berre, M., and de Kinkelin, P.,** Viral hemorrhagic septicemia: comparative susceptibility of rainbow trout *(Salmo gairdneri)* and hybrids *(S. gairdneri X Oncorhynchus kisutch)* to experimental infection, *J. Fish. Res. Board Can.,* 33, 1205, 1976.

142. **de Kinkelin, P., Le Berre, M., Meurillon, A., and Calmels, M.,** Septicemie hemorrhagique virale: Demonstration de l etat refractaire du saumon coho *(Oncorhynchus kisutch)* et de la truit fario *(Salmo trutta), Bull. Fr. Pisci.,* 253, 166, 1974.

143. **Ahne, W. and Thomsen, I.,** Occurrence of VHS in wild white fish *(Coregonus* sp.), *J. Vet. Med.,* 32, 73, 1985.

144. **Meier, W. and Jørgensen, P. E. V.,** Isolation of VHS virus from pike fry *(Esox lucius)* with hemorrhagic symptoms, in *Fish Diseases,* Ahne, W. Ed., Springer-Verlag, Berlin, 1980, 8.

145. **Meier, W. and Wahli, T.,** Viral haemorrhagic septicemia (VHS) in grayling *(Thymallus thymallus)* L., *J. Fish Dis.,* 11, 481, 1970.

146. **Jørgensen, P. E. V.,** Indirect fluorescent antibody techniques for demonstration of trout viruses and corresponding antibodies, *Acta Vet. Scan.,* 15, 198, 1974.

147. **Ahne, W., Jørgensen, P. E. V., Olesen, N. J., Schafer, W., and Steinhagen, P.,** Egtved virus: Occurrence of strains not clearly identifiable by means of virus neutralization tests, *J. Appl. Ichthyol.,* 4, 187, 1986.

148. **de Kinkelin, P., Dorson, M., and Renault, T.,** Interferon and viral interference in viruses of salmonid fish, in *Salmonid Diseases,* Kimura T., Ed., Hokkaido University Press, Hakodat, Japan, 1992, 241.

149. **Ristow, S. S., Lorenzen, N., and Jørgensen, P. E. V.,** Monoclonal-antibody-based immunodot assay distinguishes between viral hemorrhagic septicemia virus (VHSV) and infectious hematopoietic necrosis virus (IHNV), *J. Aquat. Anim. Health,* 3, 176, 1991.

150. **Olesen, N. J., Lorenzen, N., and Jørgensen, P. E. V.,** Detection of rainbow trout antibody to Egtved virus by enzyme-linked immunosorbent assay (ELISA), immuno-fluorescence (IF), and plaque neutralization tests (50% PNT), *Dis. Aquat. Org.,* 10, 31, 1991.

151. **Jørgensen, P. E. V., Olesen, N. J., Lorenzen, N., Winton, J. R., and Ristow, S. S.,** Infectious hematopoietic necrosis virus (IHN) and viral hemorrhagic septicemia (VHS): detection of trout antibodies to the causative viruses by means of plaque neutralization, immunofluorescence, and enzyme-linked immunosorbent assay, *J. Aquat. Anim. Health,* 3, 100, 1991.

152. **Olesen, N. J. and Jørgensen, P. E. V.,** Rapid detection of viral haemorrhagic septicemia virus in fish by ELISA, *J. Appl. Ichthyol.,* 7, 183, 1991.

153. **Zwellenberg, L. O., Jensen, M. H., and Zwellenberg, H. H. L.,** Electron microscopy of the virus of viral haemorrhagic septicemia of rainbow trout (Egtved virus), *Arch. Gesamte Virusforsch.,* 17, 1, 1965.

154. **Ahne, W.,** Vergleichende untersuchungen uber die stabilitat von vier fischpathgenen viren (VHSV, PFR, SVCV, IPNV), *Z. Veter.,* (B), 29, 457, 1982.

155. **Le Berre, M., de Kinkelin, P., and Metzger, A.,** Identification serologique des rhabdovirus de salmonides, *Bull. Off. Int. Epizool.,* 87(5–6), 391, 1977.

156. **Olesen, N. J., Lorenzen N., and Jørgensen, P. E. V.,** Serological differentiation of Egtved virus (VHSV) using monoclonal and polyclonal antibodies: V, International EAFP Conference, Budapest, Hungary, 1991, 79 (Abstr.).

157. **de Kinkelin, P. and Scherrer, R.,** Le virus D'Egtved, *Ann. Rech. Vet.,* 1, 17, 1970.

158. **Bellet, R.,** Viral hemorrhagic septicemia (VHS) of the rainbow trout bred in France, *Ann. N.Y. Acad. Sci.,* 126, 461, 1965.

159. **Winton, J. R., Batts, W. N., Deering, Brunson, R. E., Hooper, K., Nislizarea, T., and Stehr, C.,** Characteristics of the first North American isolates of viral hemorrhagic septicemia virus, *Proceedings of the Second International Symposium on Viruses of Lower Vertebrates,* Oregon State University Press, Corvallis, OR, 1991, 43.

160. **Bernard, J., Bremont, M., and Winton, J.,** Sequence homologies between the N genes of the 07-71 and Makah isolates of viral hemorrhagic septicemia virus, *Proceedings of the Second International Symposium on Viruses of Lower Vertebrates,* Oregon State University Press, Corvallis, OR, 1991, 109.

161. **Jørgensen, P. E. V.,** The survival of viral hemorrhagic septicemia (VHS) virus associated with trout eggs, *Ital. Pisci. Riv. Ittiopat.,* 5, 13, 1970.

162. **de Kinkelin, P., Chilmonczyk, S., Dorson, M., Le Berre, M., and Baudouy, A, M.,** Some pathogenic facets of rhabdoviral infection of salmonid fish, in *Mechanisms of Viral Pathogenesis and Virulence,* Bachmann, P., Ed., Munich 1979, 357.

163. **Ahne, W.,** Viral infection cycles in pike (*Esox lucius* L.), *J. Appl. Ichthyol.,* 2, 90, 1985.

164. **Castric, J. and de Kinkelin, P.,** Occurrence of viral hemorrhagic septicemia in rainbow trout *Salmo gairdneri* Richardson reared in sea water, *J. Fish Dis.,* 3, 21, 1980.

165. **Jørgensen, P. E. V.,** Artificial transmission of viral haemorrhagic septicemia (VHS) of rainbow trout, *Ital. Pisci. Riv. Ittiopat.,* 8, 101, 1973.

166. **de Kinkelin, P.,** Viral haemorrhagic septicemia, in *Antigens of Fish Pathogens: Development and Production for Vaccines and Serodiagnostics,* Anderson, D. P., Dorson, M., and Dubourget, Ph., Eds., Collection Fondation Marcel Merieux, Lyon, France, 1983, 51.

167. **Rasmussen, C. J.,** A biological study of the Egtved disease (IHNV), *Ann. N.Y. Acad. Sci.,* 126, 427, 1965.

168. **Horlyck, V., Mellergard, S., Dalsgaard, E., and Jørgensen, P. E. V.,** Occurrence of VHS in Danish maricultured rainbow trout, *Bull. Eur. Assoc. Fish Pathol.,* 4, 11, 1984.

169. **Meier, W., Ahne, W., and Jørgensen, P. E. V.,** Fish viruses: Viral haemorrhagic septicemia in white fish (*Coregonus* sp.), *J. Appl. Ichthyol.,* 4, 181, 1986.

170. **Giorgetti, G.,** Rhabdoviruses in salmonids, infectious pancreatic necrosis in salmonids, spring viremia of carp, *Bull. Int. Off. Epizool.,* 93(9–10), 1017, 1980.

171. **Jørgensen, P. E. V.,** Egtved virus: Occurrence in inapparent infections with virulent virus in free-living rainbow trout, *Salmo gairdneri* Richardson, at low temperature, *J. Fish Dis.,* 5, 251, 1982.

172. **Wolf, K.**, Lecture notes from Fish Disease Long Course, Eastern Fish Disease Laboratory, U.S. Fish and Wildlife Service, Leetown, WV, 1966.

173. **Eskildsen, U. K. and Jørgensen, P. E. V.**, On the possible transfer of trout pathogenic viruses by gulls, *Ital. Pisci. Riv. Ittiopat.*, 8(4), 104, 1973.

174. **Eaton, W. D., Hulett, J., Brunson, R., and True, K.**, The first isolation in North America of infectious hematopoietic necrosis virus (IHNV) and viral hemorrhagic septicemia virus (VHSV) in coho salmon from the same watershed, *J. Aquat. Anim. Health*, 3, 114, 1991.

175. **Neukirch, M.**, An experimental study of the entry and multiplication of viral haemorrhagic septicemia virus in rainbow trout, *Salmo gairdneri* Richardson, after water-born infection, *J. Fish Dis.*, 7, 231, 1984.

176. **de Kinkelin, P.**, Vaccination against viral haemorrhagic septicemia, in *Fish Vaccination*, Ellis, A. E., Ed., Academic Press, London, 1988, 171.

177. **Ghittino, P., Schwedler, H., and de Kinkelin, P.**, The principal infectious diseases of fish and their general control measures, in *Symposium on Fish Vaccination*, de Kinkelin, P. and Michel, C., Eds., Office of International des Epizooties, Paris, 1984, 5.

178. **Kaastrup, P., Horlyuck, V., Olesen, N. J., Lorenzen, N., Jørgensen, P. E. V., and Berg, P.**, Paternal association of increased susceptibility to viral haemorrhagic septicemia (VHS) in rainbow trout *(Oncorhynchus mykiss)*, *Can. J. Fish. Aquat. Sci.*, 48, 1188, 1991.

179. **Jørgensen, P. E. V.**, Partial resistance of rainbow trout *(Salmo gairdneri)* to viral haemorrhagic septicemia (VHS) following exposure to nonvirulent Egtved virus, *Nord. Veterinaermed.*, 28, 570, 1976.

180. **Lorenzen, N., Olesen, N. J., and Jørgensen, P. E. V.**, Neutralization of Egtved virus pathogenicity to cell cultures and fish by monoclonal antibodies to the viral G protein, *J. Gen. Virol.*, 71, 561, 1990.

181. **Ghittino, P.**, Viral hemorrhagic septicemia (VHS) in rainbow trout in Italy, *Ann. N.Y. Acad. Sci.*, 126, 468, 1965.

182. **Holt, R. and Rohovec, J.**, Anemia of coho salmon in Oregon, *Fish Health Sect. Am. Fish. Soc. News Lett.*, 12, 4, 1984.

183. **Leek, S. L.**, Viral erythrocytic inclusion body syndrome (EIBS) occurring in juvenile spring chinook salmon *(Oncorhynchus tshawytscha)* reared in fresh water, *Can. J. Fish. Aquat. Sci.*, 44, 685, 1987.

184. **Rodger, H. D., Dinan, E. M., Murphy, T. M., and Lunder, T.**, Observation of erythrocytic inclusion body in Ireland, *Bull. Eur. Assoc. Fish Pathol.*, 11, 108, 1991.

185. **Takahashi, K., Okamoto, N., Kumagai, A., Maita, Y., and Rohovec, J. S.**, Epizootics of erythrocytic inclusion body syndrome in coho salmon in seawater in Japan, *J. Aquat. Anim. Health*, 4, 174, 1992.

186. **Lunder, T., Thorud, K., Pappe, T. T., Holt, R. A., and Rohovec, J. S.**, Particles similar to the virus of erythrocytic inclusion body syndrome, EIBS, detected in Atlantic salmon *(Salmo salar)* in Norway, *Bull. Eur. Assoc. Fish Pathol.*, 10, 21, 1990.

187. **Okamoto, N., Takahashi, K., Maita, M., Rohovec, J. S., and Ikeda, Y.**, Erythrocytic inclusion body syndrome: Susceptibility of selected sizes of salmonid fish, *Fish Pathol.*, 27, 153, 1992.

188. **Piacentini, J. S., Rohovec, J. S., and Fryer, J. L.**, Epizootiology of erythrocytic inclusion body syndrome. *J. Aquat. Anim. Health*, 1, 173, 1989.

189. **Arakawa, C. K., Hursh, D. A., Lannan, C. N., Rohovec, J. S., and Winton, J. R.**, Preliminary characterization of a virus causing infectious anemia among stocks of salmonid fishes in the western United States, in *Viruses of Lower Vertebrates*, Ahne, W. and Kurstak, E., Eds., Springer-Verlag, Berlin, 1988, 442.

190. **Takahashi, K., Okamoto, N., Maita, M., Rohovec, J. S., and Ikeda, Y.**, Progression of erythrocytic inclusion body syndrome in artificially infected coho salmon, *Fish Pathol.*, 27, 89, 1992.

191. **Hedrick, R. P., McDowell, T., Eaton, W. D., Kimura, T., and Sano, T.**, Serological relationships of five herpesviruses isolated from salmonid fishes, *J. Appl. Icthyol.*, 3, 87, 1987.

192. **Wolf, K. and Taylor, W. G.**, Salmonid viruses: a syncytium-forming agent from rainbow trout, *Fish Health News*, 4, 3, 1975.

193. **Hedrick, R. P., McDowell, T., Eaton, W. D., Chan, L., and Wingfield, W.**, *Herpesvirus salmonis* (HPV): first occurrence in anadromous salmonids, *Bull. Eur. Assoc. Fish Pathol.*, 6, 66, 1986.

194. **Eaton, W. D., Wingfield, W. H., and Hedrick, R. P.,** Prevalence and experimental transmission of the steelhead herpesvirus in salmonid fishes, *Dis. Aquat. Org.,* 7, 23, 1989.

195. **Wolf, K. and Smith, C. E.,** *Herpesvirus salmonis:* Pathological changes in parenterally infected rainbow trout, *Salmo gairdneri* Richardson, fry, *J. Fish Dis.,* 4, 445, 1981.

196. **Hedrick, R. P. and Sano, T.,** Herpesvirus of Fishes, in *Viruses of Lower Vertebrates,* Ahne, W. and Kurstak, E., Eds., Springer-Verlag, Berlin, 1989, 161.

197. **Wolf, K., Darlington, R. W., Taylor, W. G., Quimby, M. C., and Nagabayashi, Y.,** *Herpesvirus salmonis:* characterization of a new pathogen of rainbow trout, *J. Virol.,* 27, 659, 1978.

198. **Kimura, T., Yoshimizu, M., and Tanaka, M.,** Fish viruses: tumor induction in *Oncorhynchus keta* by the herpesvirus, in *Phyletic Approaches to Cancer,* Dawe, C. J., Ed., Japan Scientific Society Press, Tokyo, 1981, 59.

199. **Kimura, T., Yoshimizu, M., Tanaka, M., and Sannohe, H.,** Studies on a new virus (OMV) from *Oncorhynchus masou* I. Characteristics and pathogenicity, *Fish Pathol.,* 15, 143, 1981.

200. **Kimura, T., Yoshimizu, M., and Tanaka, M.,** Studies on a new virus (OMV) from *Oncorhynchus masou.* II. Oncogenic nature, *Fish Pathol.,* 15, 149, 1981.

201. **Kimura, T., Yoshimizu, M., and Tanaka, M.,** Susceptibility of different fry stages of representative salmonid species to *Oncorhynchus masou* virus (OMV), *Fish Pathol.,* 17, 251, 1983.

202. **Yoshimizu, M., Tanaka, M., and Kimura, T.,** *Oncorhynchus masou* virus (OMV): incidence of tumor development among experimentally infected representative salmonid species, *Fish Pathol.,* 22, 7, 1987.

203. **Gou, D. F., Kubota, H., Onuma, M., and Kodoma, H.,** Detection of salmonid herpesvirus (*Oncorhynchus masou* virus) in fish by southern-blot technique, *J. Vet. Med. Sci.,* 53, 43, 1991.

204. **Tanaka, M., Yoshimizu, M., and Kimura, T.,** Ultrastructures of OMV infected RTG-2 cells and hepatocytes of chum salmon *Oncorhynchus keta, Nippon Suis. Gak.,* 53, 47, 1987.

205. **Sano, T., Fukuda, H., Okamoto, N., and Kaneko, F.,** Yamame tumor virus: lethality and oncogenecity, *Bull. Jpn. Soc. Sci. Fish.,* 49, 1159, 1983.

206. **Eaton, W. D., Wingfield, W. H., and Hedrick, R. P.,** Comparison of the DNA homologies of five salmonid herpesviruses, *Fish Pathol.,* 26, 183, 1991.

207. **Sano, T.,** Characterization, pathogenicity, and oncogenicity of herpesvirus in fish, *Int. Fish Health Conf.,* Vancouver, B.C., Canada, 1988, 157.

208. **Yoshimizu, M., Nomura, T., Awakura, T., Ezura, Y., and Kimura, T.,** Prevalence of pathogenic fish viruses in anadromous masou salmon *(Oncorhynchus masou)* in the northern part of Japan, 1976–1987, *Phys. Ecol. Jpn.,* 559, 1989.

209. **Tanaka, M., Yoshimizu, M., and Kimura, T.,** *Oncorhynchus masou* virus: Pathological changes in masou salmon *(Oncorhynchus masou),* chum salmon *(O. keta)* and coho salmon *(O. kisutch)* fry infected with OMV by immersion method, *Bull. Jpn. Soc. Sci. Fish.,* 50, 431, 1984.

210. **Yoshimizu, M., Tanaka, M., and Kimura, T.,** Histopathological study of tumors induced by *Oncorhynchus masou* virus (OMV) infection, *Fish Pathol.,* 23, 133, 1988.

211. **Kimura, T. and Yoshimizu, M.,** Salmon herpesvirus: OMV, *Oncorhynchus masou* virus, in *Viruses of Lower Vertebrates,* Ahne, W. and Kurstak, E., Eds., Springer-Verlag, Berlin, 1989, 171.

212. **Yoshimizu, M., Takizawa, H., and Kimura, T.,** UV susceptibility of some fish pathogenic viruses, *Fish Pathol.,* 21, 47, 1986.

213. **Kimura, T., Suzuki, S., and Yoshimizu, M.,** *In vitro* antiviral effect of 9-(2-hydroxyethoxymethyl) guanine on the fish herpesvirus, *Oncorhynchus masou* virus (OMV), *Antiviral Res.,* 3, 93, 1983.

214. **McAllister, P. E. and Herman, R. L.,** Epizootic mortality in hatchery-reared lake trout *Salvelinus namaycush* caused by a putative virus possibly of the herpesvirus group, *Dis. Aquat. Org.,* 6, 113, 1989.

215. **Bradley, T. M., Medina, D. J., Chang, P. W., and McClain, J.,** Epizootic epitheliotropic disease of lake trout *(Salvelinus namaycush):* history and viral etiology, *Dis. Aquat. Org.,* 7, 195, 1989.

216. **Horner, R.,** Illinois Department of Conservation, Wolf Lake State Fish Hatchery, personal communication, 1991.

217. Unpublished data, Tallapoosa River, Southeastern Cooperative Fish Disease Laboratory, Auburn University, Alabama, 1986.

Chapter IX

Sturgeon (Acipenseridae)

The number of white sturgeon *(Acipenser transmontanus)* being cultured in northern California (U.S.) and northern Italy is steadily increasing, and with this intensive culture, diseases that affect the species are beginning to emerge. Four virus diseases have thus far been associated with mortality in cultured white sturgeon: white sturgeon adenovirus (WSAV),[1] the white sturgeon iridovirus (WSIV),[2] and two white sturgeon herpesviruses (WSHV-1 and WSHV-2).[3-4]

A. WHITE STURGEON ADENOVIRUS

White sturgeon adenovirus (WSAV) was identified from diseased juvenile white sturgeon between 1984 and 1986,[1] but has since precipitated no serious disease problems.[5] The virus was initially observed among 0.5-g sturgeon at a fish farm in northern California. Affected fish appeared lethargic, anorexic, emaciated, livers were pale, and intestines were void of food.

Histologically, epithelial cells lining the straight intestine and spiral valve exhibited nuclear hypertrophy. Nuclei of some infected cells (up to one third of the cells in some fish) were five times larger than were noninfected cells. The infected cells continued to enlarge until they ruptured and released their contents into the gut lumen. Electron microscopy examination of affected tissues revealed large numbers of electron-dense hexagonal virions in the nuclei. The virions averaged 74 nm in diameter. Methyl green-pyronin stains of infected cells showed nuclear inclusions that contained DNA. Hedrick et al.[1] tentatively placed the virus in the family Adenoviridae, based on DNA, morphology, and the absence of an envelope. Attempts to isolate the virus in white sturgeon spleen (WSS-1) and white sturgeon heart (WSH-1) cell cultures were not successful.

The initial WSAV epizootic was not explosive, but cumulative mortality approached 50% over a 4-month period. Studies involving transmission of the disease to noninfected sturgeon were only partially successful. Intraperitoneal injections of filtrates into naive fish resulted in enlargement of cell nuclei in the gut epithelium, but the reaction was not as severe as that observed in naturally infected fish. However, the experimentally infected fish were larger than the naturally infected sturgeon and this may have affected the results.

When WSAV was first isolated and diagnosed the disease caused significant mortality, but whether or not it will adversely affect sturgeon seed production in the future is unknown.

B. WHITE STURGEON IRIDOVIRUS

White sturgeon iridovirus (WSIV) was first isolated from cultured juvenile white sturgeon at several fish farms in northern California in 1988.[2] The disease has been observed in Oregon and Idaho and in all California white sturgeon farms where adequate numbers of fish have been virologically assayed.[5] WSIV has been found in juvenile sturgeon that ranged in length from 7 to 46 cm. Affected fish appear weak, experience weight loss, drop to the bottom of the tank, cease swimming, and eventually die. Gills are pale, display hyperplasia and necrosis of the pillar cells lining the lamellar vascular channels, and some petechiae are present. Internally there is no body fat, livers are pale, and the gut is void of food. Histologically, the gills and skin have numerous hypertrophic and occasionally basophilic cells with swollen nuclei in the epithelium and epidermis.

By use of electron microscopy, abundant numbers of virions have been observed in the cytoplasm of degenerating gill tissue cells. Complete icosahedral virions are about 260 nm in diameter.

Hedrick et al.[6] reported isolation of WSIV in white sturgeon spleen (WSS-2) cells, where it induced cell enlargement and slow but progressive degeneration. Virus replication occurred at 10, 15, and 20°C (optimum temperature) but not at 5 or 25°C. WSIV has been placed in the iridovirus-like group because of its DNA genome and its large size of 262 nm.

WSIV is highly virulent. In the original epizootic 1 farm lost 95% of 200,000 juvenile sturgeon in 4 months.[2] Experimental transmission studies by Hedrick et al.[6] produced an 80% mortality in 50 d at 15°C. Initial signs of disease began 10 d postexposure. No deaths occurred in experimentally infected channel catfish, striped bass, or chinook salmon, but lake sturgeon (*A. fulvescens*) did suffer mild mortality.

Because of its pathogenic potential WSIV could be a serious threat to white sturgeon culture. Also, there is the possibility that the virus may appear in other areas where California white sturgeon have been introduced.[5]

C. WHITE STURGEON HERPESVIRUS

Two herpesviruses have been isolated from white sturgeon: white sturgeon herpesvirus-1 (WSHV-1) in 1989[3] and white sturgeon herpesvirus-2 (WSHV-2) a year later.[5]

WSHV-1 was first reported in fish less than 10 cm in length.[3] There are no specific external clinical signs associated with the disease and affected fish continue to feed until death. Internally, the stomach and intestines are filled with fluid but other tissues appear normal. During histopathological examination, hematoxylin-eosin stained tissues exhibit focal or diffused dermatitis and epidermal lesions are characterized by acanthosis and intercellular edema. Hydropic degeneration and hypertrophy of Malpighian cells with loss of intercellular junctions were common. Chromatin margination was associated with flocculent nonmembrane-bound intranuclear inclusion.

The WSHV-1 virion nucleocapsids, 110 nm in diameter, were found within the nucleus and cytoplasm. Mature particles were hexagonal and enveloped in cytoplasmic vacuoles and possessed all the characteristics of a herpesvirus. The virus was isolated from epithelial tissue only in white sturgeon skin (WSSK-1) cells where syncytia was detected 3 d after inoculation.[3] Total CPE occurred 5 to 7 d after inoculation at 15°C. An identical virus was observed in inoculated WSSK-1 cells as in the cells of infected fish. The virus grew in WSSK-1 cells at 10, 15, and 20°C, but not at 5 or 25°C. Anti-WSHV-1 serum can be used to distinguish WSHV-1 from WSHV-2.

The original sturgeon population from which WSHV-1 was isolated suffered 97% mortality after being moved into the laboratory. However, the disease process was complicated by the presence of *Flexibacter columnaris*. Experimental virus transmission from cell cultures resulted in 35% mortality after fish were exposed to $10^{5.3}$ $TCID_{50}$/ml of water (immersion) for 30 min at 15°C. Virus was recovered for 2 weeks postinfection, but not at 4 weeks.

WSHV-2[5] was isolated from internal organs of captive adult brood stock and later from juvenile fish on commercial farms. Preliminary results indicate that the more recently isolated herpesviruses are different than WSHV-1. WSHV-2 appears to be more pathogenic under experimental conditions than does WSHV-1 and can be found in both skin and internal organs. It is more lytic in cell culture and has a broader host range because it will replicate in several white sturgeon cell lines. Further biochemical and serological investigations of the relationships between WSHV-1 and WSHV-2 are underway.[5]

Viral diseases which affect the white sturgeon would have to be considered significant, due to the fact that the fish have a relatively short history of culture and already four viruses that appear specific for the species have surfaced. Also, the virulence of the adenovirus and herpesviruses adds to the disease significance. To date, no management procedures have been established relative to controls of viral diseases in cultured white sturgeon.

REFERENCES

1. **Hedrick, R. P., Speas, J., Kent, M. L., and McDowell, T.,** Adenovirus-like particles associated with a disease of cultured white sturgeon, *Acipenser transmontanus, Can. J. Fish. Aqua. Sci.,* 42, 1321, 1985.
2. **Hedrick, R. P., Groff, J. M., McDowell, T., and Wingfield, W. H.,** An iridovirus infection of the integument of the white sturgeon *Acipenser transmontanus, Dis. Aquat. Org.,* 8, 39, 1990.
3. **Hedrick, R. P., McDowell, T. S., Groff, J. M., Yun, S., and Wingfield, W. H.,** Isolation of an epitheliotropic herpesvirus from white sturgeon *Acipenser transmontanus Dis. Aquat. Org.,* 11, 49, 1991.

4. **Hedrick, R. P., McDowell, T. S., Groff, J. M., and Yun, S.,** Characteristics of two viruses isolated from white sturgeon *Acipenser transmontanus,* in *Proc. 2nd Int. Symp. Viruses Lower Vertebrates,* Oregon State University, Corvallis, 1991, 165.

5. **Hedrick, R. P.,** Department of Medicine, University of California, Davis, CA, personal communication, 1992.

6. **Hedrick, R. P., McDowell, T. S., Groff, J. M., Yun, S., and Wingfield, W. H.,** Isolation and some properties of an iridovirus-like agent from white sturgeon *Acipenser transmontanus, Dis. Aquat. Org.,* 12, 75, 1992.

Chapter X

Walleye (Percidae)

The walleye *(Stizostedion vitreum)* is not a widely cultured fish, however, interest in its commercialization in the northern U.S. justifies inclusion of diseases associated with the species. Known viral diseases of walleye are (1) epidermal hyperplasia, (2) walleye dermal sarcoma, (3) diffuse epidermal hyperplasia (walleye herpesvirus), and (4) lymphocystis (Table 1). Lymphocystis is discussed elsewhere and will be referred to in this section only for comparative purposes.

A. EPIDERMAL HYPERPLASIA

Epidermal hyperplasia, synonymous with walleye epidermal hyperplasia and discrete epidermal hyperplasia,[1] was first suspected of being caused by a virus by Walker.[2] It was not until 15 years later that Yamamoto et al.[3] described the C-type retrovirus particles in electron micrographs of these lesions. Epidermal hyperplasia has been found only in adult walleye in Saskatchewan and Manitoba in Canada and Lake Oneida in New York.

Epidermal hyperplasia lesions are gently raised, translucent, mucoid-like patches that are more discrete and less granular than lymphocystis or dermal sarcoma lesions (Figure 1). The lesions appear singly or in groups and may cover large areas of the skin and/or fins. Size of the lesions may vary in diameter from a few millimeters to several centimeters and may be 1 to 2 mm in height. The demarcation between the normal epidermis and the tumor is sharply defined.

Epidermal hyperplasia is confined to the epidermis where the growths are distinct from the lesions associated with diffuse epidermal hyperplasia, dermal sarcoma, and lymphocystis (Table 1), and a presumptive diagnosis can be made in the field. However, positive confirmation is made by histological procedures and/or electron microscopy. Histopathologically, the lesion contains predominately cuboidal cells with mucus cells at the surface (Figure 2). The C-type retrovirus associated with the lesions is pleomorphic and measures 120 nm in diameter and viral particles are randomly distributed in microvilli-like extensions of the cell.[4] Virions are not found in the cytoplasmic matrix, but occur at the periphery of the cell. The virus has not been isolated in tissue culture.

B. WALLEYE DERMAL SARCOMA

According to Wolf,[1] walleye dermal sarcoma was first described by Walker[5] who recognized that the lesions were different from classical lymphocystis which is also found in walleye. Bowser et al.[6] stated that on a seasonal basis up to 27% of adult walleye throughout North America are affected with dermal sarcoma, noting that the number of dermal sarcoma-infected fish in Lake Oneida was higher in spring and fall than during the summer. Walleye dermal sarcoma lesions have also been reported from adult walleye in Lakes Oneida and Champlain and a number of other lakes and rivers in New York, the Great Lakes, and in central Canada.

Histopathological examination can be used to definitively identify the different virus-induced growths of walleye (Figure 2). Dermal sarcoma lesions may be confused with lymphocystis, but probably not with epidermal hyperplasia (retrovirus) or diffuse epidermal hyperplasia (Table 1 and Figure 1). However, lymphocystis and sarcoma tumor cells may occur within the same lesions. The sarcoma lesions consist of irregularly shaped tumors with normal size cells which are often arranged in whorls (Figure 2).[3] These vascularized tumors are smooth textured, usually variably firm, and have a pinkish to white color. There is some disagreement about the diameter of the retrovirus: Walker[2] reported a diameter of 100 nm, Yamamoto et al.[4] gave a slightly larger size of 135 nm, and Martineau et al.[7] described a 90-nm retrovirus.

Martineau et al.[7] homogenized 20 tumors and after centrifugation found the C-type retrovirus by electron microscopy in the tissue pellets. Viral RNA, presumably from the walleye dermal sarcoma virus (WDSV), was electrophoresed, and the cDNA synthesized from the viral RNA was hybridized with viral DNA in the walleye tumors, suggesting that WDS is the result of an infection caused by a

Table 1 Comparison of virus-caused epidermal lesions on walleye (Stizostedion vitreum) and the prevalence of each type of disease on 25 fish examined from a single population in Canada[a]

Disease	Number positive	Lesion characteristic	Virus family	Virus diameter	Isolated
Epidermal hyperplasia	10	Clear, raised, distinct margin epithelium	Retroviridae	120 nm	No
Dermal sarcoma	7	Smooth, rounded	Retroviridae	135 nm	No
Difuse epidermal hyperplasia	4	Slimy, indistinct, flat, translucent	Herpesviridae	200 nm (enveloped)	Yes
Lymphocystis	12	White, redish, large spherical cells, grape-like clusters	Iridoviridae	260 nm	Yes

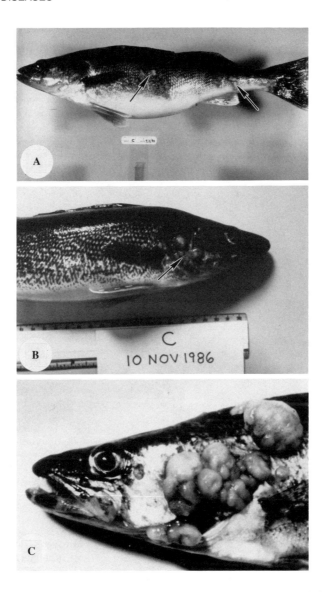

Figure 1 Skin virus diseases of walleyes. (A) Walleye dermal sarcoma (arrows); (B) difuse epidermal hyperplasia (arrow); and (C) lymphocystis disease. (Photographs by P. R. Bowser.)

unique exogenous retrovirus of which the large genome is predominantly unintegrated in tumor cells. The virus has not been isolated in cell cultures but Martineau et al.[8] characterized its molecular structure.

Experimental transmission of walleye dermal sarcoma has rarely been documented, but Martineau et al.[9] successfully transmitted the disease to healthy 4-month-old walleye by intramuscular injection with homogenized tumors from adult fish. Tumors developed in the recipient fish 4 months later. In a subsequent study, transmission of the tumor-producing retrovirus was greater at 15°C than at 10 or 20°C.[19] Tumors developed in 8 weeks at 15°C. Bowser and Wooster[11] showed that tumors regressed either totally or partially in walleye during an 18-week study period in which water temperature rose from 7 to 29°C. Total regression was observed only in females.

C. DIFFUSE EPIDERMAL HYPERPLASIA

A herpesvirus was described in diffuse epidermal hyperplasia lesions on the skin of spawning walleye in central Canada by Kelly et al.[12] After further characterization, the virus was named *Herpesvirus*

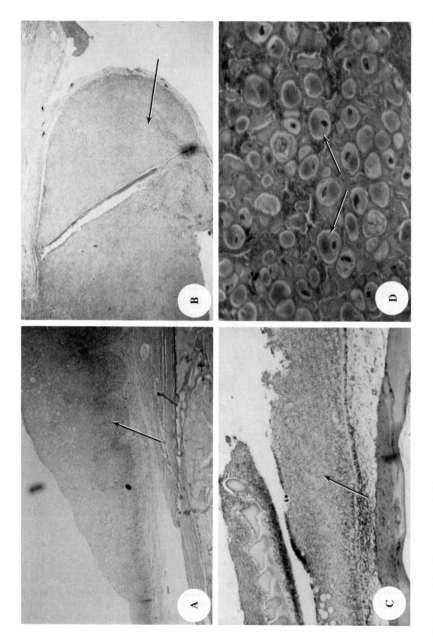

Figure 2   Histopathology of the four types of skin tumors (arrows) on walleye. (A) Epidermal hyperplasia; (B) walleye dermal sarcoma; (C) diffuse epidermal hyperplasia; and (D) lymphocystis. (Photographs A, B, and C by P. R. Bowser.)

vitreum.[13] The herpesvirus infection was named "diffuse epidermal hyperplasia" by Yamamoto et al.[3] because of the flat, diffused appearance of the lesion on the skin (Figure 1). Walleye herpesvirus causes epidermal lesions which resemble thick areas of slime. They are flat, translucent growths with soft, swollen underlying tissue.[3] These lesions may measure several centimeters in diameter, and spread laterally on the surface of the fish. They are not as obvious as lymphocystis or dermal sarcoma of walleye.[3] The transient lesions appear during spawning, but have no apparent ill effects on the host.

The virus, presumably with an envelope, measures 200 nm in diameter. *H. vitreum* replicates in walleye ovary (WO), and walleye embryo (We-2) cells, but not in CHSE-214, BB, FHM, or RTG-2. Syncytium formation is the primary cytopathic effect, which is followed by cell lysis. Replication of walleye herpesvirus occurs at 4 to 15°C in cell cultures, with maximum replication at 15°C. No replication occurs at 20°C, but at 15°C the maximum titer in cell cultures is 10^5 pfu/0.1 ml after 10 to 13 d of incubation.

Cells in diffuse epidermal hyperplasia lesions are disorganized with slightly enlarged nuclei which contain granular inclusions. Electron microscopic observation shows the presence of typical herpesvirus particles within the nucleus and mature enveloped virions near the periphery of the cell.

The primary significance of epidermal hyperplasia, walleye dermal sarcoma, and diffuse epidermal hyperplasia in walleye is that their appearance renders the fish unacceptable to fishermen and consumers. Epidermal growths and tumors of fish and their relationship to environmental carcinogens and virus-caused cancers have received attention in recent years because of increased interest in tumors and cancers in general, even though there is no evidence that the tumors themselves, or the viruses that are involved, are harmful to humans. Because they do not appear to induce mortality, the viruses and the associated lesions have generally been ignored by traditional fish pathologists, but have attracted the attention of scientists interested in tumors.

In some walleye populations the frequency of these neoplastic lesions can be high (Table 1) and an individual fish may have more than one type of tumor. Yamamoto et al.[3] examined 25 walleye from 1 population and found that 12 had lymphocystis, 7 had dermal sarcoma, 4 had diffuse epidermal hyperplasia (walleye herpesvirus), and 10 had epidermal hyperplasia. These tumors have not been reported from cultured walleyes, but the culinary quality of walleye makes it an excellent candidate for aquaculture exploitation. When, and if, this occurs the virus-induced tumors could become more important. There are no known controls or management procedures for tumors caused by viruses of walleyes.

REFERENCES

1. **Wolf, K.,** *Fish Viruses and Fish Viral Diseases,* Cornell University Press, Ithaca, NY, 1988, 476.
2. **Walker, R.,** Virus associated with epidermal hyperplasia in fish, *Natl. Cancer Inst. Monogr.* 31, 195, 1969.
3. **Yamamoto, T., Kelly, R. K., and Nielsen, O.,** Morphological differentiation of virus associated skin tumors of walleye *(Stizostedion vitreum vitreum), Fish Pathol.,* 20, 361, 1985.
4. **Yamamoto, T., MacDonald, R. D., Gillespie, D. C., and Kelly, R. K.,** Viruses associated with lymphocystis disease and dermal sarcoma of walleye *(Stizostedion vitreum vitreum), J. Fish. Res. Board Can.,* 33, 2408, 1976.
5. **Walker, R.,** Lymphocystis disease and neoplasia in fish, *Anal. Rec.,* 99, 559, 1947.
6. **Bowser, P. R., Wolfe, M. J., Forney, J. L., and Wooster, G. A.,** Seasonal prevalence of skin tumors from walleye *(Stizostedion vitreum)* from Oneida Lake, New York, *J. Wildl. Dis.,* 24, 292, 1988.
7. **Martineau, D., Renshaw, R., Williams, J. R., Casey, J. W., and Bowser, P. R.,** A large unintegrated retrovirus DNA species present in a dermal tumor of walleye *(Stizostedion vitreum), Dis. Aquat. Org.,* 10, 153, 1991.
8. **Martineau, D., Bowser, P. R., Renshaw, R. R., and Casey, J. W.,** Molecular characterization of a unique retrovirus associated with a fish tumor, *J. Virol.,* 66, 596, 1992.
9. **Martineau, D., Bowser, P. R., and Wooster, G. A.,** Experimental transmission of dermal sarcoma in fingerling walleyes *(Stizostedion vitreum), Vet. Pathol.,* 27, 230, 1990.
10. **Bowser, P. R., Martineau, D., and Wooster, G. A.,** Effects of water temperature on experimental transmission of dermal sarcoma in fingerling walleyes, *J. Aquat. Anim. Health,* 2, 157, 1990.

11. **Bowser, P. R. and Wooster, G.,** Regression of dermal sarcoma in adult walleyes *(Stizostedion vitreum), J. Aquat. Anim. Health,* 3, 147, 1991.

12. **Kelly, R. K., Nielsen, O., and Yamamoto, S. C.,** A new herpes-like virus (HLV) of fish *(Stizostedion vitreum vitreum), In Vitro,* 16, 249, 1980.

13. **Kelly, R. K., Nielsen, O., Mitchell, S. C., and Yamamoto, T.,** Characterization of *Herpesvirus vitreum* isolated from hyperplastic epidermal tissue of walleye, *Stizostedion vitreum vitreum* (Mitchill), *J. Fish Dis.,* 6, 249, 1983.

Chapter XI

Other Viral Diseases

A. LYMPHOCYSTIS

Lymphocystis, a hypertrophic disease of cells primarily in the skin and fins of fish, is the oldest and perhaps best known of all viral diseases of fish. Although lymphocystis has been recognized since 1874, and theorized to be of a viral nature in 1920,[1] the viral etiology was not proven until Wolf[2] did so in the early 1960s. The etiological agent of lymphocystis is lymphocystis virus. Rivers postulates were fulfilled for lymphocystis in 1964.[3]

Since its early description, lymphocystis has been a much studied disease. Investigations have involved geographical range, host susceptibility, histopathology, and electron microscopy; a trend which continues today even though the virus can be isolated in cell cultures. However, since 1980 greater attention has been given to the biophysical and biological properties of the virus.

1. GEOGRAPHICAL RANGE AND SPECIES SUSCEPTIBILITY

Lymphocystis virus is the most widely distributed of the fish viruses. The virus affects freshwater fish in most areas of the world including North and South America, Europe, Africa, Australia, and Asia.[4-6] Lymphocystis is also found in fish from nearly all marine waters including the eastern and western Pacific and Atlantic Oceans, Gulf of Mexico, Red Sea, Bering Sea, and the Mediterranean, to name a few specific areas.[5-7]

Anders[6] named 11 orders, 45 families, and 141 species of fish that have been associated with lymphocystis virus infections and the list grows annually. The most prominent orders are the Perciformes and Pleuronectiformes which, combined, contribute 85% of the susceptible species. The Centrarchidae (sunfishes), Percidae (perch), Sciaenidae (seatrout and drum), Chaetodontidae (butterflyfish and angelfish), Cichlidae (cichlids), and Pleuronectidae (flat fishes) are the families that contain the most frequently infected species. Susceptible fishes include cultured and wild fishes and traditional aquarium species in both fresh and salt water. The disease has been reported in tropical fishes that were imported into the U.S. and New Zealand for ornamental purposes.[8-9]

Interestingly, salmonid and cyprinids do not appear to be susceptible to the virus.

2. CLINICAL SIGNS AND FINDINGS

Lymphocystis manifests itself as a series of greatly hypertrophied cells occurring in the connective tissue below the epidermis. The cells are large, gray or whitish in color, and may measure up to 2 mm in diameter. These cells, which are easily seen with the unaided eye, may occur singly or grouped together in "grape-like" clusters giving a tumorous appearance (Figure 1). Lesions appear at any location on the fish, but are more prevalent on the fins, head, and lateral surfaces of the body. They have also been noted on the eye and gills.[10] Rarely, lymphocystis lesions have been found in the spleen, liver, and mesenteries of juvenile marine fish.[11] There are no behavioral abnormalities associated with lymphocystis virus infections.

3. DIAGNOSIS AND VIRUS CHARACTERISTICS

Lymphocystis virus infection can be identified by gross clinical signs coupled with the presence of typical enlarged cells containing Feulgen-positive cytoplasmic inclusions in stained histologically sectioned lesions (Chapter X, Figure 2). Lymphocystis virus has been isolated in BF-2 cells and cells from the largemouth bass, but generally, isolation of the virus is not easily accomplished.[3] CPE develops in the infected cell cultures.[12] At 23 to 25°C, *in vitro* infected cells increase in size and become basophilic. Feulgen-positive, basophilic inclusions can be seen in the cytoplasm 6 d after infection followed by the formation of a hyaline capsule at 10 d. The cells reach maturity in 3 to 4 weeks and are identical in appearance to those seen in infected fish, however they are smaller, measuring 40 to 50 μm in diameter. The nucleus of infected cells is also enlarged, but not to the same degree as *in vivo*.

Figure 1 Lymphocystis virus. (A) Large lymphocystis virus-induced lesions (arrows) on fins of a 2-month-old Kentucky spotted bass; (B) lymphocystis lesions on a 40-cm smallmouth bass from a wild population (photograph by J. Hooper); (C) lymphocystis lesion on an 8-cm croaker taken from a wild population and held in captivity where lesions developed (formalin preserved).

Lymphocystis virus belongs to the family Iridoviridae. It is an icosahedral (hexagonal) particle with a diameter of 145 to 330 nm[13-14] (Figure 2). The virus has a double-stranded DNA genome that replicates in the nucleus, but the virus is assembled in the cytoplasm of infected cells.

4. EPIZOOTIOLOGY

Lymphocystis is a chronic, benign condition that seldom results in the death of infected fish. The primary problem resulting from the disease is that infected fish are unattractive and less marketable to the consumer. However, studies in California indicate that the lesions may become so large that the bucal cavity is occluded and the fish dies of starvation,[15] but less severely affected wild fish may

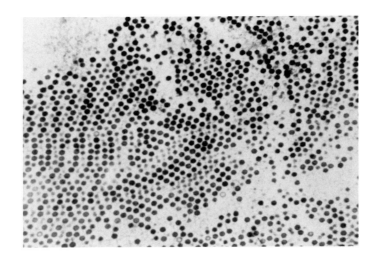

Figure 2 Electron micrograph of lymphocystis virus. (Photograph by R. Walker.)

experience a loss of weight. Infected walleyes from Lake Michigan weighed less than noninfected fish[16] and other studies have shown similar reductions in weight.[17-18] The presence of large lesions on the skin or fins may also render smaller fish more vulnerable to gill nets.

Transmission of lymphocystis virus is influenced by several factors including host species, fish density, environmental conditions, injury, and levels of pollution or parasite load. Lymphocystis can be transmitted by cohabitation, or by spraying infected material on scarified skin.[3] In natural waters, the virus is presumably transmitted to healthy fish when infected cells rupture and release virus into the water. Ryder[19] found that injuries sustained during spawning activities led to a higher incidence of lymphocystis. Handling, netting, tagging, etc., may also increase the incidence of the disease because lesions tend to develop where scales are disturbed or fins are damaged.[20-21] Even some external parasites (copepods or isopods) that disrupt the protective mucus layer may increase efficiency of transmission.[22] Lawler et al.[11] postulated that parasitic isopods *(Linoreca ovalis)* may be instrumental in the infection of silver perch *(Bairdiella chrysura)* with lymphocystis virus by irritating the skin, thus allowing the virus to invade the tissue.

It has been suggested that an increased occurrence of lymphocystis is influenced by pollution.[23] In Norway, Reierson and Fugelli[24] described a higher incidence of lymphocystis in flounder *(Platichthys flesus)* taken from polluted waters than in those taken from less polluted waters, but cautioned against drawing any definitive conclusions because of the complexity of the disease-environment relationship.

Incubation of lymphocystis depends upon the host species, virus, and temperature. At 10 to 15°C, incubation may take up to 6 weeks for lesions to develop compared to 5 to 12 d at 20 to 25°C.[25-26] Also, lymphocystis virus from one species of fish may not be infective for a different species.[27]

Lymphocystis occurs in all sizes and ages of fish, but Yamamoto et al.[28] did report a higher incidence of lymphocystis in small walleye (250 to 600 g) than in medium (600 to 1200 g) or larger fish. They also noted that the infection rate in females was three times greater than it was in males. Lorenzen et al.,[29] in a population dynamics study of flounder *(Platichthys flesus),* determined that a higher incidence of lymphocystis in young fish than in older fish was due to acquired immunity and lower density of older fish.

Incidence of lymphocystis infections may reach 100% in some crowded fish populations.[8] In contrast, the incidence in wild populations of marine fish may range from 4 to 57% depending on the season of the year, with the highest incidence usually occurring in the spring and summer.[30-31] On the other hand, Reierson and Fugelli[24] reported increased incidence of the disease in the winter. Lymphocystis lesions usually do not cover much of the skin or fins, but Overstreet[32] reported that naturally infected croakers had lymphocystis lesions over 60 to 80% of their body surface. However, this degree of infection is unusual. Experimental infections or infections that develop under confined conditions tend to cover a larger portion of the body surface.

5. PATHOLOGICAL MANIFESTATIONS

Hypertrophied lymphocystis cells may be 20 to 1000 times larger than the normal 10- to 100-μm cell.[6] These easily visible infected cells are generally round to oval. Histologically, they contain an enlarged, centrally located nucleus and Feulgen positive, basophilic, ribbon-shaped cytoplasmic inclusions consisting of DNA[33] (Chapter X, Figure 2). A hyaline capsule surrounds each cell and when the cells are clustered the hyaline material becomes a matrix.[34] The morphology of the inclusions may vary with fish species, but they are considered sites of viral maturation.[12]

There is little inflammation associated with lymphocystis, but the infected cells are often imbedded in a matrix of collagenous tissues. Electron micrographs show that the inclusion bodies are dense areas of hexagonally shaped virus particles. Within the fish the virus is transported to different sites in the bloodstream. In most instances, the lymphocystis lesions regress after a period of time or the cells lyse or slough, and the infected area heals with connective tissue, leaving little evidence of the infection.

6. SIGNIFICANCE

Lymphocystis virus-infected fish generally do not die as a result of the infection, but they do have a reduced economic value. Entire populations of fingerlings or juveniles may be rejected if they are heavily infected with the disease. Also, infected fish are not retained for sale by fishermen, although the lesions are transient.

7. MANAGEMENT

Control and management of lymphocystis is difficult. Removal of infected fish from the population before cells rupture is recommended so as to prevent the release of virus into the water. In aquaculture, avoidance of known carrier populations is recommended. In aquaria, the replacement of 50% of the water every 2 to 3 d will dilute the virus and possibly reduce infection incidence. Sanitation and disinfection of equipment and containers is essential for prevention of transmission of the virus to noninfected populations.

B. VIRAL ERYTHROCYTIC NECROSIS

Viral erythrocytic necrosis (VEN) is an infection of erythrocytes that affects a wide variety of marine and anadromous fish species. The disease was known as piscine erythrocytic necrosis (PEN), until Walker and Sherburne[35] suggested that the new name would better reflect the nature of the disease if the emphasis was placed on blood instead of fish, thus "viral erythrocytic necrosis". A viral etiology of VEN as proposed by Laird and Bullock,[36] and Walker[37] provided evidence from electron microscopy that VEN was associated with a viral infection. The etiological agent of VEN is erythrocytic necrosis virus (ENV). In his book, Wolf[4] cites literature from the early 1900s that described the disease but attributed its cause to intracellular parasites.

1. GEOGRAPHICAL RANGE AND SPECIES SUSCEPTIBILITY

VEN was first reported in the Atlantic cod, sea snail *(Liparis atlanticus),* and shorthorn sculpin *(Myxocephalus scropius)* in the coastal waters of the North Atlantic of Canada and the U.S. by Laird and Bullock.[36] In later studies, VEN was reported from waters of the U.K.,[38] in the coastal waters of Portugal,[39] and the Pacific region of the U.S.[40-41] Essentially, VEN is confined to marine environments, but has been reported in eels in Taiwan[42] and freshwater stocks of anadromous salmon.[41]

Walker and Sherburne[35] listed 12 species of fish from the Atlantic coast of North America that were affected by VEN, 5 species that were questionable, and 25 that appeared refractive. The primary species infected were alewife *(Alosa pseudoharangus)* (53%), rainbow smelt *(Osmerus mordax)* (64%), and sea raven *(Hemitripterus americanus)* (93%). The most economically important species of fish infected with ENV include the Atlantic cod, chum salmon, coho salmon, chinook salmon, pink salmon, steelhead trout, Atlantic herring *(Clupea harengus),* alewife, and mullet *(Mugil cephalus).*[39-41,43-44]

2. CLINICAL SIGNS AND FINDINGS

Stricken fish will lie listless or whirl. The principle clinical sign of VEN is a severe anemia characterized by pale gills, watery colorless blood, and discolored livers.[45] Pacific herring infected with ENV do show some hemorrhage in the epithelium (Figure 3). Hematocrits will be as low as 2 to 10%.[46] It was reported by Walker and Sherburne[36] that captured infected cod were somewhat emaciated and were

Figure 3 Pacific herring with viral erythrocytic necrosis virus. Note the hemorrhage in the epithelium (arrows). (Photograph by J. R. Winton.)

darker than noninfected fish. Infected erythrocytes contain pathognomonic small to large, rounded to lobate, eosinophilic inclusion bodies in the cytoplasm (Figure 4). Depending on the stage and severity of infection, 9 to 100% of the erythrocytes contain these inclusions. Pinto et al.[47] found from 2.7 to 25% of erythrocytes in ENV-infected Mediterranean seabass (*Dicentrarchus labrax*) had inclusions.

3. DIAGNOSIS AND VIRUS CHARACTERISTICS

ENV can only be confirmed by electron microscopy because the virus cannot be propagated in established cell lines;[46] however, Reno and Nicholson[48] demonstrated ENV could replicate in cod erythrocytes *in vitro* at 4°C. ENV infections can be detected by staining blood smears and observing the cytoplasmic inclusion bodies in erythrocytes. Nuclei of infected cells may be round, bilobed, or "U" shaped with a rounded, eosinophilic (Giemsa stain) cytoplasmic inclusion (Figure 4).[35] The nucleus is pyknotic or karyorrhectic. The inclusion body most likely contains the replicating pool of viral nucleic acid that is Feulgen positive and fluoresces bright green with acridine orange. Electron microscopy of ENV-infected erythrocytes reveal hexagonal virions in the cytoplasm, which confirms the presence of the virus infection (Figure 4). Although serological methods for identification of ENV are not available, Pinto et al.[47] purified viral material from virus-infected erythrocytes from sea bass and developed an immunofluorescent reagent that allowed differentiation from other similar viral infections of fish.

The ENVs described to date are icosahedral virions with DNA genome. They are classified as members of the iridovirus family. The size of the ENV virion varies with the species of fish afflicted; the virus described in the erythrocytes of Atlantic herring and salmonids is 140 to 190 nm,[46,49] but a larger virus, 300 to 350 nm, was reported to infect Atlantic cod.[35] Without the benefit of serological procedures and *in vitro* culture techniques it is impossible to determine if all of these viruses are the same but, with the exception of particle size, they are very similar.

4. EPIZOOTIOLOGY

An ENV infection may manifest itself differently depending on the species of fish affected, the degree of infection, and the presence of associated clinical signs (Table 1). Mean hematocrits decline to less than 10% from a normal of near 40% in chum salmon.[50] It is proposed that newly infected circulating erythrocytes are destroyed in the hematopoietic centers of the spleen and anterior kidney, thus facilitating virus release. The host compensates for this anemia by erythroblastosis, but the additional virus infection of the erythroblasts may prolong the infection.

Smail and Egglestone[38] recorded a 32% incidence of VEN in young-of-the-year Atlantic cod compared to an almost zero incidence in 3- to 4-year-olds, but the older fish may have recovered from an earlier infection. Similarly, MacMillan and Mulcahy[49] detected a 43% incidence of the virus in young Pacific herring compared to a 4% incidence in 4- to 8-year-old fish, while Meyers et al.[45] found a prevalence

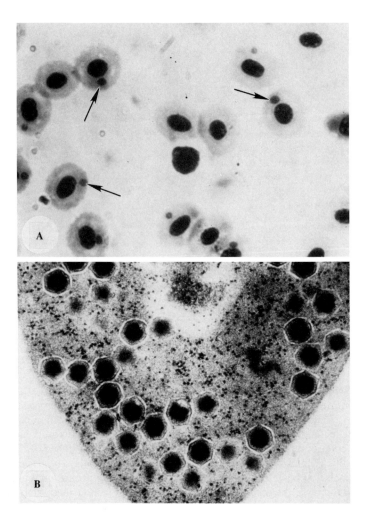

Figure 4 Viral erythrocytic necrosis: (A) erythrocytes with singular cytoplasmic inclusion bodies (arrows); (B) electron micrograph of the hexagonal viruses in Atlantic cod erythrocytes. (Photographs by B. Nicholson.)

of 56 to 100% VEN in 1+-year-old and 17 to 80% in older Pacific herring in Alaska. Captive Mediterranean sea bass had a 20 to 100% VEN infection rate compared to only 6% in wild members of the same species.[47] These data indicate that younger and confined fish are more susceptible to VEN.

The route of ENV infection is most likely from fish to fish, through water. Experimentally, ENV can be transmitted by injection of recipient fish with filtrates derived from tissue of infected fish. Eaton[50] transmitted ENV from Pacific herring to pink, chum, and sockeye salmon, with chum salmon being the most susceptible. These studies show that ENV from one species of fish will infect another. Blood-sucking parasites such as the salmon louse (*Lepophtheirus* sp.) and gill maggot (*Salmonicola* sp.) may also serve as vectors.[51]

Evelyn and Traxler[46] found that chum and pink salmon experimentally infected with cell-free filtrates became clinically infected in 12 d (chum) and 3 weeks (pink). The infection rate in the chum salmon increased from 50% at 12 d to 100% at 48 d postinoculation. They also reported that 7 months postinoculation, 4 of the infected fish were negative for ENV, which would indicate that recovery from VEN is possible. Rohovec and Amandi[41] reported ENV infection in freshwater fish. The virus was found in juvenile coho salmon (20 g each) which were experiencing mortalities. The fish had no clinical signs other than hematocrits of less than 10%.

Table 1 **Summary of pathology of viral erythrocytic necrosis infections in some fish**[36,45-48,52]

Host	Percent erythrocytes	Pathology
Atlantic cod	0.01–99% Mature	Skin darkening, no mortality
Common blenny	3–60% Mature	None
Alewife	0.17% Mature	None
Atlantic herring	90% Mature and immature	Hyperemic liver hemolyzed blood
Chum salmon	80% Mature and immature	Two types of inclusions 0.3% Mortality, prone to vibriosis, BKD, anemia
Pacific herring	60–80% Mature and immature	Erythroblastosis

Reno et al.[52] reported that Atlantic cod and Atlantic herring exhibited lower hematocrits and erythrocyte counts, while there was reduced hemoglobin concentration in cod but not herring. There was no effect on plasma electrolyte or protein concentration in either species. Chum salmon that were artificially infected with ENV developed changes in their physiology and hematology including lowered RBC counts, hematocrits, and hemoglobin concentration.[53] ENV-infected fish also had higher white blood cell counts, and less fragile erythrocytes. Infection progressed through 4 weeks before the fish began to recover, therefore, the physiological and hematological changes were transient. However, infected fish recovered more slowly than control fish from stress and had increased osmoregulatory difficulties. Meyers et al.[45] also found evidence that ENV-infected Pacific herring suffered osmoregulatory stress that may precipitate death, particularly in young fish.

The survival of ENV-infected fish may vary. MacMillan and Mulcahy[49] and Rohovec and Amandi[41] assayed 35 stocks of sea run salmonids and found 25% to be positive for erythrocytic inclusions. The percentage of fish that were ENV positive in the populations ranged from 1.0 to 15.1%. When evaluating the effect of VEN on released salmon, it was found that mortality directly due to the viral infection was limited, but that environmental stressors caused a higher rate of mortality in infected fish than in uninfected controls.[54] There is some evidence that VEN is directly responsible for deaths in natural and cultured stocks, as the disease renders the fish more susceptible to environmental insults and secondary infections.

5. PATHOLOGICAL MANIFESTATIONS
Appy et al.[55] described the pathology in the erythrocytes of cod due to VEN. The nucleus became more rounded and densely stained with the presence of magenta-colored cytoplasmic inclusions with a diameter of 1 to 4 μm (Figure 4). In some cells the inclusions appeared to possess a small central vacuole which increased in size with subsequent degeneration. Infected cells appeared more rounded and irregular than the uninfected cells. Some cells fractured, which resulted in the formation of fragments containing nuclear remains of the inclusion. The cytoplasmic inclusions of infected cells are composed of a pool of icosahedral viral particles (Figure 4) with incomplete viral particles associated with viroplasm at the edge of the pool.

6. SIGNIFICANCE
The prevalence of VEN in certain fish populations can be very high, but its actual importance to these fish stocks is unknown. However, this virus does pose a threat to mariculture populations. While limited mortality in cultured populations does occur from the disease, its direct impact on other populations is not clear. VEN does alter blood parameters and thus may make infected populations more susceptible to opportunistic bacterial diseases (i.e., *Vibrio anguillarum*) and to dissolved oxygen depressions and handling.[52-53]

7. MANAGEMENT
No control measures for VEN exist. However, since fry and young-of-the-year fish seem to be more susceptible to the virus they should be reared in water that is free of infected fish. Adverse environmental conditions, such as low dissolved oxygen and other stressors, should be avoided.

REFERENCES

1. **Weissenberg, R.,** Fifty years of research on the lymphocystis virus disease of fishes (1914–1964), *Ann. N.Y. Acad. Sci.,* 126, 362, 1965.
2. **Wolf, K.,** Experimental propagation of lymphocystis disease of fishes, *Virology,* 18, 249, 1962.
3. **Wolf, K., Gravell, M., and Malsberger, R. G.,** Lymphocystis virus: isolation and propagation in centrarchid fish cell lines, *Science,* 151, 1004, 1966.
4. **Wolf, K.,** *Fish Viruses and Fish Virus Diseases,* Cornell University Press, Ithaca, NY, 1988, 486.
5. **Nigrelli, R. F. and Ruggieri, G. D.,** Studies on virus diseases of fishes. Spontaneous and experimentally induced cellular hypertrophy (lymphocystis disease) in fishes of the New York Aquarium with a report of new cases and an annotated bibliography (1874–1965), *Zoologica,* 50, 83, 1965.
6. **Anders, K.,** Lymphocystis disease of fishes, in *Viruses of Lower Vertebrates,* Ahne, W. and Kurstak, E., Eds., Springer-Verlag, New York, 1989, 141.
7. **Faisal, M.,** Lymphocystis in the Mediterranean golden grouper *Epinephelus alexandrinus* Valenciennes 1828 (Pisces, Serranidae), *Bull. Eur. Assoc. Fish Pathol.,* 9, 17, 1989.
8. **Durham, P. J. K. and Anderson, C. D.,** Lymphocystis disease in imported tropical fish, *N.Z. Vet. J.,* 29, 88, 1978.
9. **Huizinga, H. W.,** Lymphocystis disease in the green terror, *Trop. Fish Hobbyist,* 25, 47, 1977.
10. **Dukes, T. W. and Lawler, A. R.,** The ocular lesions of naturally occurring lymphocystis in fish, *Can. J. Comp. Med.,* 39, 406, 1975.
11. **Lawler, A. R., Howse, H. D., and Cook D. W.,** Silver perch, *Bairdiella chrysura:* new host for lymphocystis, *Copeia,* 1974, 266, 1974.
12. **Wolf, K. and Carlson, C. P.,** Multiplication of lymphocystis virus in the bluegill *(Lepomis macrochirus), Ann. N.Y. Acad. Sci.,* 126, 414, 1965.
13. **Kelly, D. C. and Robertson, J. S.,** Icosahedral cytoplasmic deoxyriboviruses, *J. Gen. Virol.,* Suppl. 20, 17, 1973.
14. **Robin, J. and Bertholimue, L.,** Purification of lymphocystis disease virus (LDV) grown in tissue culture, evidences for the presence of two types of viral particles, *Rev. Can. Biol.,* 40, 323, 1981.
15. **McCosker, J. E.,** A behavioral correlate for passage of lymphocystis disease in three blennioid fish, *Copeia,* 1969, 636, 1969.
16. **Hile, R.,** Fluctuations in growth and year-class strength of the walleye in Saginaw Bay, *U.S. Fish Wildl. Serv. Fish. Bull.,* 56, 7, 1954.
17. **Witt, A., Jr.,** Seasonal variation in the incidence of lymphocystis in the white crappie from the Niangua arm of the Lake of the Ozarks, Missouri, *Trans. Am. Fish. Soc.,* 85, 271, 1957.
18. **Petty, L. L. and Magnuson, J. J.,** Lymphocystis in age 0 bluegills *(Lepomis macrochirus)* relative to heated effluent in Lake Monona, Wisconsin, *J. Fish. Res. Board Can.,* 31, 189, 1974.
19. **Ryder, R. A.,** Lymphocystis as a mortality factor in a walleye population, *Prog. Fish Cult.,* 23, 183, 1961.
20. **Clifford, T. J. and Applegate, R. L.,** Lymphocystis disease in tagged and untagged walleyes in a South Dakota lake, *Prog. Fish Cult.,* 32, 177, 1970.
21. **Olson, D. E.,** Statistics of a walleye sport fishery in a Minnesota lake, *Trans. Am. Fish. Soc.,* 87, 52, 1958.
22. **Nigrelli, R. F.,** Lymphocystis disease and ergasilid parasites in fishes, *J. Parasitol.,* 36, 36, 1950.
23. **Alpers, C. E., McCain, B. B., Myers, M. S., and Wellings, S. R.,** Lymphocystis disease in yellowfin sole *(Limanda aspera)* in the Bering Sea, *J. Fish. Res. Board Can.,* 34, 611, 1977.
24. **Reiersen, L. O. and Fugelli, K.,** Annual variation in lymphocystis infection frequency in flounder, *Platichthys flesus* (L.), *J. Fish Biol.,* 24, 187, 1984.
25. **Roberts, R. J.,** Experimental pathogenesis of lymphocystis in the plaice *(Pleuronectes platessa),* in *Wildlife Diseases,* Page, L. A., Ed., Plenum Press, New York, 1975, 431.
26. **Cook, D. W.,** Experimental infection studies with lymphocystis virus from Atlantic croaker, in *Proceedings of the 3rd Annual Workshop, World Mariculture Society,* St. Petersburg, FL, Avault, J. W., Boudreaux, E., and Jaspers, E., Eds., 1972, 329.
27. **Overstreet, R. M. and Howse, H. D.,** Some parasites and diseases of estuarine fishes in polluted habitats of Mississippi, *Ann. N.Y. Acad. Sci.,* 298, 427, 1977.

28. **Yamamoto, T., Kelly, R. K., and Nielsen, O.,** Morphological differentiation of virus associated skin tumors of walleye *(Stizostedion vitreum vitreum), Fish Pathol.,* 20, 361, 1985.

29. **Lorenzen, K., Clers, S. A., and Anders, K.,** Population dynamics of lymphocystis disease in estuarine flounder, *Platichthys flesus* (L.), *J. Fish Biol.,* 39, 577, 1991.

30. **Shelton, R. G. J. and Wilson, K. W.,** On the occurrence of lymphocystis, with notes on other pathological conditions, in the flatfish stocks of the north-east Irish Sea, *Aquaculture,* 2, 395, 1973.

31. **Amin, O. M.,** Lymphocystis disease in Wisconsin fishes, *J. Fish Dis.,* 2, 207, 1979.

32. **Overstreet, R. M.,** Aquatic pollution, Southeastern U.S. coasts: histopathologic indicators, *Aquat. Toxicol.,* 11, 213, 1988.

33. **Dunbar, C. E. and Wolf, K.,** The cytological course of experimental lymphocystis in bluegill, *J. Infect. Dis.,* 116, 466, 1966.

34. **House, H. D. and Christmas, J. Y.,** Lymphocystis tumors: histochemical identification of hyaline substances, *Trans. Am. Micros. Soc.,* 89, 276, 1970.

35. **Walker, R. and Sherburne, S. W.,** Piscine erythrocytic necrosis virus in Atlantic cod, *Gadus morhua,* and other fish: ultrastructure and distribution, *J. Fish. Res. Board Can.,* 34, 1188, 1977.

36. **Laird, M. and Bullock W. L.,** Marine fish hematozoa from New Brunswick and New England, *J. Fish. Res. Board Can.,* 26, 1075, 1969.

37. **Walker, R.,** PEN, a viral lesion of erythrocytes, *Am. Zool.,* 11, 707, 1971.

38. **Smail, D. A. and Egglestone, S. I.,** Virus infections of marine fish erythrocytes: prevalence of piscine erythrocytic necrosis in cod *Gadus morhua* L. and blenny *Blennius pholis* L. in coastal and off shore waters of the United Kingdom, *J. Fish Dis.,* 3, 41, 1980.

39. **Eiras, J. C.,** Virus infection of marine fish: prevalence of viral erythrocytic necrosis (VEN) in *Mugil cephalus* L., *Blennius pholis* L. and *Platichthys flesus* L. in coastal waters of Portugal, *Bull. Eur. Assoc. Fish Pathol.,* 4, 52, 1984.

40. **Bell, G. R. and Traxler, G. S.,** First record of viral erythrocytic necrosis and *Ceratomyxa shasta* Noble, 1950 *(Myxozoa: Myxosporea)* in feral pink salmon *(Oncorhynchus gorbuscha* Walbaum), *J. Wildl. Dis.,* 21, 169, 1985.

41. **Rohovec, J. S. and Amandi, A.,** Incidence of viral erythrocytic necrosis among hatchery reared salmonids of Oregon, *Fish Pathol.,* 15, 135, 1981.

42. **Chen, S. N., Kou, G. H., Hedrick, R. P., and Fryer, J. L.,** The occurrence of viral infections of fish in Taiwan, in *Fish and Shellfish Pathology,* Ellis, A. E., Ed., Academic Press, Orlando, FL, 1985, 313.

43. **Sherburne, S. W.,** Occurrence of piscine erythrocytic necrosis (PEN) in the blood of the anadromous alewife, *Alosa pseudoharengus* from Maine coastal streams, *J. Fish. Res. Board Can.,* 34, 281, 1977.

44. **MacMillan, J. R., Mulcahy, D., and Landolt, M.,** Viral erythrocytic necrosis: some physiological consequences of infection in chum salmon *(Oncorhynchus keta), Can. J. Fish. Aquat. Sci.,* 37, 799, 1980.

45. **Meyers, T. R., Hauck, A. K., Blandenbeckler, W. D., and Minicucci, T.,** First report of viral erythrocytic necrosis in Alaska, USA, associated with epizootic mortality in Pacific herring, *Clupea harengus pallasi* (Valenciennes), *J. Fish Dis.,* 9, 479, 1986.

46. **Evelyn T. P. T. and Traxler, G. S.,** Viral erythrocytic necrosis: natural occurrence in Pacific salmon and experimental transmission, *J. Fish. Res. Board Can.,* 35, 903, 1978.

47. **Pinto, R. M., Alvarez-Pellitero, P., Bosch, A., and Jafre, J.,** Occurrence of viral erythrocytic infection in the Mediterranean sea bass, *Dicentrarchus labrax* (L.), *J. Fish Dis.,* 12, 185, 1989.

48. **Reno, P. W. and Nicholson, B. L.,** Viral erythrocytic necrosis (VEN) in Atlantic cod *(Gadus morhua):* in vitro studies, *Can. J. Fish. Aquat. Sci.,* 37, 2276, 1980.

49. **MacMillan, J. R. and Mulcahy, D.,** Artificial transmission to and susceptibility of Puget Sound fish to viral erythrocytic necrosis (VEN), *J. Fish. Res. Board Can.,* 36, 1097, 1979.

50. **Eaton, W. D.,** Artificial transmission of erythrocytic necrosis virus (ENV) from Pacific herring in Alaska chum, sockeye, and pink salmon, *J. Appl. Ichthyol.,* 6, 136, 1990.

51. **Smail, D. A.,** Viral erythrocytic necrosis in fish: a review, *Proc. R. Soc. Edinburgh (B),* 81, 169, 1982.

52. **Reno, P. W., Serreze, D. V., Hellyer, S. K., and Nicholson, B. L.,** Hematological and physiological effects of viral erythrocytic necrosis (VEN) in Atlantic cod and herring, *Fish Pathol.,* 20, 353, 1985.

53. **Haney, D. C., Hursh, D. A., Mix, M. C., and Winton, J. R.,** Physiological and hematological changes in chum salmon artificially infected with erythrocytic necrosis virus, *J. Aquat. Anim. Health,* 4, 48, 1992.

54. **Anon.,** Impact of Disease on Survival of Wild and Propogated Fish, Seattle National Fisheries Research Center, Sport Fisheries Research, U.S. Fish and Wildlife Service, 1980, 109.

55. **Appy, R. G., Burt, M. D. B., and Morris, T. J.,** Viral nature of piscine erythrocytic necrosis (PEN) in the blood of Atlantic cod *(Gadus morhu), J. Fish. Res. Board Can.,* 33, 1380, 1976.

PART THREE
BACTERIAL DISEASES

The aquatic environment contains numerous species of bacteria, many of which are beneficial to the balance of nature and are of no consequence to fish. However, about 60 to 70 bacterial species are capable of causing diseases of aquatic animals and on rare occasions these fish disease-producing bacteria will also cause infections in humans.

The aquatic environment, especially aquacultural and eutrophic waters, provide a natural habitat for the growth and proliferation of bacteria because of the availability of nutrient-producing organic materials that enhance bacterial growth. The bacterial flora of water is influenced by the availability of nutrients, pH, temperature, and other environmental characteristics — factors that affect the growth pattern of bacteria and, in turn, may affect their virulence and pathogenicity. Some bacteria will grow and thrive if there is any organic matter available to provide nutrients, while others are fastidious and survive in the environment only in a host. Also, the salinity in water, or culture media, may affect the growth and survival of some bacteria. Generally, pH 6 to 8 is desirable for growth of most bacteria, while many are killed above pH 11 or below pH 5. Bacteria generally have an optimum growth temperature of 20 to 42°C, but some will grow at temperatures above 50°C (thermophiles) while others are capable of growth at 0°C (psychrophiles). Bacteria that are most prolific at temperatures of 18 to 45°C are mesophiles. Bacteria that are pathogenic to fish are mesophiles and usually require optimum temperatures of 10 to 30°C.

Three basic bacterial cell morphologies are spherical (coccus), rod shaped (bacillus), or spiral shaped (spirillium). Bacteria have a particular staining characteristic referred to as Gram positive (blue) or Gram negative (red or pink). Gram positive organisms may also be acid fast, which is related to the presence or absence of mycotic acid in the cell wall. Generally, bacteria that cause disease problems in fish are Gram negative rods, but some pathogens are Gram positive rods or cocci while a few are acid fast.

There are basically two disease-producing types of bacteria that cause problems in fish: (1) obligate pathogens, and (2) nonobligate or facultative pathogens. Although there are few true obligate pathogenic bacteria that cause infections in fish, *Renibacterium salmoninarum,* the etiological agent of bacterial kidney disease, and *Mycobacterium* are two examples of obligate pathogens. Facultative bacteria are able to survive in water, but under certain conditions, usually environmentally induced stress, they cause infectious disease in fish. *Aeromonas hydrophila,* one of the primary bacteria involved in the motile *Aeromonas* septicemia complex, is one of the better known examples of facultative bacteria.

The role that environmentally induced stress plays in the development of bacterial diseases in fish cannot be overemphasized. Most bacterial fish diseases result directly from an environmental stressor such as water quality, or from handling, or nonlethal parasites. Many bacterial infections are "secondary" even when obligate pathogens are involved. Carrier fish of some obligate bacteria (e.g., *A. salmonicida*) may not be adversely affected by the presence of that bacteria until stress responses of the fish reach a point that resistance and immunity are compromised, causing a dormant infection to become an active, debilitating, clinical infection. When a disease is caused by a facultative organism, it is often tempting to classify the infection as "secondary" and not to consider the organism as a serious cause of disease, but this is not always true. Secondary is often a misleading term because many facultative bacteria are the actual cause of death and must be treated accordingly.

There are some general, recognizable clinical signs of bacterial infections in fish, few of which are pathognomonic. General clinical signs are fish "go off feed", behavior and swimming becomes erratic or lethargic, excessive mucus is produced on gills and skin, necrotic lesions develop on the integument, fins become frayed, a bloody fluid can be present in the abdominal cavity, and internal organs may be hemorrhagic and swollen. Gills can be pale, swollen, or with necrotic lesions.

Bacterial infections of fish can occur as a bacteremia or septicemia. A bacteremia implies that the organisms are simply present in the bloodstream and the fish's defense mechanisms keep their numbers low enough that clinical infection does not occur. Septicemia indicates a more severe condition in which bacteria and the toxins they produce are actually present in the circulatory system. Inflammation, hemorrhage, and necrosis are clinical signs associated with septicemia. Pathogenic bacteria can cause disease-producing exotoxins, a characteristic generally, but not exclusively, of Gram positive organisms. Gram negative bacteria may produce either exotoxins or endotoxins. These toxins may consist of proteolytic enzymes which kill host cells and cause necrosis or they may make the blood vessels more porous and cause hemorrhage.

Detection of bacterial disease is accomplished by recognition of clinical signs, conventional isolation of bacteria from lesions or internal organs on culture media, and proceeding through determination of morphological and biochemical characterization of these organisms for confirmation.[1] Bacterial pathogens of fish can be presumptively identified using a limited number of characteristics,[2-4] but definitive identification requires application of a relatively large number of media and biochemical tests to determine specific reactions on specific substrates,[1,5-6] or utilizing one of the available commercial identification systems (API, MiniTek, etc.).[7] These commercial systems were not designed to identify bacterial pathogens of fish, therefore, some adaptation is often required before they can be used for bacteria isolated from fish.

The classical immunological method of identifying bacterial organisms is by agglutination using antigen specific antisera. In light of new technology and the need to more quickly identify bacterial pathogens, time-consuming biochemical procedures, and in some cases culturing, have been totally eliminated[8] having been replaced by serological methods which include fluorescent antibody techniques (FAT), enzyme-linked immunosorbent assay (ELISA), and the use of nucleic acid probes that are specific for the target organism. When available, these highly sensitive, accurate, and rapid procedures can be used to identify either isolated bacterial organisms, or bacteria in smears made directly from diseased fish.

Bacterial diseases may be categorized several ways: according to the species of fish they infect; staining reaction of the cell; or by family, genus, and species of the bacteria. However, in keeping with the organizational rationale of this book bacterial diseases are categorized according to the family of cultured fishes that they most severely affect. It should not be construed that the diseases discussed with a particular group of fish are the only ones which infect that group. When a particular pathogen causes serious disease in more than one group or family of fishes, the reader will be referred to the most in-depth discussion of the disease.

REFERENCES

1. **Austin, B. and Austin, D. A., Eds.,** *Methods for the Microbiological Examination of Fish and Shellfish,* Ellis Horwood, Chichester, 1989.
2. **Shotts, E. B. and Bullock, G. L.,** Bacterial diseases of fishes: diagnostic procedures for Gram negative pathogens, *J. Fish. Res. Board Can.,* 32, 1243, 1975.
3. **Shotts, E. B. and Bullock, G. L.,** Rapid diagnostic approaches in the identification of Gram-negative bacterial diseases of fish, *Fish Pathol.,* 10, 187, 1976.
4. **Shotts, E. B. and Teska, J. D.,** Bacterial pathogens of aquatic vertebrates, in *Methods for the Microbiological Examination of Fish and Shellfish,* Austin, B. and Austin, D. A., Eds., Ellis Horwood, Chichester, 1989, 8.
5. **Lewis, D. H.,** *Predominant Aerobic Bacteria of Fish and Shellfish,* Texas A & M University Sea Grant College, TAMY-SG-73-401, Galveston, 1973, 102.
6. **Thoesen, J., Ed.,** *Blue Book Version 1: Suggested Procedures for the Detection and Identification of Certain Finfish and Shellfish Pathogens,* 4th ed., American Fisheries Society/Fish Health Section, in press, 1994.
7. **Teska, J. H., Shotts, E. B., and Hsu, T. C.,** Automated biochemical identification of bacterial pathogens using the Abbott Quantum II, *J. Wildl. Dis.,* 25, 103, 1989.
8. **Schill, W. B., Bullock, G. L., and Anderson, D. P.,** Serology, in *Methods for the Microbiological Examination of Fish and Shellfish,* Austin, B. and Austin, D. A., Eds., Ellis Horwood, Chichester, 1989, 6.

Chapter XII

Catfish (Ictaluridae)

Catfish, more specifically channel catfish *(Ictalurus punctatus)*, are the most extensively cultured fish in the U.S. There are three major bacterial infections of cultured catfish: (1) columnaris *(Flexibacter columnaris)* is one of the most common diseases; (2) enteric septicemia of catfish *(Edwardsiella ictaluri)* is most often found in channel catfish but occasionally occurs in other species; and (3) motile *Aeromonas* septicemia *(Aeromonas hydrophila* and related motile aeromonads).

A. COLUMNARIS

Columnaris is an acute to chronic infectious skin disease of fish, especially channel catfish. Synonyms of columnaris include cotton wool and mouth fungus. The disease was first described by Davis,[1] but it was not until 22 years later that the organism was isolated and characterized.[2] In freshwater fish columnaris is caused by *Flexibacter columnaris*. A very similar disease of marine fish in brackish and salt water is caused by *Flexibacter maritimus*.[3]

The causative agent of columnaris has gone through several taxonomic reclassifications and name changes since its original designation, *Bacterium columnaris*. Subsequently the organism has been named *Chondrococcus columnaris*,[2] *Cytophaga columnaris*,[4] *Flexibacter columnaris*, and again *Cytophaga columnaris*.[5] Each of these nomenclatures was supported by convincing taxonomic criteria, but Bernardet and Grimont[6] presented justification to retain *Flexibacter columnaris* based on DNA relatedness and phenotypic characterization. They also include the "yellow pigmented group" *Flexibacter maritimus*[3] (saltwater columnaris) and *Flexibacter (Cytophaga) psychrophila*[7] as the etiological agent of "cold water disease" of salmonids. For purposes of convenience, *F. columnaris* and *F. maritimus* will be discussed here and *F. psychrophila* will be discussed in the salmonid section (Chapter XV).

1. GEOGRAPHICAL RANGE AND SPECIES SUSCEPTIBILITY

Columnaris disease exists worldwide in fresh and brackish water habitats, especially in the U.S., Europe, and Asia. Channel catfish and other ictalurids are most severely affected by the disease.[8] However, cultured eels (fresh and brackish water) are also highly susceptible to *F. columnaris*.[9] Salmonids, particularly hatchery populations of trout and migrating adult salmon,[10-11] cultured centrarchids, common carp *(Cyprinus carpio)* in Europe,[12] and golden shiners *(Notemigonus chrysoleucas)*, fathead minnows *(Pimephales promelas)*, and gold fish *(Carasius auratus)* in the U.S. are particularly susceptible. Although some groups of fishes appear more susceptible than others, no wild or cultured freshwater fish, including ornamental fish in aquaria, are totally resistant to it.

The marine counterpart to freshwater columnaris disease, *F. maritimus*, occurs only in saltwater fishes. It was first reported in cultured marine fishes in Japan[3,13] and has since been reported in Dover sole *(Solea solea)* in Scotland, where it caused a condition known as "black patch necrosis".[14] A columnaris-like disease has also been described in juvenile cod *(Gadus morhua)* in estuaries of the Elbe and Weser Rivers of Germany.[15] The complete species susceptibility of marine columnaris is not yet fully known.

2. CLINICAL SIGNS AND FINDINGS

Clinical signs of columnaris are easily recognized and differ little between species. However, location of lesions will vary from outbreak to outbreak.[11,16-17] Disease severity, type and location of lesions, and virulence may correspond to the strain of *F. columnaris* involved in the infection.[18]

Although clinical signs of columnaris are nearly pathognomonic for the disease, one must always be aware that it can be complicated by dual infections and that another bacteria or protozoan parasite may also be involved. In catfish, the disease begins as an external infection on the fins, body surface, or gills. Fins become frayed (necrotic) with grayish to white margins. Initial lesions on the skin appear as small, discrete bluish-gray areas that enlarge into depigmented necrotic lesions causing infected catfish to lose their metallic sheen (Figure 1). The lesions have yellowish or pale margins which are

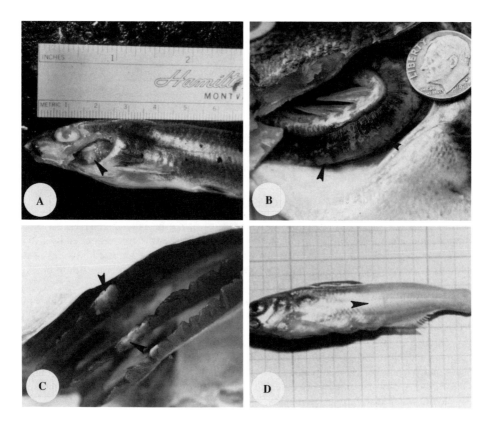

Figure 1   Columnaris lesion (arrows) on (A) gill of channel catfish; (B) gill of largemouth bass; (C) gill of largemouth bass; (D) body and caudal peduncle of channel catfish.

accompanied by mild inflammation. The mouth of infected fish may be covered with a yellowish mucoid material. These same types of lesions occur on eel, trout, cyprinids, and centrarchids.

Gill lesions are white to brown, depending on the presence of debris and secondary fungus (Figure 1). Lesions may develop exclusively on the gills, in which case the disease and morbidity tends to advance rapidly.

In some instances columnaris becomes systemic with few pathological changes occurring in the visceral organs. Whether or not the bacteria isolated from internal organs are taxonomically *F. columnaris* is not clear, but they may be isolated from the kidneys of more than 50% of catfish necropsied with epidermal *F. columnaris*.[19]

F. maritimus results in development of similar lesions and disease in marine fishes, including mouth erosion and tail rot in juvenile fish. Initially, lesions appear as gray to white areas on the fins and body surfaces and progress into ulcers on older fish.[3,13]

3. DIAGNOSIS AND BACTERIAL CHARACTERISTICS

Columnaris is normally detected by the recognition of typical lesions on the skin, fins, and gills of diseased fish along with the presence of long, slender rods in wet mounts made from lesions (Figure 2). These nonflagellated bacteria display a gliding motility and form "hay stacks" or columns, which is essentially confirmatory of the disease (Figure 2).

For isolation of *F. columnaris,* a moist media with a low nutrient level is required. The organism grows poorly on conventional media such as brain heart infusion (BHI) agar or tryptic soy agar (TSA). Two good primary media for isolation of *F. columnaris* are cytophaga (Ordal's)[20] and Hsu-Shotts.[21] The media can also be prepared as broth by eliminating the low concentration of agar. Hawke and Thune[19] found that a modification of the media described by Fijan[22] proved best for isolation of *F. columnaris,* especially when other bacteria may be present. However, Song et al.[23] found that the media

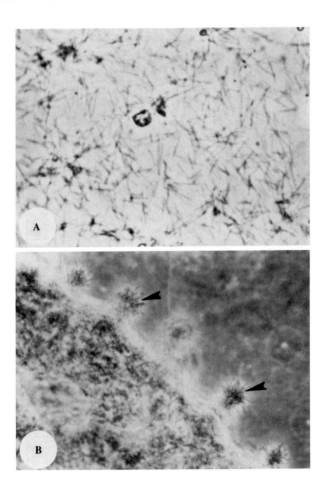

Figure 2 Wet mounts of *F. columnaris* from channel catfish showing (A) long slender, flexing rods; (B) the hay-stacking (arrows) typical of virulent *F. columnaris.*

of Shieh[24] was superior to any other. After incubation at 25 to 30°C for 48 h, growth will appear as spreading, rhizoid, discrete colonies with yellow centers that adhere tightly to the media (Figure 3). In broth culture *F. columnaris* forms a distinct yellow, mucoid pellicle at the meniscus. Isolation may be enhanced by addition of polymyxin B (10 IU/ml) and neomycin (5 μg/ml) in media because they inhibit growth of noncolumnaris organisms.[22] Columnaris is confirmed by slide agglutination using *F. columnaris* specific antisera or by the direct fluorescent antibody test. Baxa et al.[25] reported positive identification of *F. maritimus* by FAT in experimentally infected black sea bream fry.

F. columnaris* is a Gram-negative rod that measures about 0.4 μm in diameter and 3 to 10 μm in length. There are few morphological or biochemical differences between *F. columnaris, F. maritimus,* and *F. psychrophila* (Table 1). The major distinguishing factors of these three organisms involves length of cell, acid production from glucose, H$_2$S production, catalase, optimum growth temperature, salinity tolerance, and the presence of chondroitinase. Griffin[26] devised a simplified method of identifying *F. columnaris* using the five following characteristics that separate it from other yellow pigment-producing aquatic bacteria:

1. Ability to grow in the presence of neomycin sulfate and polymyxin B
2. Typical thin, rhizoid, yellowish colonies
3. Ability to degrade gelatin
4. Bind congo red
5. The production of chondroitin lyase

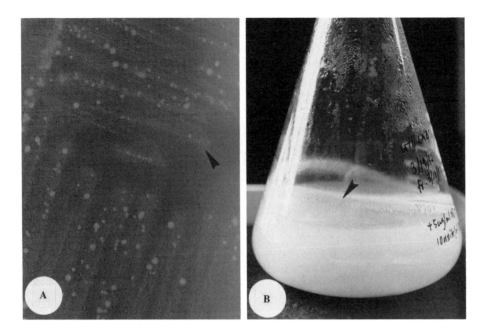

Figure 3 (A) *F. columnaris* (arrow) culture on Hsu-Shotts media from gill of largemouth bass. Solid, entire, white colonies are waterborne contaminants. (B) *F. columnaris* growing in Hsu-Shotts broth illustrating the mucoid pelicle (arrow) growing at the meniscus.

Pyle and Shotts,[27] using DNA homology, suggested that the organisms infecting salmonids and warmwater fish were indeed different strains, and that there appeared to be three distinct groups within the coldwater isolates. Song et al.[28] found three distinct groups among 26 *F. columnaris* isolates from Canada, Chile, Japan, Korea, the Republic of China, and the U.S. based on DNA homology, although there was diversity in colony morphology and some biochemical characteristics; 20 of the isolates in 1 homologous group included representatives from each country.

4. EPIZOOTIOLOGY

Columnaris disease, one of the most common bacterial diseases of fish, has an interesting epizootiological pattern. It may occur as a primary infection without any significant stress to the host or, more commonly, it may occur as a secondary pathogen as the result of stressful environmental conditions or trauma. In either case, the disease can develop as an acute infection concomitant with a rapidly developing mortality.

Columnaris most often occurs as an external infection of skin, fins, or gills but may also become systemic. Hawke and Thune[19] found that in 53 cases involving *F. columnaris* 11% were solely external, 17% were solely internal, but in total, external and systemic infection occurred in 86% of the cases.

Columnaris often appears as a mixed infection in association with one or more other pathogens and is possibly secondary to a more primary but less lethal pathogen (e.g., external protozoan parasites). Of the 53 *F. columnaris* infections of channel catfish studied by Hawke and Thune,[19] 46 involved mixed infections of other bacteria, especially *Edwardsiella ictaluri* and *Aeromonas* spp. Marks et al.[29] could not experimentally induce an *F. columnaris* infection unless a *Corynebacterium* sp. was also present, and Chowdhury and Wakabayashi[30] reported that *F. columnaris* invaded fish when several other bacteria were present but not when *A. hydrophila* or *Pseudomonas fluorescens* were present. They also showed that *F. columnaris* did not survive well *in vitro* when the density of *A. hydrophila* was approximately 100 times higher than that of *F. columnaris*. In view of these conflicting reports, the role of *F. columnaris* in primary or secondary infections and its relationship with other pathogens is unclear.

In cultured channel catfish populations where there are usually no other species present, particularly in ponds filled with well water, the catfish itself would be the source of *F. columnaris*. Fish in stream

Table 1 **Biophysical and biochemical characteristics of
Flexibacter columnaris and *Flexibacter maritimus,* two
species of yellow pigmented bacteria that produce disease
in fish[3,5-6]**

Characteristic	*Flexibacter columnaris*	*Flexibacter maritimus*
Colony morphology	Flat, thin, rhizoid, sticks to agar	Flat, irregular
Length of cell (μm)	3–10	2–30
Width of cell (μm)	0.3–0.5	0.5
Motility	Gliding	Gliding
Flexiruben pigments	+	−
Resistant to neomycin sulfate, polymyxin B	+	+
Produce chondroitin lyase	+	−
ONPG	−	−
Nitrogen sources		
Peptones	+	+
Casamino acids	+	+
Growth on peptone	+	+
Glucose sole source carbon	−	−
Acid from carbohydrates	−	+
Degradation of		
Gelatin	+	+
Casein	−	+
Starch	−	−
Tyrosine	−	+
Urease	?	+
H_2S	+	−
Nitrate reduced	−	+
Catalase	+	+
Cytochrome oxidase	+	+
Temperature (growth)		
Optimum (°C)	25–30	30
Tolerance (°C)	10–37	15–34
Highest NaCl (%)	0.5	Sea water
Habitat	Fresh water	Marine

or reservoir water supplies would also serve as sources for the bacteria. It has been proposed by Bullock et al.[17] and Bottomley and Holland[31] that course fish (suckers, carp, etc.) are reservoirs of infection.

Transmission of columnaris is generally by fish to fish via the water, but numerous factors affect the actual transmission and contraction of the disease. As a secondary infection in channel catfish, handling, seining, transportation, temperature shock, water quality (low dissolved or supersaturation of gases — oxygen, nitrogen, etc.), and other infectious diseases are the most common precursors to columnaris.[17,32-33] Channel catfish are susceptible to columnaris at temperatures from 15 to 30°C and young fish are more severely affected by the disease than are larger fish. Centrarchids are especially susceptible when held in abnormally cool water during the summer and juvenile rainbow trout and other salmonids are more susceptible to columnaris than are older fish, particularly when they are held in water at temperatures above 15°C. Adult migrating salmon become more susceptible to columnaris the farther they move upstream toward spawning areas. *Flexibacter columnaris* infections in eels are enhanced by low levels of dissolved oxygen and elevated ammonia,[34] and this probably holds true in other cultured fish species as well.

Trauma, mechanical injury, and stressful environmental conditions often predispose channel catfish to *F. columnaris*. Salmonid fingerlings may suffer up to 100% mortality if they are injured and held at 18 to 20°C, but a very low mortality occurs to noninjured fish. Hussain and Summerfelt[35]

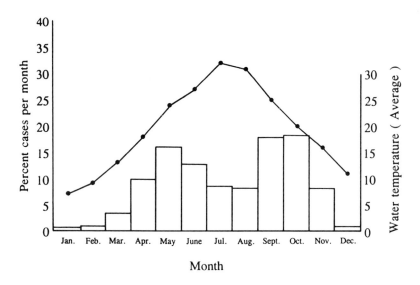

Figure 4 Monthly distribution of *F. columnaris* infections (all fish species) in Alabama from 1980 through 1990 (N = 1289). Diagnostic data are from the Southeastern Cooperative Fish Disease Laboratory, Auburn University, AL and the Alabama Fish Farming Center, Greensboro, AL. Water temperatures are from Boyd.[48]

experimentally induced columnaris infections in a group of 7- to 9-cm walleye *(Stizostedion vitreum)*. Up to 70% of the mechanically injured fish contracted the disease compared to no infections in noninjured fish.

An *F. columnaris* infection can be chronic and cause a lingering, gradually accelerating mortality, but more often it appears suddenly and accelerates to acute mortality in a matter of days — 90% mortality in tank-held fingerlings is not uncommon during optimum disease conditions. In pond populations, mortalities are usually lower but may still reach 50 to 60%, and several researchers have reported over 90% mortality during epizootics in salmonids and eels.[32,35] In a recent columnaris infection of 3-week-old channel catfish fry, about 50% were killed in a 24-h period.

While columnaris does occur in every month of the year it tends to be seasonal, especially in temperate climates where it exhibits two peak periods (Figure 4). Infections begin to increase in late March through April, present a low incidence during the summer, and an increase again in the fall. This pattern is probably dependent upon the presence of optimum water temperatures and a greater number of susceptible size and age fish in the spring, and to movement and transport of fingerlings in the fall. Bowser[37] found that in black bullheads *(Ameirus melas)*, in Clear Lake, Iowa, widespread columnaris infections occurred in May and June with no reported incidences after July. *Flexibacter columnaris* infections in Taiwan were reported by Kuo et al.[38] to be highest in tilapia *(Tilapia* spp.) and eel populations during September and October, lower in March through June, and very low January through February and June through August. These peak infectious patterns are very similar to those reported in the U.S.

The ability of *F. columnaris* to survive in the aquatic environment has been examined several times. Fijan[22] showed that survival of the pathogen decreased at pH 7 or less, in waters with hardness less than 50 mg/l $CaCO_3$, and/or in waters with low organic matter content. Chowdhury and Wakabayashi[39] found that survival of *F. columnaris* decreased very little over 7 d in chemically defined water containing 0.03% NaCl, 0.01% KCl, 0.002% $CaCl_2 \cdot H_2O$, and 0.004% $MgCl_2 \cdot 6H_2O$, but survival was reduced in water containing higher concentrations of these ions. When sterile mud was seeded with *F. columnaris*, 62% of the cells survived after 77 h at 10°C but only 35% survived at 20°C.[11] These data imply that the organism does not survive well for extended periods of time under normal conditions without a fish host.

5. PATHOLOGICAL MANIFESTATIONS

Initial infections of *F. columnaris* are usually the result of mechanical or physiological injury or environmental stress, but may appear independent of these stressors. It has been proposed that lesion severity depends on strain virulence and that necrotic lesions probably result from the presence of proteolytic enzyme activity. Necrotic gills become congested and the epithelium separates from the lamellae. Gill lesions start at the margins of the filaments and necrosis progresses toward the gill arch (Figure 1).

When lesions occur on the body surface the dermis and the underlying musculature becomes necrotic.[40] Capillaries become congested and are destroyed, allowing blood to appear at the margins of the ulceration. Once the integument is compromised by the bacterium, systemic infections may occur. Generally, there is very little pathology associated with systemic infections of channel catfish with *F. columnaris* although the organism can be isolated from a large percentage of infected animals. However, Hawke and Thune[19] found swelling of the trunk kidney in some cases.

6. SIGNIFICANCE

According to Thune,[41] columnaris was the most frequently reported infectious disease in the catfish industry from 1987 to 1989, accounting for 58% of the bacterial cases. Its broad geographic range and extensive species susceptibility adds to its significance. However, because its presence is often considered to be of a secondary nature *F. columnaris* has not received as much attention from researchers as it deserves. Nevertheless, columnaris is probably responsible for killing as many cultured fish as any other bacterial organism.

7. MANAGEMENT

Management of columnaris is accomplished through prevention, maintenance of an optimum environment, proper handling of fish, judicious use of prophylactic treatments, and implementation of good health management procedures. Water temperature control appears to be an important environmental management tool, particularly in tanks, raceways, and aquaria. Overstocking, which leads to excessive feed and organic material in the water, causes water quality deterioration. When running water is available, flows must be sufficient to flush away metabolic waste (ammonia) while oxygen concentration and other water quality parameters must be maintained.

Columnaris generally responds to chemotherapy. Fijan and Voorhees[42] found that the organism is sensitive to oxytetracycline, tetracycline, and several other drugs that are no longer available. Potassium permanganate and copper sulfate, neither of which are FDA-approved for disease treatment of fish, are most commonly used for treatment of columnaris. Potassium permanganate is used at 2 to 4 mg/l in ponds as an indefinite treatment and up to 10 mg/l in tanks for up to 1 h.[43] Organic load in the water will dictate the concentration of potassium permanganate to be used; the higher the organic load the higher the chemical concentration required. The amount of permanganate which is quickly reduced to manganese dioxide in a particular body of water is called the potassium permanganate demand. Tucker[44] showed how this potassium permanganate demand can be measured colorimetrically, and Jee and Plumb[45] showed that 2 mg/l above the potassium permanganate demand was necessary to control columnaris. Copper sulfate should be used with caution because in soft water (<20 mg/l $CaCO_3$) copper is highly toxic to fish. The best therapeutic concentration is 0.5 mg/l or higher. Oxytetracycline can be used as an antibiotic feed additive for 10 d (50 mg/kg of fish per day), however, this treatment is not approved by the FDA. The most successful treatment for columnaris on catfish in ponds is a combined therapy of potassium permanganate and medicated feed.

Vaccination of fish against *F. columnaris* showed some promise as early as 1972[46] when high antibody titers were achieved by injecting channel catfish. Moore et al.[47] vaccinated fingerling channel catfish by immersion in a formalin-killed bacterin of *F. columnaris*. Survival was better in vaccinated fish and they required less antibiotic treatment than did the control. In another vaccination study, Maas and Bootsma[48] found that carp will absorb *F. columnaris* by immersion and indicated that this could be a means of vaccinating carp against columnaris.

B. ENTERIC SEPTICEMIA OF CATFISH

The genus *Edwardsiella* includes two species of bacteria which cause major disease problems in fish: *E. tarda*[50] and *E. ictaluri*.[51] *Edwardsiella tarda* is the causative agent in a disease with several common names: fish gangrene, emphysematous putrifactive disease of catfish, red disease or hepatonephritis of eels, and Edwardsiella septicemia.[52-53] Because *E. tarda* and *E. ictaluri* cause distinctly different diseases they will be discussed separately. *E. tarda* is included in the eel disease section (Chapter XIV).

Edwardsiella ictaluri causes enteric septicemia of catfish (ESC), also known as "hole-in-the-head".[51,54] ESC is one of the two most important infectious diseases of cultured catfish in the U.S. It was the most frequently reported infectious fish disease in the southeastern U.S. from 1986 through 1991, causing losses in the millions of dollars.

1. GEOGRAPHICAL RANGE AND SPECIES SUSCEPTIBILITY

E. ictaluri has been confirmed only in the U.S., Thailand,[55] and Australia.[56] In the U.S. it has been isolated from channel catfish primarily across the south from Florida to Texas and north to Missouri and Kentucky. However, the bacterium has been reported among cultured channel catfish in other states including Arizona, California, Idaho, Indiana, Kansas, New Mexico, and Virginia. As the propagation of channel catfish continues to spread, dissemination of *E. ictaluri* into new geographical areas will likely continue and, in fact, there are unconfirmed reports of its presence in countries where the channel catfish has been introduced.

Edwardsiella ictaluri has a more narrow host range than most other warmwater fish disease-producing bacteria. Cultured channel catfish are much more susceptible than other ictalurids, but white catfish *(I. melas)*, blue catfish *(I. furcatus)*, and brown bullhead *(I. nebulosus)* occasionally have been found to be infected with *E. ictaluri*. Natural infections have been reported in walking catfish *(Clarias batrachus)* in Thailand,[55] and in two aquarium species: Bengal danio *(Danio devario)*[57] and the glass knife fish *(Eigemannia virescens)*.[58] Experimental infections were established in chinook salmon *(Oncorhynchus kisutch)* and rainbow trout *(O. mykiss)*.[59] European catfish (sheatfish) *(Silurus glanis)* are only slightly susceptible[60] while several commonly cultured warmwater species were shown to be refractive.[61]

2. CLINICAL SIGNS AND FINDINGS

Enteric septicemia is a chronic to acute disease with almost pathognomonic clinical signs in channel catfish.[54] Diseased fish hang listlessly at the surface with a "head-up-tail-down" posture, sometimes spinning in circles followed quickly by morbidity and death. Petechial hemorrhage or inflammation in the skin under the jaw and on the operculum and belly often becomes so extensive that the skin will become bright red (paint brush hemorrhage). Hemorrhage may also occur at the base of fins. Affected fish have pale gills, exophthalmia, and occasionally enlarged abdomens. Small depigmented lesions of 1 to 3 mm in diameter appear on the dark skin of the flanks and back of infected fish. The lesions then progress into similar sized inflamed cutaneous ulcers (Figure 5). In chronically ill fish an open lesion may develop along the central line of the skull between the eyes at the insertion of the two frontal bones, thus the name "hole-in-the-head" disease (Figure 5). Internally, the body cavity may contain a cloudy or bloody fluid and on rare occasions a clear yellow fluid; the kidney and spleen are hypertrophied, and the spleen is dark red; inflammation occurs in adipose tissue, peritoneum, and intestine; and the liver is either pale or mottled with congestion.

3. DIAGNOSIS AND BACTERIAL CHARACTERISTICS

Generally, *E. ictaluri* is isolated from clinically infected fish on BHI or TSA agar, a method that requires 36 to 48 h for punctate colonies to form when incubated at 28 to 30°C. When using BHI agar to detect the presence of *E. ictaluri* careful observation of primary culture plates is essential, as there is the possibility that other more rapidly growing bacteria, such as *Aeromonas* spp., will overgrow the *E. ictaluri*. Shotts and Waltman[62] developed *Edwardsiella ictaluri* isolation media (EIM), a selective media that distinguishes *E. ictaluri* from other fish bacterial pathogens (Figure 6). On EIM, *E. ictaluri* produces clear greenish colonies and at the same time inhibits growth of Gram-positive bacteria. Other Gram-negative organisms will grow on the media, but can be separated by colony morphology and color. *E. tarda* colonies will appear small with black centers, *Aeromonas hydrophila* colonies are brownish and larger than *E. ictaluri*, and *Pseudomonas fluorescens* colonies will appear blackish and

Figure 5 Channel catfish infected with *E. ictaluri*. (A) Lower fish is not infected and the upper fish shows the early stages of disease development with depigmented areas (arrow). The middle fish has petechial hemorrhage (arrow) in the skin. (B) Channel catfish exhibiting open lesions in the cranial region (large arrow), inflamed nares (small arrows), typical of chronic infection and exophthalmia.

more punctate. When the biochemical characteristics of *E. ictaluri* (Table 2) were described very little diversity among the many isolates was noted.[51,63-64] The organism is a Gram-negative, short rod (0.8 μm × 1 to 3 μm) that tends to be longer in actively growing cultures. It is cytochrome oxidase negative and weakly motile at 25 to 28°C, but not at 30°C or above. *E. ictaluri* grows very poorly or not at all at 37°C. At 20 to 30°C *E. ictaluri* ferments and oxidizes glucose while producing gas. It will not tolerate a NaCl level higher than 1.5% in the media. *E. ictaluri* can easily be separated from *E. tarda* by its indole negative reaction and lack of H_2S production on TSI.

Serological tests can be used to confirm the identity of *E. ictaluri*. Methods developed by Rogers[65] utilized indirect immunofluorescent antibody (FA) or enzyme-linked immunoassay (ELISA) using polyvalent rabbit-anti-*E. ictaluri* sera. In these tests no cross reactivity was detected with *E. tarda, Salmonella,* or *A. hydrophila*. The lack of serological cross-reactivity of *E. ictaluri* with *E. tarda* or *A. hydrophila* was also demonstrated by Klesius et al.[66] using ELISA and monoclonal antibody. Both techniques successfully identified the bacterium in necropsied tissues of channel catfish. Ainsworth et al.[67] also used monoclonal antibodies in an indirect FAT for diagnosing *E. ictaluri*. When the FAT was compared to cultured isolates from the brain, liver, spleen, and anterior and posterior kidney it was noted that 90.3% of culture-positive fish were also FA-positive compared to 85% positive by conventional culture. As in similar studies of other pathogens, these studies emphasize the time-saving advantage of using immunoassay by giving results in 2 h vs. 48 h required for culture. Klesius[68] further refined the ELISA system by developing FAST-ELISA to detect the presence of antibodies against *E. ictaluri* exoantigen in 30 min. Utilization of monoclonal antibody specific for the outer membrane antigen of *E. ictaluri* should be beneficial in bacterial identification. These techniques could be valuable in detecting covert carrier fish or predicting overt infection in advance. Baxa-Antonio et al.[69] suggested that detection of humoral antibody in channel catfish could be used to identify populations that have been exposed to *E. ictaluri*. However, the absence of antibody may not indicate they had not been exposed to the pathogen because all fish in a population do not produce antibody following exposure.[70]

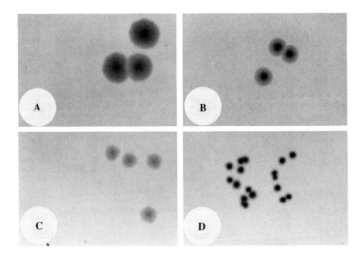

Figure 6   Colonies of four bacterial pathogens commonly isolated from channel catfish growing on *Edwardsiella ictaluri* identification media (EIM). (A) *Aeromonas hydrophila* colonies are larger and brownish; (B) *E. tarda* are smaller and have black centers; (C) *E. ictaluri* colonies are a pale green with darker green centers; (D) *Pseudomonas fluorescens* are very small and black. (Photographs by D. Earlix.)

DNA plasmids may offer help in detecting *E. ictaluri* carrier fish. Lobb and Rhoades[71] and Newton et al.[72] described one to three plasmids in *E. ictaluri* isolates and indications are that all *E. ictaluri* isolates, regardless of origin, possess plasmids. The plasmids are specific enough that Speyerer and Boyle[73] suggested the possibility of using them as a probe to detect fish infected with *E. ictaluri.*

4. EPIZOOTIOLOGY

When ESC was first described[54] it was thought that *E. ictaluri* was an obligate pathogen that could not survive for any extended period outside of the host. It has since been found that the organism can survive in pond bottom mud for over 90 d at 25°C, and in water for under 15 d at 25°C or less.[74] Once *E. ictaluri* is introduced into a particular body of water it appears that it will remain there for an extended period of time in either the water, mud, or carrier fish. There is some question concerning the presence of *E. ictaluri* in epizootic survivor fish but Mgolomba and Plumb[75] found significant numbers of bacteria present in the blood and all organs of exposed fish 81 d after initial exposure to *E. ictaluri,* and Klesius[76] isolated bacteria from channel catfish 280 d after initial detection of the pathogen.

    E. ictaluri is transmitted through water via carrier channel catfish which, in all probability, sequester the bacterium in their intestines, especially in summer months and during clinical infections. Transmission from adult to offspring at spawning is likely, but as yet not proven. *E. ictaluri* is readily phagocytized but indications are that the bacteria are not destroyed in these cells (Figure 7),[77-78] therefore, phagocytes may contribute to the longevity of bacteria in carrier fish.

    During an *E. ictaluri* epizootic, a significantly higher concentration of the pathogen was found in pond water in areas with a large number of dead fish than in areas with no carcasses. This indicates that removal of dead fish may reduce the level of bacteria to which noninfected fish are exposed.[79] Taylor[80] detected *E. ictaluri* by fluorescent antibody in the lower intestines of fish-eating birds (cormorants and herons) and later demonstrated that the bacteria were viable, suggesting the possibility that these birds could serve as sources of infection.

    Although ESC has been diagnosed during every month of the year and in a wide range of water temperatures, it is considered to be a seasonal disease. ESC occurs primarily when water temperatures range from 18 to 28°C in late spring to early summer and again in the fall (Figure 8). Mortality in *E. ictaluri* experimentally infected channel catfish fingerlings was highest at 25°C, slightly lower at 23 and 28°C, and no deaths occurred at 17, 21, or 32°C.[81] However, there has been a noted increase of ESC outbreaks during July and August, and from late fall to early spring, which would indicate an

Table 2 **Biochemical characteristics of *Edwardsiella tarda*, *Edwardsiella ictaluri*, and *Edwardsiella hoshinae*[51,63-64,99]**

Characteristic	*Edwardsiella tarda*	*Edwardsiella ictaluri*	*Edwardsiella hoshinae*
Motility			
25°C	+	+	+
37°C	+	−	+
Growth at 40°C	+	−	+
Tolerance to NaCl			
1.5%	+	+	+
4.0%	+	−	+
Indole	+	−	±
Methyl red	+	−	+
Citrate (Christensen's)	+	−	+
H₂S production			
Triple sugar iron	+	−	−
Peptone iron agar	+	−	+
Lysine decarboxylase	+	+	+
Ornithine decarboxylase	+	+	+
Malonate utilization	−	−	+
Gas from glucose	+	+	±
Acid production from			
D-Mannose, maltose	+	+	+
D-Mannitol, sucrose	−	−	+
Trehalose	−	−	±
L-Arabinose	−	−	±
Jordans tartrate	±	−	−
Nitrate reduced to nitrite	+	+	+
Tetrathionate reductase	+	?	+
E. ictaluri media (EIM)	Black centers	Green translucent	?
Mol% G + C of DNA	55–58	56–57	53

Note — Symbols: +, positive for 90–100% of strains; −, negative for 90–100% of strains; ±, mixed reactions. All strains of *E. tarda, E. ictaluri,* and *E. hoshinae* tested are negative for Voges-Proskauer, Simmons citrate, urea, phenylalanine deaminase, arginine dihydrolase, gelatin hydrolysis, growth in KCN; acid production from glycerol, salacin, adonitol, D-arabitol, cellobiose, dulcitol, erythritol, lactose, inositol, melobiose, α-methyl-D-glucoside, raffinose, L-rhamnose, and D-xylose; acid from mucate; esculin hydrolysis, acetate utilization, deoxyribonuclease, lipase, β-galactosidase (ONPG), pectate hydrolysis, pigment production, tyrosine clearing, and oxidase test (Kovacs).

expanding temperature tolerance of the pathogen. Baxa-Antonio et al.[69] found that channel catfish experimentally infected by immersion experienced 47 and 25% mortality when the fish were held at 20 and 30°C, respectively. These data indicate a wider temperature range of ESC than the normal optimum of 22 to 28°C.

The mortality rate in *E. ictaluri*-infected catfish populations varies from less than 10% to over 50%. The pathogen infects fingerling as well as production-sized fish and occurs in all types of cultural conditions: ponds, raceways, recirculating systems, and cages. Most fish diseases are accompanied by some type of environmental stressor that precedes infection; however, ESC is an exception and does develop independently of extrinsic influences. It is not unusual for ESC to occur in catfish populations during favorable environmental conditions, however, adverse environmental circumstances may intensify the severity of infection. The precise relationship between environmental quality and outbreaks of *E. ictaluri* has not been fully elucidated, but during experimental infections Wise et al.[82] established a relationship between confinement-induced stress and increased susceptibility of channel catfish to

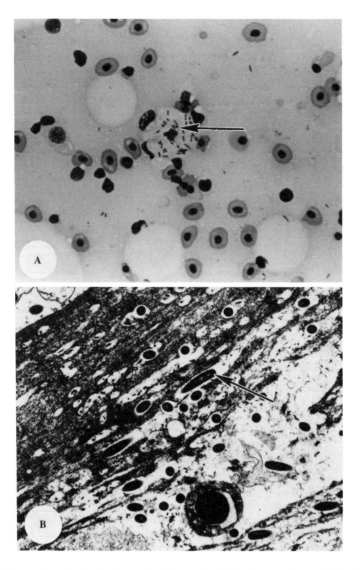

Figure 7 (A) Smear from lesion in the skull of channel catfish showing intracellular *E. ictaluri* (arrow) in macrophages (H & E). (Photograph by T. Miyazaki.) (B) Electron micrograph of olefactory organ from channel catfish illustrating *E. ictaluri* (arrow) invading the tissue. (Original magnification × 15,000; photograph by E. Morrison.)

E. ictaluri. They also concluded that stress induced by handling and hauling will likely result in increased mortality of channel catfish that are infected with *E. ictaluri*.

5. PATHOLOGICAL MANIFESTATIONS

The most severely damaged organs in an *E. ictaluri*-infected catfish are the trunk kidney and spleen. Both organs develop necrosis, and the liver becomes edematous.[83] Further, *E. ictaluri* studies by others revealed that interlamellar tissue of the gills proliferate and skin epidermis is destroyed. A mild focal infiltration takes place in the musculature underlying the epidermis where bacteria apparently colonize capillaries, causing depigmentation and then necrosis.[77,84-85] Ulcerative lesions of the head are necrotic and hemorrhaged, whereas systemic infection is associated with necrosis of hepatocytes and pancreatic cells. Intact, apparently dividing *E. ictaluri* cells are seen within macrophages (Figure 7). This was also seen in electron micrographs of macrophages in tissues in the olfactory sac.[86]

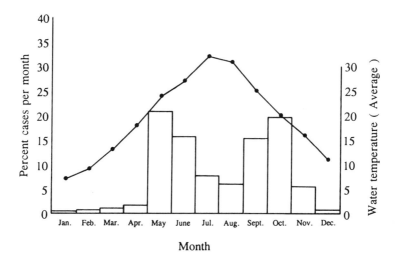

Figure 8 Seasonal occurrence of *E. ictaluri* (N = 1041) in Alabama from 1980 through 1990 showing greatest incidence of disease in May, June, September, and October when average water temperatures are 20 to 28°C. Water temperatures from Boyd.[48]

Following experimental exposure via immersion to 5×10^8 CFU/ml, 93% of affected channel catfish developed acute ESC and 7% developed chronic disease.[87] Acute disease was characterized grossly by hemorrhage and ulceration, and microscopically by enteritis, olfactory sacculitis at 2 d postexposure, followed by hepatitis and dermatitis. Chronic ESC, most commonly observed at 3 to 4 weeks postexposure was characterized by dorsocranial swelling, ulceration, and granulomatous inflammation and meningeoencephalitis of the olfactory bulbs, tracts, and lobes of the brain.

Fish can become infected with *E. ictaluri* by waterborne exposure or ingestion. Waterborne bacteria will invade the olfactory organ via the nasal opening, migrate into the olfactory nerve, enter the brain[78,85] and then spread from the meninges to the skull and skin, creating a "hole-in-the-head" condition. Morrison and Plumb[86] showed that *E. ictaluri* can indeed attach to the olfactory epithelium and migrate into the submucosa (Figure 7). During the infection process the celia on the epithelium are destroyed and the mucosa becomes secretory.

When ingested, *E. ictaluri* enters the blood through the intestine and results in a septicemia.[83,87] Catfish exposed orally to *E. ictaluri* developed enteritis, hepatitis, interstitial nephritis, and myositis within 2 weeks of infection. Baldwin and Newton[88] showed that at 0.25 h postintestinal infection, *E. ictaluri* appears in kidneys, indicating a rapid transmucosal passage. They further suggested that *E. ictaluri* may have invasion and survival strategies in the host that are similar to other invasive Enterobacteriaceae. Francis-Floyd et al.[81] described petechia or ecchymoses in the mucosa of the gastrointestinal tract and intestinal distension associated with gas production.

6. SIGNIFICANCE

Enteric septicemia is the most economically important infectious disease of cultured channel catfish. *E. ictaluri* infections cost the aquaculture industry millions of dollars annually in killed fish and/or expenditures for preventive measures and chemotherapeutic treatments. The effect of ESC on reduced growth and higher feed conversions is only speculative, but because channel catfish depend on their olfactory functions in feeding, any injury to these vital tissues most likely affects feeding during and after infections, and therefore helps explain why affected fish will not feed once the disease has reached a certain level of severity.

Relatively few cases of ESC were detected immediately following its discovery, but in the early 1980s the number of *E. ictaluri* isolates began to climb at an alarming rate. In the southeastern U.S. 47 cases of ESC were reported in 1981, 1042 cases in 1985, and 1605 cases in 1988. Of all reported fish disease outbreaks in the southeast, ESC accounted for 28% of them in 1985 and 30.4% in 1988. However, reported outbreaks from 1990 to 1992 have leveled off.[89]

7. MANAGEMENT

Successful therapy of ESC depends on immediate treatment with medicated feed, therefore early diagnosis is imperative. Only two drugs for treating bacterial diseases of fish are presently approved by the FDA. Both drugs, oxytetracycline (Terramycin®) and the potentiated sulfonamide combination of sulfadimethoxine and ormetoprim (Romet-30®) are effective against ESC,[90-91] but only Romet-30® includes *E. ictaluri* on its label. Terramycin® is fed at 50 mg of oxytetracycline per kilogram of fish per day for 12 to 14 d, followed by a 21-d withdrawal period. Romet-30® is also fed at 50 mg of drug per kilogram of fish per day, but for only 5 d, followed by a required 3-d withdrawal period before slaughter. To be effective, drug therapy must be initiated before fish stop feeding. Unpublished reports of *E. ictaluri* developing resistance to Terramycin® and Romet-30® have occurred, with as high as 10 to 15% of the isolates being resistant to one or both drugs. Waltman et al.[92] found that the resistance of *E. ictaluri* to Romet® was mediated by an R plasmid and Starliper et al.[93] showed that the plasmid could be transferred to nonresistant isolates of *E. ictaluri,* thus transferring resistance. Cooper et al.[94] suggested that agricultural runoff is a potential source for the plasmid responsible for the emergence of Romet-30® resistance of *E. ictaluri,* and the use of feed containing this drug enhances the selection for *E. ictaluri* that have received the R plasmid. While this resistance can be plasmid or genetically induced, the problem is likely exacerbated by improper use of the antibiotics. Too often, fish farmers use medicated feed when it is not necessary, apply it at improper rates, and/or continue treatment longer than recommended, all of which promotes an increased rate of conversion from sensitive to resistant.

E. ictaluri is a strong immunogen when injected, which makes it an excellent candidate for vaccine development against ESC. The organism is also antigenic when the fish is immersed in a solution containing *E. ictaluri* or when it is incorporated into feed.[95] Antibody titers against *E. ictaluri* are easily measured but the antibody level is yet to be proven as a measure of protection. However, channel catfish that survived a natural epizootic of *E. ictaluri* and had titers greater than 1:512 were protected against *E. ictaluri.*[70]

Vaccination studies of *E. ictaluri* have used whole-cell bacterins and cell extracts such as LPS and purified "immunodominant" antigens. Several studies have shown that fractionating *E. ictaluri* and then using a purified immunodominant antigen can provide protection against the disease.[96-98] A protein with a molecular mass of 36 kDa is the primary immunodominant antigen in the cell membrane and it is retained by the cells when cultured for up to 30 passages on media.[95]

Commercial vaccines for *E. ictaluri* have already undergone successful laboratory and field testing and show an effective immunity against mass mortalities due to *E. ictaluri* in most instances. These vaccines are generally applied by an initial immersion of fry (10 to 14 days old) or fingerlings, followed by an oral booster in feed 1 to 2 months later, or vaccination by immersion or the oral route only.

While vaccination of channel catfish against *E. ictaluri* infections is becoming a management tool, this method of prevention should not be perceived as absolute. However, if the disease does develop, mortalities will be reduced in vaccinated populations. A manager will have to analyze the benefit to cost ratio to determine if vaccination is feasible. It must also be considered that, generally, vaccinated fish grow faster and show a lower feed conversion rate in addition to 10 to 30% better survival than the nonvaccinated channel catfish. Aquaculturists must not be lulled into thinking that vaccines are "cure-alls" and use their availability to justify increased stocking rates and continued abuse of the culture environment.

Hatchery and farm sanitation is very important in curtailing the spread of *E. ictaluri* within a farm or between farms; therefore, all seines, nets, and fish hauling units, etc., should be disinfected or thoroughly air dried after use and any failure to do so increases the potential of spreading the pathogen to naive fish.

C. MOTILE *AEROMONAS* SEPTICEMIA

Motile *Aeromonas* septicemia (MAS) is associated with infections caused by motile members of the genus *Aeromonas.* Synonyms of the syndrome are hemorrhagic septicemia, infectious dropsy, infectious abdominal dropsy, red pest, red disease, red sore, rubella, and others. The disease, presently known as MAS became part of modern fish health in the 1930s when it was named infectious dropsy (of carp). In North America the syndrome was known as hemorrhagic septicemia until the mid 1970s, when the name was changed to "motile *Aeromonas* septicemia" because it was realized that more than one member of the *Aeromonas* genus could cause the same disease syndrome, and that each bacterial species was motile. The principal motile species of *Aeromonas* that affect fish are *A. hydrophila (punctata, liquefaciens), A. sobria,* and *A. caviae.*[100]

1. GEOGRAPHICAL RANGE AND SPECIES SUSCEPTIBILITY

MAS is an ubiquitous disease found in warm, cool, and cold freshwater fish and occasionally in brackish- and saltwater fishes around the world. Europe, North and South America, and Asia are most significantly affected by the disease but no area is free of MAS.

MAS is generally, but not exclusively, associated with warmwater fish. Channel catfish, other ictalurids, silurids, clariads, carp and other cyprinids, eels, centrarchids, and the true basses (e.g., striped bass) are susceptible to the disease. Trout and salmon are also susceptible, but are usually affected only when water temperatures reach the fish's upper tolerance (i.e., stress) limits.[101-102] As far as is known, no species of fish is totally immune or resistant to the organisms that cause MAS. Motile aeromonads infect other aquatic animals as well, causing "red-leg" disease in frogs and fatal disease in reptiles.[103] They can also produce fatal septicemias and localized infections in humans.[104-105]

2. CLINICAL SIGNS AND FINDINGS

Clinical signs of MAS are as varied and diverse as are the species of fishes affected. Generally, clinical observations are categorized as either behavioral or external, but internal lesions also occur.

MAS-infected fish will lose their appetite, become lethargic, and swim lazily at the surface. When diseased fish first appear at the surface, they usually dive when disturbed but eventually lose their equilibrium, return to the surface, and move into shallow water. When in flowing water they will float downstream after losing their equilibrium.

Slight differences in manifestation of the disease can be noted between scaled and scaleless fish and external signs are varied; none are specific but are often similar to those associated with other septicemic bacterial infections. In catfish and other scaleless fishes, fins become frayed, hemorrhaged or hyperemic, and congested. Epidermal lesions will begin as irregularly shaped depigmented areas that eventually develop into necrotic skin which sloughs off, leaving open, ulcerated lesions with exposed muscle (Figure 9). Margins of the lesion are whitish or hemorrhaged. External lesions may occur at any location on the fish: caudal peduncle, dorsally, ventrally, laterally, or on top of the head. On scaled fish, the lesions which begin as hemorrhages at the base of the scales have a red (inflamed) central area surrounded by whitish (necrotic) tissue from which the scales have been lost (Figure 9). In both scaleless and scaled fish, fungus (e.g., *Saprolegnia* spp.) will often attack necrotic tissue and give the lesion a fuzzy, brownish appearance. MAS-infected fish may also have exophthalmia accompanied by hemorrhages or opaqueness of the eye, enlarged abdomens with ascites, pale gills indicative of anemia, edematous musculature, and scales will show lepidorthosis before being lost.

Internally, organs are friable and have a generalized hyperemic appearance; the kidney and spleen are swollen; and the liver is often mottled with hemorrhage interspersed with light areas. The body cavity may contain a clear fluid (ascites) but more often the fluid is bloody and cloudy. The intestine is flaccid, hyperemic, contains yellowish mucus, and is void of food.

3. DIAGNOSIS AND BACTERIAL CHARACTERISTICS

Diagnosis of MAS in fish cannot be based solely on clinical signs associated with the disease because other bacterial organisms (e.g., *Pseudomonas* sp.) and protozoan parasites (e.g., *Epistylis* sp. and *Ichthyobodo*) often produce similar clinical signs and identical external lesions. A definitive diagnosis of MAS can only be made by a complete necropsy of diseased fish in conjunction with isolation and identification of the causative organism. Primary isolation of the motile aeromonads (i.e., *A. hydrophila, A. sobria,* or *A. caviae*) can be made on either BHI or TSA. The cultures should be incubated at 25 to 30°C for 24 to 48 h, at which time white to yellowish, mucoid, entire, slightly convex colonies will be obvious. *A. hydrophila* can be isolated and identified on Rimler-Shotts (RS) selective media and will form an orange-yellow colony when incubated at 35°C.[106] Once the bacteria is isolated it should be determined if it is a Gram-negative, short, motile rod that is cytochrome oxidase positive and ferments glucose. These bacteria measure 0.8 to 0.9 μm $\times$ 1.5 μm, are polar flagellated, produce no soluble pigments (Table 3), and are resistant to vibriostat (0/129) (2,4-diamino-6,7-disopropyl pteridine phosphate) and Novobiocin. For definitive identification additional biochemical characteristics should be determined (Table 3).

Figure 9   Motile *Aeromonas* septicemia *(Aeromonas hydrophila)* infected fish. (A) Channel catfish with necrotic epithelial lesions (arrows) with exposed musculature; (B) striped bass with hemorrhaged, necrotic lesion (arrow).

The genus *Aeromonas* is presently in the family Vibrionaceae, however, Colwell et al.[107] have proposed that the family Aeromonodaceae, which would include *Aeromonas,* be recognized. Carnahan et al.[108] listed seven motile species of the genus, but only three have been associated with diseased fish with any regularity. The species *A. hydrophila (liquefaciens, punctata)* and *A. sobria* are most commonly reported, particularly in cultured channel catfish, with *A. caviae* being only occasionally isolated from diseased fish.

A. hydrophila, A. sobria, and *A. caviae* can be distinguished by several biochemical tests such as gas from glucose, esculin hydrolysis, and acid from arabinose (Table 4). Serological identification of motile aeromonads is not a common diagnostic tool because of antigenic diversity and genetic complexity.[109] This is particularly true of the ubiquitous *A. hydrophila,* which has as many as 12 O-antigens and 9 H-antigens.[110] Historically, monovalent antisera to a particular *A. hydrophila* strain agglutinates only a small percentage of heterologous *A. hydrophila.*

4. EPIZOOTIOLOGY

According to data compiled by the Fish Disease Committee of the Southern Division of the American Fisheries Society,[89] MAS has been one of the most frequently diagnosed bacterial diseases of fish since 1972 and was the most severe disease problem encountered by catfish farmers until the mid 1980s. The disease accounted for as many as 60% of the total bacterial cases reported in some years during this period.

Motile *Aeromonas* septicemia is generally a seasonal disease, but has been diagnosed in every month of the year across the southern U.S. where it peaks in the spring (Figure 10). As one moves northward, the peak of motile *Aeromonas* infections tends to occur later in spring or early summer.[8] This pattern of susceptibility can be attributed to fish having a lowered resistance to *Aeromonas* infections following

Table 3 **Characteristics and reactions frequently used to identify the motile Aeromonas (Aeromonas hydrophila, Aeromonas caviae, and Aeromonas sobria)**[140]

Characteristic	Reaction or feature
Morphology	Short, Gram-negative, single or paired rods
Motility	+
Cytochrome oxidase	+
Catalase	+
Growth in nutrient broth (37°C)	+
Ornithine decarboxylase	−
Indole production	+
Fermentation	
Glucose, sucrose, mannitol	+
Dulcitol, rhamnose, xylose	−
Raffinose, inositol, adonitol	−
NO_3 reduction to NO_2	+
Growth in peptone without NaCl	+
0/129 Resistance	+
Starch, gelatine, ONPG, RNA, and DNA hydrolysis	+
Tween® 80 esterase	+

winter and spring environmental stress periods, to the presence of large numbers of susceptible young fish at this time of year, and to the fact that water temperatures in the spring are more conducive to bacterial infections. Spawning activities also increase susceptibility to *Aeromonas* sp. As summer progresses, fish tend to develop an immunity and/or natural resistance to the *Aeromonas* sp. organisms and disease outbreaks decrease, only to rise again in the fall when water temperatures are more favorable to the bacterium and during which time juvenile fish must be handled and transported.

Fish mortality associated with motile *Aeromonas* infections usually follows a chronic pattern with obscure peaks of deaths per day; however, cumulative mortality can be significant. If a highly virulent bacterial strain is involved, high mortality may result, but total losses will normally be below 50%. Infections of *A. hydrophila* can occur in very young fish to adults. Epizootics of young fish may be acute, whereas in older fish the die-off is more chronic. Kage et al.[111] found that 90 to 100% of 7-day-old Japanese catfish *(Silurus asotus)* died as result of an *A. hydrophila* infection.

Infections may be external, internal (systemic), or more commonly, both. In an external infection, skin lesions from which bacteria can be isolated are the only obvious clinical signs of disease. During this type of infection, bacteria cannot be isolated from internal organs and it can be argued that this phase of the syndrome should not be considered motile *Aeromonas* septicemia because there is no septicemia. Systemic infections are characterized by a septicemia and the causative organism can easily be isolated from any internal organ as well as from skin lesions.

A. hydrophila is found in most natural freshwater ponds, streams, or reservoirs and bottom muds[112] where it is a facultative organism that can utilize any available organic material as a nutrient source; therefore, it is environmentally adaptable.

Infections of *A. hydrophila* in fish are often associated with some type of predisposing stress such as temperature shock, exposure to low oxygen (Chapter I, Figure 5), high ammonia and other adverse water quality problems, trauma from improper handling or hauling, and presence of other disease organisms.[113-114] Peters et al.[115] found that even social stress of juvenile rainbow trout enhanced ventilation, elevated glucose and leukocyte volume, and increased susceptibility to *A. hydrophila*. Motile aeromonad infections in most animals including man usually occur as a secondary disease and are the result of a debilitating condition, but once an infection of *A. hydrophila* becomes established it is capable of causing morbidity and death.[116]

Table 4   **Biochemical reactions that separate the three
species of bacteria (*Aeromonas hydrophila, Aeromonas
caviae,* and *Aeromonas sobria*) involved in motile
Aeromonas septicemia**[106,140-141]

Substrate	Reaction		
	Aeromonas hydrophila	*Aeromonas caviae*	*Aeromonas sobria*
[a]Aerogenic (gas from glucose)	+	−	+
Salacin	+	+	−
β-δ Xylosidase	+	+	−
Arbutine	+	+	−(14%+)
β-Glucosidase	+	+	−(29%+)
[a]Acid from arabinose	+	+	−
Voges-Proskauer	+	−	+
Elastase	+	−	−
[a]Esculin	+	+	−
Growth on KCN	+	+	−
L-Histidine utilization	+	+	−
L-Arginine utilization	+	+	−
H2S from cystine	+	−	+

[a]Characteristics most often used to separate the three species if they were isolated from fish.

A. hydrophila infections are associated with eutrophication of lakes and ponds that receive large amounts of organic enrichment,[103] similarly to conditions in catfish culture ponds that receive large daily rations of feed. Excess feed not only supplies nutrients for microbial flora and phytoplankton, but has a detrimental effect on oxygen concentrations, which can cause stress in fish populations and enhance susceptibility to infection. The relationship between water quality-induced stress and an *A. hydrophila* infection in a highly eutrophic channel catfish culture pond was shown by Plumb et al.[113] and later duplicated in the laboratory.[117] An abrupt decrease in oxygen concentration in the pond resulted in a drop in pH and an increase in concentration of ammonia and carbon dioxide, and was followed in 6 d by an *A. hydrophila* infection (see Chapter I, Figure 3). Levels of pH and concentrations of O_2, CO_2, and ammonia are interdependent factors and constitute a major part of the low dissolved oxygen syndrome (LODOS) proposed by Schmottou.[118]

Susceptibility of some species of fish to *A. hydrophila* is also linked to specific water temperatures. Groberg et al.[101] reported that when exposed to *A. hydrophila* coho and chinook salmon and steelhead trout suffered 64 to 100% mortality at 18°C compared to 0% mortality at 9.4°C. Rainbow trout demonstrated an increased susceptibility to *A. hydrophila* when water temperature was raised from 5.5 to 8°C to more than 11°C.[102] Algae in unclean holding tanks can contribute to higher *A. hydrophila* concentrations and incidence of infection.[119]

There is evidence that the motile *Aeromonas* complex is a secondary and opportunistic pathogen, but the ability of *A. hydrophila* to cause disease and death of fish should not be overlooked. The organism does have the ability to cause infection and high mortality among cultured fish without the presence of severe external (i.e., stressful) influences, although this is a rare event and may be caused by *A. hydrophila* strains that possess specific virulent or pathogenic characteristics. Regardless of whether or not motile *Aeromonas* serves as the primary or secondary invader of stressed fish, in many instances it is the final insult that leads to death.

In Asia *A. hydrophila* has been closely associated with epizootic ulcerative syndrome (EUS), a disease that has plagued both wild and cultured fish populations from Indonesia to India since the early 1980s.[120] The definitive etiology of EUS is uncertain because a virus, a fungus, and *A. hydrophila* have been associated with the lesions. While clinical signs of EUS are almost identical to those of MAS, EUS investigators feel that *A. hydrophila* is totally secondary to the primary cause, but this bacterium may be what kills the fish.

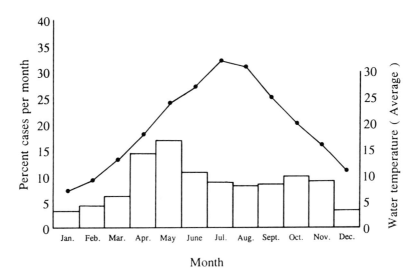

Figure 10 Monthly distribution of motile *Aeromonas* septicemia infections in fish in Alabama from 1980 through 1991 (N = 1040). Data are from disease case histories at the Southeastern Fish Disease Laboratory, Auburn, AL and the Alabama Fish Farming Center, Greensboro, AL. Temperature curve was taken from Boyd.[48]

Often different bodies of water or watersheds have a unique strain of *A. hydrophila* because of the serological diversity and widespread facultative nature of the bacterium. If fish which have become resistant to a particular strain of the bacterium are moved to a new body of water, they will in all probability be exposed to a strain of *A. hydrophila* to which they have no immunity and due to handling and transport stressors these fish may become infected by the new strain of bacteria.

While *A. hydrophila* primarily affects fish, other aquatic animals and humans can be infected by the bacterium. These infections can become quite severe in humans, causing enteritis, meningitis, and localized infections on the extremities.[121] Puncture wounds inflicted by catfish spines should not be taken lightly because several such instances have led to serious *A. hydrophila* infections.[122-123] Several deaths due to *A. hydrophila* have been reported, but in such instances the patient had been debilitated by another ailment.[104] Also, Goncalves et al.[105] attributed pneumonia to an *A. hydrophila* septicemia in a healthy male 3 d after swimming in the ocean and suggested that this may have been the source of infection. King et al.[124] reported an *Aeromonas* spp. isolation rate of 10.6 cases per 1,000,000 population in California, and though 2% of the patients died, all had serious underlying medical conditions apart from the *Aeromonas* infection. They concluded that *Aeromonas* spp. infection in humans is not an important public health problem and is largely unpreventable. Nevertheless, with an increase in incidence of *A. hydrophila* infections in humans, handling infected fish with caution is advisable.

The individual pathogenic capabilities of *A. hydrophila*, *A. sobria*, and *A. caviae* are still unclear. Gray et al.[125] found that in 61 isolates of motile *Aeromonas* spp. taken from pig and cow feces and other environmental sources (none were from fish), 96.4% of the *A. hydrophila*, 36.4% of the *A. sobria*, and only 13.6% of the *A. caviae* were cytotoxic. The conclusion drawn from this study was that *A. hydrophila* was highly pathogenic, *A. caviae* was not pathogenic, and *A. sobria* was only moderately pathogenic. However, this may not hold true in channel catfish. In isolates taken from MAS-diseased channel catfish in Mississippi, *A. sobria* was found more frequently than *A. hydrophila*.[126] In view of this, it is possible that *A. sobria* may be the primary cause of MAS infection in cultured channel catfish, but is often confused with *A. hydrophila*.

5. PATHOLOGICAL MANIFESTATIONS

The pathology of MAS is not distinctly different from most other septicemic bacterial infections of fish.[127-129] Lesions are typical of those caused by bacterially produced proteases and hemolysins. Epidermal infections are characterized by necrotic lesions which have spongy centers and hemorrhagic margins. The epidermis adjacent to lesions is edematous, and the dermis becomes hemorrhagic with macrophage and lymphocyte accumulation accompanied by severe inflammation. Scale pockets become edematous, causing lepidorthosis. In systemic infections most internal organs are edematous with diffused necrosis of the liver, kidney, and spleen. Inflammation is not apparent in diseased internal organs, but hemorrhage and/or erythemia does occur.

The virulence of *A. hydrophila* may be influenced by the presence of specific biochemical characteristics or by production of extracellular products that independently produce pathological effects when injected into fish.[130,131] It was reported by Wakabayashi et al.[132] that the virulence of *A. hydrophila* was related to proteolytic casein and elastin hydrolysis. Correlation of extracellular enzymatic activity of 127 strains of *A. hydrophila* showed that elastase-positive strains produced lesions and mortality when injected into channel catfish. It has also been suggested that other extracellular substances such as hemolysins and proteases may be involved with the pathogenic mechanism of *A. hydrophila*.[133] Additional factors that may influence the pathogenesis of the organism are its ability to adhere to tissue surfaces; its resistance to phagocytosis; or its production of surface proteins, siderophores, LPS; the presence of pili, S-layer, or outer membrane proteins; or the bacteriostatic activity of host serum. Also, water isolates of the bacterium may be avirulent or less pathogenic to fish than are isolates from diseased fish.[134] What these potentially pathogenic factors means is as yet not clear; therefore, in the final analysis the only accurate measure of a microorganism's virulence and pathogenicity is whether or not it kills fish *in vivo*.

6. SIGNIFICANCE

Considerable disagreement exists among fish pathologists as to the significance of MAS. Although the frequency of its appearance cannot be disputed, the fact that MAS is usually a secondary pathogen reduces its importance to many fish pathologists. Nevertheless, when MAS, especially *A. hydrophila* and *A. sobria,* are present in less than ideal aquaculture environments and the potential for stress is high, the disease cannot be ignored.

7. MANAGEMENT

Management of MAS is best achieved in fish populations by proper handling, maintaining a quality environment, and reducing environmental stressors. Ideally, fish should never be handled when in a weakened condition or when environmentally stressed, but if handling cannot be avoided great care should be taken to ensure that the mucus membrane of the skin, fins, and gills, that serves as a biological and physical barrier to potential waterborne pathogens, is not disturbed.

Prophylactic bath treatments of 1 to 3% salt (NaCl) or 2 to 4 mg/l of potassium permanganate will reduce the incidence of posthandling infections. If disease occurs, prolonged bath treatments with potassium permanganate at 2 to 4 mg/l will be effective as long as the infection is confined to the skin. In fish that are being fed commercial feed, a medicated ration containing 2 to 4 g of oxytetracycline (Terramycin®) per kilogram of feed (50 to 100 mg/kg of fish) is fed for 14 d. Medicated feed should be fed as soon as possible before fish become so severely infected that they stop feeding.

Indiscriminant use of antibiotics, including Terramycin®, should be avoided to reduce the potential for antibiotic resistance. Recently, fish disease diagnostic laboratories have reported that over 45% of the *A. hydrophila* isolates are resistant to Terramycin®. This acquired resistance is due, in part, to many years of exposure to drugs and their improper application. Also, plasmid-mediated R factors are involved in antibiotic resistance of *A. hydrophila*.[135]

Research to develop a vaccine for *A. hydrophila* dates back to the mid 1950s. Very little effort has gone into development of motile *Aeromonas* vaccines for channel catfish, however, studies have explored the application for other fish species. Experiments have used heat- or formalin-killed bacterins and cell extracts such as LPS with some success. Carp immunized by immersion in LPS from *A. hydrophila* were protected by the cell-mediated system.[136] Rainbow trout fry were protected against challenge following injection of heat-killed *A. hydrophila*,[137] and Japanese eel were protected from the bacterium by injection with attenuated, heat- or formalin-killed bacteria.[138] Neither of the latter studies, however, used heterologous antigen for challenge. Tiecco et al.[139] vaccinated eel with formalin- and Ampicillin-inactivated *A. hydrophila* by injection and immersion and both methods provided some protection; however, the formalin-inactivated bacterin was not as effective as the Ampicillin preparation. In spite of the limited success with immunization against MAS, no vaccines are presently available because of the serological and genetic diversity of the motile aeromonads.

REFERENCES

1. **Davis, H. S.**, A new bacterial disease of fresh-water fishes, *U.S. Bur. Fish. Bull.*, 38, 261, 1922.
2. **Ordal, E. J. and Rucker, R. R.**, Pathogenic myxobacteria, *Soc. Exp. Biol. Med. Proc.*, 56, 15, 1944.
3. **Wakabayashi, H., Hikida, M., and Masumura, K.**, *Flexibacter maritimus* sp. nov., a pathogen of marine fishes, *Int. J. Syst. Bacteriol.*, 36, 396, 1986.
4. **Garnjobst, L.**, *Cytophaga columnaris* (Davis) in pure culture. A myxobacterium pathogenic to fish, *J. Bacteriol.*, 44, 113, 1945.
5. **Larkin, J. M.**, Nonphotosynthetic, nonfruiting gliding bacteria, Sect. 23, in *Bergeys Manual of Determinative Bacteriology*, 9th ed., Williams & Wilkins, Baltimore, 1984, 2010.
6. **Bernardet, J.-F. and Grimont, P. A. D.**, Deoxyribonucleic acid relatedness and phenotypic characterization of *Flexibacter columnaris* sp. nov., nom. rev., *Flexibacter psychrophilus* sp. nov., nom. rev., and *Flexibacter maritimus* Wakabayashi, Hikida, and Masumura 1986, *Int. J. Syst. Bacteriol.*, 39, 346, 1989.
7. **Borg, A. F.**, Studies on myxobacteria associated with diseases in salmonid fishes, *J. Wildl. Dis.*, 8, 1, 1960.
8. **Meyer, F. P.**, Seasonal fluctuations in the incidence of diseases on fish farms, in *A Symposium on Diseases of Fishes and Shellfishes*, Snieszko, S. F., Ed., Spec. Publ. No. 5, American Fisheries Society, Washington, D.C., 1970, 21.
9. **Wakabayashi, H., Kira, K., and Egusa S.**, Studies on columnaris of eels. I. Characteristics and pathogenicity of *Chondrococcus columnaris* isolated from pond-cultured eels, *Bull. Jpn. Soc. Sci. Fish.*, 36, 147, 1970.
10. **Rucker, R. R., Earp, B. J., and Ordal, E. J.**, Infectious diseases of Pacific salmon, *Trans. Am. Fish. Soc.*, 83, 297, 1953.
11. **Becker, C. D. and Fujihara, M. P.**, The bacterial pathogen *Flexibacter columnaris* and its epizootiology among Columbia River fish, *Am. Fish. Soc. Monogr.*, No. 2, 1978, 92.
12. **Bernardet, J. F.**, *Flexibacter columnaris:* first description in France and comparison with bacterial strains from other origins, *Dis. Aquat. Org.*, 6, 37, 1989.
13. **Hikida, M., Wakabayashi, H., Egusa, S., and Masumura, K.**, *Flexibacter* sp., a gliding bacterium pathogenic to some marine fishes in Japan, *Bull. Jpn. Soc. Sci. Fish.*, 45, 421, 1979.
14. **Bernardet, J. F., Campbell, A. C., and Buswell, J. A.**, *Flexibacter maritimus* is the agent of black patch necrosis in Dover sole in Scotland, *Dis. Aquat. Org.*, 8, 233, 1990.
15. **Hilger, I., Ullrich, S., and Anders, D.**, A new ulcerative flexibacteriosis-like disease (yellow pest) affecting young Atlantic cod *Gadus morhua* from the German Wadden Sea, *Dis. Aquat. Org.*, 11, 19, 1991.
16. **Bullock, G. L., Conroy, D. A., and Snieszko, S. F.**, Bacterial diseases of fishes, Book 2A, in *Diseases of Fishes*, Snieszko, S. F. and Axelrod, H. R., Eds., T.F.H. Publications, Neptune City, NJ, 1971, 151.
17. **Bullock, G. L., Hsu, T. C., and Shotts, E. B., Jr.**, Columnaris disease of fishes, U.S. Department of the Interior, *Fish Dis. Leaf l.*, 72, 9, 1986.
18. **McCarthy, D. H.**, Columnaris disease, *J. Inst. Fish. Manage.*, 6, 44, 1975.
19. **Hawke, J. P. and Thune, R. L.**, Systemic isolation and antimicrobial susceptibility of *Cytophaga columnaris* from commercially reared channel catfish, *J. Aquat. Anim. Health*, 4, 109, 1992.
20. **Anacker, R. L. and Ordal, E. J.**, Studies on the myxobacterium *Chondrococcus columnaris*. I. Serological typing, *J. Bacteriol.*, 78, 25, 1959.
21. **Shotts, E. B.**, Selective isolation methods for fish pathogens, *J. Appl. Bacteriol. Symp. Suppl.*, 70, 75S, 1991.
22. **Fijan, N. N.**, The survival of *Chondrococcus columnaris* in waters of different quality, Symp. Comm. de L'office Int. Episooties Pour L'Etude des Maladies des Poissons, Stockholm, 1968.
23. **Song, Y.-L., Fryer, J. L., and Rohovec, J. S.**, Comparison of six media for the cultivation of *Flexibacter columnaris, Fish Pathol.*, 23, 91, 1988.
24. **Shieh, H. S.**, Studies on the nutrition of a fish pathogen, *Flexibacter columnaris, Microb. Lett.*, 13, 129, 1980.
25. **Baxa, D. V., Kawai, K., and Kusuda, R.**, Detection of *Flexibacter maritimus* by fluorescent antibody technique in experimentally infected black sea bream fry, *Fish Pathol.*, 23, 29, 1988.

26. **Griffin, B. R.,** A simple procedure for identification of *Cytophaga columnaris, J. Aquat. Anim. Health,* 4, 63, 1992.

27. **Pyle, S. W. and Shotts, E. B., Jr.,** DNA homology studies of selected flexibacteria associated with fish diseases, *Can. J. Fish. Aquat. Sci.,* 38, 146, 1981.

28. **Song, Y.-L., Fryer, J. L., and Rohovec, J. S.,** Comparison of gliding bacteria isolated from fish in North America and other areas of the Pacific rim, *Fish Pathol.,* 23, 197, 1988.

29. **Marks, J. E., Lewis, D. H., and Trevino, G. S.,** Mixed infection in columnaris disease of fish, *J. Am. Vet. Med. Assoc.,* 177, 811, 1980.

30. **Chowdhury, B. R. and Wakabayashi, J.,** Effects of competitive bacteria on the survival and infectivity of *Flexibacter columnaris, Fish Pathol.,* 24, 9, 1989.

31. **Bottomley, P. R. and Holland, D. G.,** Columnaris disease in coarse fish, *Year Book of Association of River Authorities,* Westminster, London, 1966, 101.

32. **Holt, R. A., Sanders, J. E., Zinn, J. L., Fryer, J. L., and Pilcher, K. S.,** Relation of water temperature to *Flexibacter columnaris* infection in steelhead trout *(Salmo gairdneri),* coho *(Oncorhynchus kisutch)* and chinook *(O. tshawytscha)* salmon, *J. Fish. Res. Board Can.,* 32, 1553, 1975.

33. **Hanson, L. A. and Grizzle, J. M.,** Nitrite-induced predisposition of channel catfish to bacterial diseases, *Prog. Fish Cult.,* 47, 98, 1985.

34. **Chia-Reiy, L., Chung, H.-Y., and Kuo, G.-H.,** Studies on the pathogenicity of *Flexibacter columnaris.* I. Effect of dissolved oxygen and ammonia on the pathogenicity of *Flexibacter columnaris* to eel *(Anguilla japonica), CAPD Fish. Ser.,* 8, 57, 1982.

35. **Hussain, M. and Summerfelt, R. C.,** The role of mechanical injury in an experimental transmission of *Flexibacter columnaris* to fingerling walleye, *J. Iowa Acad. Sci.,* 98, 93, 1991.

36. **Fujihara, M. P., Olson, P. A., and Nakatani, R. E.,** Some factors in susceptibility of juvenile rainbow trout and chinook salmon to *Chondrococcus columnaris, J. Fish. Res. Board Can.,* 28, 1739, 1971.

37. **Bowser, P. R.,** Seasonal prevalence of *Chondrococcus columnaris* infection in black bullheads from Clear Lake, Iowa, *J. Wildl. Dis.,* 9, 115, 1973.

38. **Kuo, S.-C., Chung, H.-Y., and Kuo, G.-H.,** Studies on artificial infection of the gliding bacteria in cultured fishes, *Fish Pathol.,* 15, 309, 1981.

39. **Chowdhury, B. R. and Wakabayashi, H.,** Effects of sodium, potassium, calcium and magnesium ions on the survival of *Flexibacter columnaris* in water, *Fish Pathol.,* 23, 231, 1988.

40. **Bootsma, R. and Clerx, J. P. M.,** Columnaris disease of cultured carp *Cyprinus carpio* (L.), characterization of the causative agent, *Aquaculture,* 7, 371, 1976.

41. **Thune, R. L.,** Major infectious and parasitic diseases of channel catfish, *Vet. Human Toxicol.,* 33(Suppl. 1), 14, 1991.

42. **Fijan, N. N. and Voorhees, P. R.,** Drug sensitivity of *Chondrococcus columnaris, Vet. Arh.,* 39, 259, 1969.

43. **Phelps, R. P., Plumb, J. A., and Harris, C. S.,** Control of external bacterial infections of bluegills with potassium permanganate, *Prog. Fish Cult.,* 39, 142, 1977.

44. **Tucker, C. S.,** Potassium permanganate demand of pond waters, *Prog. Fish Cult.,* 46, 24, 1984.

45. **Jee, L. K. and Plumb, J. A.,** Effects of organic load on potassium permanganate as a treatment for *Flexibacter columnaris, Trans. Am. Fish. Soc.,* 110, 86, 1981.

46. **Schachte, J. H., Jr. and Mora, E. C.,** Production of agglutinating antibodies in the channel catfish *(Ictalurus punctatus)* against *Chondrococcus columnaris, J. Fish. Res. Board Can.,* 30, 116, 1973.

47. **Moore, A. A., Eimers, M. E., and Cardella, M. A.,** Attempts to control *Flexibacter columnaris* epizootics in pond-reared channel catfish by vaccination, *J. Aquat. Anim. Health,* 2, 109, 1990.

48. **Maas, M. G. and Bootsma, R.,** Uptake of bacterial antigens in the spleen of carp *(Cyprinus carpio* L.), *Dev. Comp. Immunol.,* Suppl. 2, 47, 1982.

49. **Boyd, C. E.,** *Water Quality in Ponds for Aquaculture,* Ala. Agric. Exp. Sta., Auburn University Press, AL, 1990, 482.

50. **Ewing, W. H., McWhorter, A. C., Escobar, M. R., and Lubin, A. H.,** *Edwardsiella,* a new genus of Enterobacteriaceae based on a new species, *Edwardsiella tarda, Int. Bull. Bact. Nom. Tax.,* 15, 33, 1965.

51. **Hawke, J. P., McWhorter, A. C., Steigerwalt, A. C., and Brenner, D. J.,** *Edwardsiella ictaluri* sp. nov., the causative agent of enteric septicemia of catfish, *Int. J. Syst. Bacteriol.,* 31, 396, 1981.

52. **Meyer, F. P. and Bullock, G. L.,** *Edwardsiella tarda,* a new pathogen of channel catfish *(Ictalurus punctatus), Appl. Microbiol.,* 25, 155, 1973.
53. **Egusa, S.,** Some bacterial diseases of freshwater fishes in Japan, *Fish Pathol.,* 10, 103, 1976.
54. **Hawke, J. P.,** A bacterium associated with disease of pond cultured channel catfish, *J. Fish. Res. Board Can.,* 36, 1508, 1979.
55. **Kasornchandra, J., Rogers, W. A., and Plumb, J. A.,** *Edwardsiella ictaluri* from walking catfish, *Clarias batrachus* L., in Thailand, *J. Fish Dis.,* 10, 137, 1987.
56. **Humphrey, J. D., Lancaster, C., Gudkovs, N., and McDonald, W.,** Exotic bacterial pathogens *Edwardsiella tarda* and *Edwardsiella ictaluri* from imported ornamental fish *Betta splendens* and *Puntius conchonius* respectively; isolation and quarantine significance, *Aust. Vet. J.,* 63, 999, 1986.
57. **Waltman, W. D., Shotts, E. B., and Blazer, V. S.,** Recovery of *Edwardsiella ictaluri* from Danio *(Danio devario), Aquaculture,* 46, 63, 1985.
58. **Kent, M. L. and Lyons, J. M.,** *Edwardsiella ictaluri* in the green knife fish, *Eigemannia virescens, Fish Health News,* 2, ii, 1982.
59. **Baxa, D. V., Groff, J. M., Wishkovsky, A., and Hedrick, R. P.,** Susceptibility of nonictalurid fishes to experimental infection with *Edwardsiella ictaluri, Dis. Aquat. Org.,* 8, 113, 1990.
60. **Plumb, J. A. and Hilge, V.,** Susceptibility of European catfish *(Silurus glanis)* to *Edwardsiella ictaluri, J. Appl. Ichthyol.,* 3, 45, 1987.
61. **Plumb, J. A. and Sanchez, D. J.,** Susceptibility of 5 species of fish to *Edwardsiella ictaluri, J. Fish Dis.,* 6, 261, 1983.
62. **Shotts, E. B. and Waltman, W. D.,** A medium for the selective isolations of *Edwardsiella ictaluri, J. Wildl. Dis.,* 26, 214, 1990.
63. **Waltman, W. D., Shotts, E. B., and Hsu, T. C.,** Biochemical characteristics of *Edwardsiella ictaluri, Appl. Environ. Microbiol.,* 51, 101, 1986.
64. **Plumb, J. A. and Vinitnantharat, S.,** Biochemical, biophysical, and serological homogeneity of *Edwardsiella ictaluri, J. Aquat. Anim. Health,* 1, 51, 1989.
65. **Rogers, W. A.,** Serological detection of two species of *Edwardsiella* infecting catfish, *Dev. Biol. Stand. Fish Biol. Serodiag. Vacc.,* 49, 169, 1981.
66. **Klesius, P., Johnson, K., Durborow, R., and Vinitnantharat, S.,** Development and evaluation of an enzyme-linked immunosorbent assay for catfish serum antibody to *Edwardsiella ictaluri, J. Aquat. Anim. Health,* 3, 94, 1991.
67. **Ainsworth, A. J., Capley, G., Waterstrat, P., and Munson, D.,** Use of monoclonal antibodies in the indirect fluorescent antibody technique (IFA) for the diagnosis of *Edwardsiella ictaluri, J. Fish Dis.,* 9, 439, 1986.
68. **Klesius, P. H.,** Rapid enzyme-linked immunosorbent tests for detecting antibodies to *Edwardsiella ictaluri* in channel catfish, *Ictalurus punctatus,* using exoantigen, *Vet. Immunol. Immunopathol.,* 36, 359, 1993.
69. **Baxa-Antonio, D., Groff, J. M., and Hedrick, R. P.,** Effect of water temperature on experimental *Edwardsiella ictaluri* infections in immersion exposed channel catfish, *J. Aquat. Anim. Health,* 4, 148, 1992.
70. **Vinitnantharat, S. and Plumb, J. A.,** Protection of channel catfish following exposure to *Edwardsiella ictaluri* and effects of feeding antigen on antibody titer, *Dis. Aquat. Org.,* 15, 31, 1993.
71. **Lobb, C. J. and Rhoades, M.,** Rapid plasmid analysis for identification of *Edwardsiella ictaluri* from infected channel catfish *Ictalurus punctatus, Appl. Environ. Microbiol.,* 53, 1267, 1987.
72. **Newton, J. C., Bird, R. C., Blevins, W. T., Wilt, G. R., and Wolfe, L. G.,** Isolation, characterization, and molecular cloning of cryptic plasmids isolated from *Edwardsiella ictaluri, Am. J. Vet. Res.,* 49, 1856, 1988.
73. **Speyerer, P. D. and Boyle, J. A.,** The plasmid profile of *Edwardsiella ictaluri, J. Fish Dis.,* 10, 461, 1987.
74. **Plumb, J. A. and Quinlan, E. E.,** Survival of *Edwardsiella ictaluri* in pond water and bottom mud, *Prog. Fish Cult.,* 48, 212, 1986.
75. **Mgolomba, T. N. and Plumb, J. A.,** Longevity of *Edwardsiella ictaluri* in the organs of experimentally infected channel catfish, *Ictalurus punctatus, Aquaculture,* 101, 1, 1992.
76. **Klesius, P.,** Carrier state of *Edwardsiella ictaluri* in channel catfish, *Ictalurus punctatus, J. Aquat. Anim. Health,* 4, 227, 1992.

77. **Miyazaki, T. and Plumb, J. A.,** Histopathology of *Edwardsiella ictaluri* in channel catfish, *Ictalurus punctatus* (Rafinesque), *J. Fish Dis.,* 8, 389, 1985.
78. **Blazer, V. S., Shotts, E. B., and Waltman, W. D.,** Pathology associated with *Edwardsiella ictaluri* in catfish, *Ictalurus punctatus* (Rafinesque), and *Danio devario* (Hamilton-Buchanan, 1922), *J. Fish Dis.,* 27, 167, 1985.
79. **Earlix, D.,** Department of Fisheries and Allied Aquacultures, Auburn University, AL, personal communication, 1993.
80. **Taylor, P.,** Fish-eating birds as potential vectors for *Edwardsiella ictaluri, J. Aquat. Anim. Health,* 4, 240, 1992.
81. **Francis-Floyd, R., Beleau, M. H., Waterstrat, P. R., and Bowser, P. R.,** Effect of water temperature on the clinical outcome of infection with *Edwardsiella ictaluri* in channel catfish, *J. Am. Vet. Med. Assoc.,* 191, 1413, 1987.
82. **Wise, D. J., Schwedler, T. E., and Otis, D. L.,** Effects of stress on susceptibility of naive channel catfish in immersion challenge with *Edwardsiella ictaluri, J. Aquat. Anim. Health,* 5, 16, 1993.
83. **Areechon, N. and Plumb, J. A.,** Pathogenesis of *Edwardsiella ictaluri* in channel catfish, *Ictalurus punctatus, J. World Maricult. Soc.,* 14, 249, 1983.
84. **Jarboe, H. H., Bowser, P. R., and Robinette, H. R.,** Pathology associated with a natural *Edwardsiella* infection in channel catfish (*Ictalurus punctatus* Rafinesque), *J. Wildl. Dis.,* 20, 352, 1984.
85. **Shotts, E. B., Blazer, V. S., and Waltman, W. D.,** Pathogenesis of experimental *Edwardsiella ictaluri* infections in channel catfish (*Ictalurus punctatus), Can. J. Fish. Aquat. Sci.,* 43, 36, 1986.
86. **Morrison, E. and Plumb, J. A.,** The chemosensory system of channel catfish, *Ictalurus punctatus,* following immersion exposure to *Edwardsiella ictaluri, Chemical Senses,* American Chemosensory Society, Orlando, FL, April 8–10, 1992 (Abstr.).
87. **Newton, J. C., Wolfe, L. G., Grizzle, J. M., and Plumb, J. A.,** Pathology of experimental enteric septicemia in channel catfish *Ictalurus punctatus* Rafinesque following immersion exposure to *Edwardsiella ictaluri, J. Fish Dis.,* 12, 335, 1989.
88. **Baldwin, T. J. and Newton, J. C.,** Early events in the pathogenesis of enteric septicemia of channel catfish caused by *Edwardsiella ictaluri:* light and electron microscopic and bacteriologic findings, *J. Aquat. Anim. Health,* 5, 189, 1993.
89. **Mitchell, A. J.,** U.S. Fish and Wildlife Service, Stuttgart, AR, personal communication, 1991.
90. **Schnick, R. A., Meyer, F. P., and Gray, D. L.,** A Guide to Approved Chemicals in Fish Production and Fishery Resource Management, U.S. Fish and Wildlife Service, MP 241, Department of the Interior, Washington, D.C., 1989, 27.
91. **Waltman, W. D. and Shotts, E. B.,** Antimicrobial susceptibility of *Edwardsiella ictaluri, J. Wildl. Dis.,* 22, 173, 1986.
92. **Waltman, W. D., Shotts, E. B., and Wooley, R. E.,** Development and transfer of plasmid-mediated antimicrobial resistance in *Edwardsiella ictaluri, Can. J. Fish. Aquat. Sci.,* 46, 1114, 1989.
93. **Starliper, C. E., Cooper, R. K., Shotts, E. B., Jr., and Taylor, P. W.,** Plasmid-mediated Romet resistance of *Edwardsiella ictaluri, J. Aquat. Anim. Health,* 5, 1, 1993.
94. **Cooper, R. K., II, Starliper, C. E., Shotts, E. B., Jr., and Taylor, P. W.,** Comparison of plasmids isolated from Romet-30-resistant *Edwardsiella ictaluri* and tribrissen-resistant *Escherichia coli, J. Aquat. Anim. Health,* 5, 9, 1993.
95. **Plumb, J. A. and Vinitnantharat, S.,** Vaccination of channel catfish, *Ictalurus punctatus* (Rafinesque), by immersion and oral booster against *Edwardsiella ictaluri, J. Fish Dis.,* 16, 65, 1993.
96. **Plumb, J. A. and Klesius, P.,** An assessment of the antigenic homogeneity of *Edwardsiella ictaluri* using monoclonal antibody, *J. Fish Dis.,* 11, 499, 1988.
97. **Klesius, P. and Horst, M. N.,** Characterization of a major outer-membrane antigen of *Edwardsiella ictaluri, J. Aquat. Anim. Health,* 3, 181, 1991.
98. **Vinitnantharat, S., Plumb, J. A., and Brown, A. E.,** Isolation and purification of an outer membrane protein of *Edwardsiella ictaluri* and its antigenicity to channel catfish (*Ictalurus punctatus), Fish Shellfish Immunol.,* 3, 401, 1993.
99. **Grimont, P. A. D., Grimont, F., Richard, C., and Sakazaki, R.,** *Edwardsiella hoshinae,* a new species of Enterobacteriaceae, *Curr. Microbiol.,* 4, 347, 1980.
100. **Austin, B. and Austin, D. A.,** *Bacterial Fish Pathogens: Diseases in Farmed and Wild Fish,* Ellis Horwood, Chichester, 1987, 364.

101. **Groberg, W. J., Jr., McCoy, R. H., Pilcher, K. S., and Fryer, J. L.,** Relation of water temperature to infections of coho salmon *(Oncorhynchus kisutch),* chinook salmon *(O. tshawytscha),* and steelhead trout *(Salmo gairdneri)* with *Aeromonas salmonicida* and *A. hydrophila, J. Fish. Res. Board Can.,* 35, 1, 1978.

102. **Nieto, T. P., Corcobado, M. J. R., Toranzo, A. E., and Barja, J. L.,** Relation of water temperature to infection of *Salmo gairdneri* with motile *Aeromonas, Fish Pathol.,* 20, 99, 1985.

103. **Shotts, E. B., Gaines, J. L., Martin, L., and Prestwood, A. K.,** *Aeromonas*-induced deaths among fish and reptiles in an eutrophic inland lake, *J. Am. Vet. Med. Assoc.,* 161, 603, 1972.

104. **Davis, W. A., Kane, J. G., and Garagusi, V. G.,** Human *Aeromonas* infections; a review of the literature and a case report of endocarditis, *Medicine,* 57, 267, 1978.

105. **Goncalves, J. R., Braum, G., Fernandes, A., Biscaia, I., Correia, M. J. S., and Bastardo, J.,** *Aeromonas hydrophila* fulminant pneumonia in a fit young man, *Thorax,* 47, 482, 1992.

106. **Shotts, E. B. and Rimler, R.,** Medium for the isolation of *Aeromonas hydrophila, Appl. Microbiol.,* 26, 550, 1973.

107. **Colwell, R. R., MacDonnell, M. T., and DeLey, J.,** Proposal to recognize the family Aeromonadaceae fam. nov., *Int. J. Syst. Microbiol.,* 36, 473, 1986.

108. **Carnahan, A. M., Behram, S., and Joseph, S. W.,** Aerokey II. A flexible key for identifying clinical *Aeromonas* species, *J. Clin. Microbiol.,* 29, 2843, 1991.

109. **Janda, J. M.,** Recent advances in the study of the taxonomy, pathogenicity, and infectious syndromes associated with the genus *Aeromonas, Clin. Microbiol. Rev.,* 4, 397, 1991.

110. **Sakazaki, R.,** Serology of mesophilic *Aeromonas* spp. and *Plesiomonas shigelloides, Experientia,* 43, 357, 1987.

111. **Kage, T., Takahashi, R., Barcus, I., and Hayashi, F.,** *Aeromonas hydrophila,* a causative agent of mass mortality in cultured Japanese catfish larvae *(Silurus asotus), Fish Pathol.,* 27, 57, 1992.

112. **Hazen, T. C., Fliermans, C. B., Hirsch, R. P., and Esch, G. W.,** Prevalence and distribution of *Aeromonas hydrophila* in the United States, *Appl. Environ. Microbiol.,* 36, 731, 1978.

113. **Plumb, J. A., Grizzle, J. M., and deFigueiredo, J.,** Necrosis and bacterial infection in channel catfish *(Ictalurus punctatus)* following hypoxia, *J. Wildl. Dis.,* 12, 247, 1976.

114. **Grizzle, J. M. and Kiryu, Y.,** Histopathology of gill, liver, and pancreas, and serum enzyme levels of channel catfish infected with *Aeromonas hydrophila* complex, *J. Aquat. Anim. Health,* 5, 36, 1993.

115. **Peters, G., Faisal, M., Lang, T., and Ahmed, I.,** Stress caused by social interaction and its effect on susceptibility to *Aeromonas hydrophila* infection in rainbow trout *Salmo gairdneri, Dis. Aquat. Org.,* 4, 83, 1988.

116. **Haley, R., Davis, S. P., and Hyde, J. M.,** Environmental stress and *Aeromonas liquefaciens* in American and threadfin shad mortalities, *Prog. Fish Cult.,* 29, 193, 1967.

117. **Walters, G. R. and Plumb, J. A.,** Environmental stress and bacterial infection in channel catfish, *Ictalurus punctatus* Rafinesque, *J. Fish Biol.,* 17, 177, 1980.

118. **Schmottou, H. R.,** Department of Fisheries and Allied Aquacultures, Auburn University, AL, personal communication, 1993.

119. **Levanon, N., Motro, B., Levanon, D., and Degani, G.,** The dynamics of *Aeromonas hydrophila* in the water of tanks used to nurse elvers of the European eel *Anguilla anguilla, Bamidgeh,* 38, 55, 1986.

120. **Boonyaratpalin, S.,** Bacterial pathogens involved in the epizootic ulcerative syndrome of fish in Southeast Asia, *J. Aquat. Anim. Health,* 1, 272, 1989.

121. **Ketover, B. P., Young, L. S., and Armstrong, D.,** Septicemia due to *Aeromonas hydrophila;* clinical and immunological aspects, *J. Infect. Dis.,* 127, 284, 1973.

122. **Hargraves, J. E. and Lucey, D. R.,** *Edwardsiella tarda* soft tissue infection associated with catfish puncture wound, *J. Infect. Dis.,* 162, 1416, 1990.

123. **Murphey, D. K., Septimus, E. J., and Waagner, D. C.,** Catfish-related injury and infection: report of two cases and review of the literature, *Clin. Infect. Dis.,* 14, 689, 1992.

124. **King, G. E., Werner, S. B., and Kizer, W.,** Epidemiology of aeromonas infections in California, *Clin. Infect. Dis.,* 15, 449, 1992.

125. **Gray, S. J., Stickler, D. J., and Bryant, T. N.,** The incidence of virulence factors in mesophilic *Aeromonas* species isolated from farm animals and their environment, *Epidemiol. Infect.,* 105, 277, 1990.

126. **Johnson, M.,** Mississippi State University, Stoneville, personal communication, 1993.

127. **Miyazaki, T. and Jo, Y.,** A histopathological study of motile aeromonad disease in ayu, *Fish Pathol.,* 20, 55, 1985.

128. **Ventura, M. T. and Grizzle, J. M.,** Lesions associated with natural and experimental infections of *Aeromonas hydrophila* in channel catfish, *Ictalurus punctatus* (Rafinesque), *J. Fish Dis.,* 11, 397, 1988.

129. **Huizinga, H. W., Esch, G. W., and Hazen, T. C.,** Histopathology of red-sore disease *(Aeromonas hydrophila)* in naturally and experimentally infected largemouth bass *Micropterus salmoides* (Lacepede) *J. Fish Dis.,* 2, 263, 1979.

130. **Hsu, T. C., Waltman, W. D., and Shotts, E. B.,** Correlation of extracellular enzymatic activity and biochemical characteristics with regard to virulence of *Aeromonas hydrophila, Dev. Biol. Stand. Fish Biol. Serodiag. Vacc.,* 49, 101, 1981.

131. **Santos, Y., Toranzo, A. E., Dopazo, C. P., Nieto, T. P., and Barja, J. L.,** Relationships among virulence for fish, enterotoxigenicity, and phenotypic characteristics of motile *Aeromonas, Aquaculture,* 67, 29, 1987.

132. **Wakabayashi, J., Kanai, K., Hsu, T. C., and Egusa, S.,** Pathogenic activities of *Aeromonas* biovar *hydrophila* (Chester) Popoff and Vernon, 1976 to fishes, *Fish Pathol.,* 15, 319, 1980.

133. **Chabot, D. J. and Thune R. L.,** Proteases of the *Aeromonas hydrophila* complex: identification, characterization and relation to virulence in channel catfish, *Ictalurus punctatus* (Rafinesque), *J. Fish Dis.,* 14, 171, 1991.

134. **de Figueiredo, J. and Plumb, J. A.,** Virulence of different isolates of *Aeromonas hydrophila* in channel catfish, *Aquaculture,* 11, 349, 1977.

135. **Shotts, E. B., Vanderwork, V. L., and Campbell, L. M.,** Occurrence of R factors associated with *Aeromonas hydrophila* isolated from aquarium fish and waters, *J. Fish. Res. Board Can.,* 33, 736, 1976.

136. **Baba, T., Immura, J., Izawa, K., and Ikeda, K.,** Immune protection in carp, *Cyprinus carpio* L., after immunization with *Aeromonas hydrophila* crude lipopolysaccharide, *J. Fish Dis.,* 11, 237, 1988.

137. **Khalifa, K. A. and Post, G.,** Immune response of advanced rainbow trout fry to *Aeromonas liquefaciens, Prog. Fish Cult.,* 38, 66, 1976.

138. **Song, Y.-L., Chen S.-N., and Kou, G.,** Agglutinating antibodies production and protection in eel *(Anquilla japonica)* inoculated with *Aeromonas hydrophila (A. liquefaciens)* antigens, *J. Fish Soc. Taiwan,* 4, 25, 1976.

139. **Tiecco, G., Sebastio, C., Francioso, E., Tantilla, G., and Corbari, L.,** Vaccination trials against "red plaque" in eels, *Dis. Aquat. Org.,* 4, 105, 1988.

140. **Joseph, S. W., Colwell, R. R., and MacDonell, M. T.,** Research on *Aeromonas.* I. Taxonomy, ecology, isolation and identification of *Aeromonas* and *Plesiomonas, Experientia,* 43, 349, 1987.

141. **Popoff, M.,** Genus III. *Aeromonas* Kluyver and Van Niel 1936, 398 AL, in *Bergey's Manual of Systematic Bacteriology,* Vol. 1, Krieg, N. R., Ed., Williams & Wilkins, Baltimore, 1984, 545.

Chapter XIII

Cyprinids (Cyprinidae)

Members of the family Cyprinidae are among the oldest cultured fishes in the world. Common carp *(Cyprinus carpio)* have been cultured in Europe since the Middle Ages and the Chinese have grown carp and goldfish *(Carrasius auratus)* in captivity for 2500 years. Presently, golden shiners *(Notemigonus chrysoleucas)* and fathead minnows *(Pimephales promelas)* are cultured extensively in the U.S. for bait fish. Two bacterial diseases that are most often associated with cyprinids are ''carp erythrodermatitis'' (CE)[1] and ''ulcer disease of goldfish'' (UDG).[2] Ironically, both of these disease syndromes are caused by the same organism; *Aeromonas salmonicida* subspecies *achromogens*, also called ''atypical nonmotile *Aeromonas*''. The following discussion includes CE, UDG, and infections caused by these pathogens in other species of fish. Other bacterial pathogens that cause disease in cyprinids, but only briefly discussed in this section, include motile *Aeromonas* septicemia (MAS) and columnaris (Chapter XII).

A. ATYPICAL NONMOTILE *AEROMONAS*

Aeromonas slamonicida achromogens has been recognized as a fish pathogen since 1963 when Smith[3] isolated the organism from brown trout *(Salmo trutta)* in Great Britain. Paterson et al.[4] later showed that the organism was identical to *Haemophilus piscium*, the etiological agent of ''ulcer disease'' of trout. Since the early 1970s atypical *A. salmonicida* (subsp. *achromogens*) has been a major disease-producing organism in cultured cyprinid fishes, causing carp erythrodermatitis and ulcer disease of goldfish, and has also been associated with similar diseases in some noncyprinid species. A more applicable common name would be ''atypical nonmotile *Aeromonas* infections'' (ANAI).

CE, once considered to be a stage of the infectious dropsy of carp (IDC) syndrome, is now considered a distinct disease.[1,5] CE is described as a subacute to chronic, contagious disease of carp and several other species of fish. Even though Bootsma and Blommaert[6] isolated a myxobacteria from carp that exhibited skin ulcerations like those present in carp erythrodermatitis, the precise etiology was not determined until Bootsma et al.[7] isolated a nonmotile *Aeromonas* spp. *(A. salmonicida achromogens)* from carp at five farms in Yugoslavia and Germany and Koch's postulates were fulfilled for the pathogen.

1. GEOGRAPHICAL RANGE AND SPECIES SUSCEPTIBILITY

Atypical nonmotile *Aeromonas* infections have been documented in Europe, Great Britain, the U.S., Asia, and Australia. In Europe and Great Britain these infections are most commonly referred to as carp erythrodermatitis.[1,8-9] CE has not been reported from carp in North America or Asia where the syndrome is best known as ulcerative disease of goldfish. In the U.S. UDG has been diagnosed in Arkansas, Missouri, Tennessee, Maryland, California, and Georgia. UDG has also been reported in Italy, Great Britain, Japan, and Australia.[2,10-12]

The common carp, including scaled and mirror varieties, is the principal species affected by atypical nonmotile *Aeromonas*, but similar types of lesions have been described in eels, goldfish, and a variety of other species of freshwater and marine fishes (Table 1). Of particular interest is the fact that several species of salmonids (i.e., Atlantic *Salmo salar*, chum *Oncorhynchus keta*, and coho *O. kisutch* salmon, and brown, rainbow *O. mykiss*, and brook *Salvelinus fontinalis* trout) and other nonsalmonid marine fish have been either naturally or experimentally infected with atypical *A. salmonicida*.[13-15]

2. CLINICAL SIGNS AND FINDINGS

Because clinical signs of atypical nonmotile *Aeromonas* infections are similar in most fish, CE and UDG cannot be separated solely on these signs.[2,8] Moribund CE fish rest near the surface, at the bottom, or close to the banks of ponds, or concentrate near the inflow of fresh water; a behavioral pattern that has not been described for UDG. Carp become darkly pigmented, with slightly extended abdomens and slight to extreme exophthalmia.

Table 1   Species of fish from which atypical *Aeromonas salmonicida* (subsp. *achromogens*) has been isolated and country in which it occurs

Fish species	Country	Ref.
Common carp *(Cyprinus carpio)*	Most European countries	7,9
Goldfish *(Carrasius auratus)*	U.S., Italy, Australia	2,12
Eels *(Anguillu japonica)*	Japan	18
Roach *(Rutilus rutilus)*	Great Britain, Israel	16-17
Silver bream *(Blicca bjoerknna)*	Great Britain	16
Atlantic cod *(Gadus morhua)*	Nova Scotia, Canada	19
Brown trout *(Salmo trutta)*	Sweden	23
Atlantic salmon *(Salmo salar)*	Nova Scotia, Canada	24
Common bream *(Abramis brama)*	Great Britain	16
Perch *(Perca fluviatilis)*	Great Britain	16
Minnow *(Phoxinus phoxinus)*	Norway	25

Deep ulcers in skin and muscle tissue are the most obvious clinical signs of CE and UDG (Figure 1). These lesions begin as small to large hemorrhage-inflamed areas in the skin and scale pockets and progress to extensive necrosis in muscle of the superficial epithelium. Hemorrhagic inflammation also occurs on the fins. Lesions associated with atypical *A. salmonicida* in other species of fish are generally very similar to those described for CE and UDG.[16-17] However, Iida et al.[18] stated that the most typical sign of ANAI in eels was a swollen head and the presence of ulcerative lesions on the jaws and cheeks. Affected goldfish have pronounced anemia, exhibit edema of the body along with lepidorthosis, and proteinacious material may collect on the lesions giving the appearance of a fungal infection (Figure 1). Internally, they have a general hemorrhagic appearance, pale liver, and inflamed intestines. Atlantic salmon (30 to 40 g) that were infected with atypical *A. salmonicida* exhibited clinical signs that bore close resemblance to those of furunculosis of salmonids.

3. DIAGNOSIS AND BACTERIAL CHARACTERISTICS

Diagnosis of atypical *A. salmonicida* infections is accomplished by noting the presence of clinical signs including ulcerative skin lesions, etc. *A. salmonicida achromogens* can be isolated from necrotic musculature, or hemorrhagic, inflamed skin lesions, and occasionally from the visceral organs of carp or goldfish. However, the pathogen is isolated more often from visceral organs of other species of fish. For example, Cornick et al.[19] isolated *A. salmonicida achromogens* from the kidneys of 22% of Atlantic cod *(Gadus morhua)* with skin lesions, and Hubert and Williams[17] working with the roach *(Rutilus rutilus),* isolated the causative organism from 95% of skin ulcers, and from 67 and 24% of blood and kidneys, respectively. Bootsma et al.[7] proposed that isolation media be supplemented with tryptone and/or serum, but Paterson et al.[4] found that blood agar produced satisfactory isolation results. The organism is fastidious, especially on primary isolation, but growth capability improves upon repeated subcultivation. Muscle lesions are often contaminated with other, less fastidious bacteria (e.g., *A. hydrophila* that grow on media more rapidly than do atypical *A. salmonicida*) thus overgrowing the target organism and making diagnosis more difficult.

Atypical *A. salmonicida* is a short bacillus that is Gram-negative, nonmotile, cytochrome oxidase positive, and produces no brown water-soluble pigment at temperatures below 25°C.[7,10] The cells are about 0.5 μm wide and 0.7 to 1.4 μm long. Optimum growth temperature is 27°C, with no growth taking place at 37°C. On culture media the bacterium forms small punctate colonies in 48 to 72 h at 25°C. Primary differences between *A. salmonicida salmonicida,* and *A. salmonicida achromogens* are summarized in Table 2.

4. EPIZOOTIOLOGY

The epizootiology of atypical nonmotile *Aeromonas* infections is similar in most species of fish, especially in terms of seasonal and temperature relationship, age of susceptibility, mortality patterns, and location of the causative organism in infected fish.

Figure 1 (A) Clinical signs of carp erythrodermatitis. Note the ulcerative lesion on the skin (arrow). (B) Goldfish with clinical signs of ulcer disease of goldfish. Note the ulcerative lesion on the skin (arrow). Both diseases are caused by atypical *Aeromonas salmonicida* (subsp. *achromogens*). (Photograph A by N. Fijan.)

Carp erythrodermatitis generally occurs in spring following stocking of production ponds, with less severe outbreaks being reported in the fall. Negligible losses from the bacteria occur in summer. An increased incidence of *A. salmonicida achromogens* infections in spring are associated with optimum water temperatures (15 to 20°C) and to spawning activities which can lead to superficial injury of the skin. During periods when water temperatures fluctuate between 6 and 20°C, losses of carp due to CE can be greater than 50%.[8] The disease in other species of fish also tends to be more prevalent in spring and fall with little evidence of its presence in the summer.[16,19] Atypical *A. salmonicida* is transmitted by cohabitation of healthy and diseased fish, inoculation of infected skin material into the dermis or scale pockets, or by rubbing infected material onto superficially scarified skin.[7,20]

UDG is a chronic disease that primarily affects cultured goldfish 1 year old or older, including brood stock. The disease usually occurs in early spring following stocking of spawning ponds with adult fish and when water temperatures range between 18 and 25°C. Losses of larger subadult and adult goldfish during this time may reach 45 to 90%. Epizootic survivors appear unaffected, with the exception that they tend to have a lower yield of eggs. Saleable sized goldfish (5 to 10 cm) become more susceptible as size, fish density, parasite load, and environmental stressors increase. Other potential pathogenic bacteria such as *F. columnaris, A. hydrophila,* and *Vibrio anguillarum* are often found in skin lesions in association with the consistently present *A. salmonicida achromogens.*

Table 2   **Selected biochemical differentiation between** *Aeromonas salmonicida salmonicida* **(typical),** *Aeromonas salmonicida achromogens* **(atypical), and** *Aeromonas salmonicida masoucida* **(atypical); these bacteria are Gram-negative, nonmotile, short rods that are cytochrome oxidase positive and ferment carbohydrates**[16,26-27]

| | *Aeromonas salmonicida* | | |
Characteristic	subspecies *salmonicida*	subspecies *achromogens*	subspecies *masoucida*
Brown soluble pigment	+	−	−
Indole in 1% peptone	−	−	−
Esculin hydrolysis	+	−	+
L-Arabinose utilization	+	−	+
Sucrose fermentation	−	+	+
Mannitol fermentation	+	−	+
Voges-Proskauer	−	−	+
Gas from glucose	+	−	+
H₂S from cysteine	−	−	+
Gelatinase	+	−	+
Digestion of casein	+	+	−
Mannose	+	?	+

5. PATHOLOGICAL MANIFESTATIONS

Infections of atypical *A. salmonicida* are generally localized in the skin where small or large hemorrhagic, inflamed lesions occur. Necrosis of the superficial skin develops at the center of an inflamed lesion.[5] The lesions may disappear but the infection remains, resulting in hydropsy complete with exophthalmia, ascites, edema, and anemia. It was noted by Bootsma et al.[7] that the bacteria responsible for CE could only be found in dermal and subdermal lesions, but nevertheless, a generalized edema occurred. As previously discussed, atypical *A. salmonicida* infections in other species of fish (e.g., cod and Atlantic salmon) may be systemic, but still the skin is most severely affected. It was theorized by Pol et al.[21] that the inflammation, tissue necrosis, and possibly the generalized edema present in CE-infected fish, could be caused by a toxic factor released by the bacterium.

6. SIGNIFICANCE

During the last 20 years, atypical nonmotile *Aeromonas* infections have assumed a more important role in aquaculture because of broader species susceptibility to the disease. Under certain conditions carp erythrodermatitis can be a very serious disease in European carp culture ponds. *A. salmonicida achromogens* isolated from goldfish were experimentally pathogenic to Atlantic salmon and brown, rainbow, and brook trout.[15] Reduced egg production and loss of yearling and adult fish have resulted from UDG infections and the problem has been exacerbated by the fact that the disease does not respond well to chemotherapeutics.

7. MANAGEMENT

Handling, crowding, or injuries to the skin of carp and goldfish should be kept to a minimum during critical periods to avoid CE and UDG infections. Normally, brood goldfish are stocked into ponds in early spring just prior to spawning, and often as a result of this potentially stressful handling, an atypical *A. salmonicida* infection may occur. If brood fish are stocked into spawning ponds in late fall or early winter, this prespawning stress can be eliminated and UDG generally will not manifest itself in the spring. Fall stocking is common on goldfish farms where atypical *A. salmonicida* is endemic. Fijan[5] recommended that general prophylactic measures be taken on culture farms to reduce the occurrence and/or severity of carp erythrodermatitis. Antibiotic baths should be applied during transport and after handling of carp of goldfish. Medicated feed containing oxytetracycline at 3.5 to 5.0 mg/kg of fish is fed daily for 4 to 6 d to diseased fish, however, success of this procedure has been limited for both CE and UDG.

Several antigen preparations from atypical *A. salmonicida* were tested for their potential immunogenic ability to protect carp against CE.[22] An injected formalin-killed whole-cell preparation offers consistent, moderate protection but a concentrated, deactivated culture supernatant affords the best protection against subsequent lethal challenge with atypical *A. salmonicida*. These data indicate that extracellular products of atypical *A. salmonicida* are involved in CE and that vaccination of carp against the disease is feasible.

B. OTHER BACTERIAL DISEASES OF CYPRINIDS

Motile aeromonads can cause disease problems in cultured carp (i.e., common, grass, silver, and bighead), golden shiner, fathead minnows, and goldfish, especially after handling, stocking, or transport. *A. hydrophila*, the principle bacterial organism in infectious dropsy of carp syndrome in Europe, is generally the etiological agent. It has also been implicated as the etiological agent in a variety of infections and mortalities in natural cyprinid populations. Clinical signs of *A. hydrophila* in cyprinids are necrotic (frayed) fins, hemorrhaged scale pockets, scale loss, and development of necrotic skin lesions. Infections become systemic and produce edema, anemia, and hyperemia of the internal organs. Mortalities in cyprinids due to *A. hydrophila* are usually chronic and seldom acute. *Pseudomonas* spp. and *P. fluorescence* produce clinical signs similar to those observed in motile *Aeromonas* septicemia in cyprinids.

The cyprinids, especially shiners and fathead minnows, are extremely susceptible to infections of *Flexibacter columnaris* (columnaris). This organism can be responsible for chronic losses in ponds during periods of stress and/or following stocking. Shiners and fathead minnows crowded into holding tanks while awaiting sale or shipment are particularly susceptible to columnaris. Mortalities among these confined populations may reach 100%. Fish infected with columnaris display frayed fins; open ulcerative skin lesions where scales have been sloughed; are grayish in color as a result of injury to the mucus layer and epithelium; and hemorrhages appear at the base of fins, around the opercle, and mouth. Margins of the columnaris lesions, particularly around the mouth, may be yellowish in color due to the presence of a large number of bacteria. Also, the lesions may contain a variety of other waterborne bacteria including *A. hydrophila* and *P. fluorescence*. Treatment of minnows infected with columnaris is usually effected by antibiotics or bactericidal chemical baths. A more detailed discussion of these infections can be found in Chapter XII.

REFERENCES

1. **Fijan, N.,** Infectious dropsy in carp — a disease complex, *Symp. Zool. Soc. London*, 30, 39, 1972.
2. **Elliott, D. G. and Shotts, E. B.,** Aetiology of an ulcerative disease in goldfish *Carassius auratus* (L): microbial examination of diseased fish from seven locations, *J. Fish Dis.*, 3, 133, 1980.
3. **Smith, I. W.,** The classification of *"Bacterium salmonicida"*, *J. Gen. Microbiol.*, 33, 263, 1963.
4. **Paterson, W. D., Douey, D., and Desautels, D.,** Relationship between selected strains of typical and atypical *Aeromonas salmonicida*, *Aeromonas hydrophila*, and *Haemophilus piscium*, *Can. J. Microbiol.*, 26, 58, 1980.
5. **Fijan, N.,** Diseases of cyprinids in Europe, *Fish Pathol.*, 10, 129, 1976.
6. **Bootsma, R. and Blommaert, J.,** Sur aetiologie der erythrodermatitis beim karpfen *Cyprinus carpio*. I, in *Neuere Erkennt. Fish Infect.*, Gustav Fischer, Stuttgart, 1978, 20.
7. **Bootsma, R., Fijan, N., and Blommaert, J.,** Isolation and preliminary identification of the causative agent of carp erythrodermatitis, *Vet. Arh.*, 47, 291, 1977.
8. **Fijan, N. and Petrinec, Z.,** Mortality in a pond caused by carp erythrodermatitis, *Riv. Ital. Pisci. Ittiopath.*, 8, 45, 1973.
9. **Buck, D., McCarthy, D., and Hill, B.,** A report of suspected erythrodermatitis in carp in Great Britain, *J. Fish Biol.*, 7, 301, 1975.
10. **Shotts, E. B., Jr., Talkington, F. D., Elliott, D. G., and McCarthy, D. H.,** Aetiology of an ulcerative disease in goldfish, *Carassius auratus* (L): characterization of the causative agent, *J. Fish Dis.* 3, 181, 1980.
11. **Elliott, D. G. and Shotts, E. B., Jr.,** Aetiology of an ulcerative disease in goldfish, *Carassius auratus* (L): experimental induction of the disease, *J. Fish Dis.*, 3, 145, 1980.

12. **Hamilton, R. C., Kalnins, H., Ackland, N. R., and Ashburner, L. D.,** An extra layer in the surface layers of an atypical *Aeromonas salmonicida* isolated from Australian goldfish, *J. Gen. Microbiol.,* 122, 363, 1981.

13. **Evelyn, T. P. T.,** An aberrant strain of the bacterial fish pathogen *Aeromonas salmonicida* isolated from a marine host, the Sable fish *(Anoplopoma finbria),* and from two species of cultured Pacific salmon, *J. Fish. Res. Board Can.,* 28, 1629, 1971.

14. **Whittington, R. J. and Cullis, B.,** The susceptibility of salmonid fish to an atypical strain of *Aeromonas salmonicida* that infects goldfish, *Carassius auratus* (L), in Australia, *J. Fish Dis.,* 11, 461, 1988.

15. **Carson, J. and Handlinger, J.,** Virulence of the aetiological agent of goldfish ulcer disease in Atlantic salmon, *Salmo salar* L., *J. Fish Dis.,* 11, 471, 1988.

16. **McCarthy, D. H.,** Fish furunculosis caused by *Aeromonas salmonicida* var. *achromogenes, J. Wildl. Dis.,* 11, 489, 1975.

17. **Hubert, R. M. and Williams, W. P.,** Ulcer disease of roach, *Rutilus rutilus* (L), *Bamidgeh,* 32, 46, 1980.

18. **Iida, T., Nakakoshi, K., and Wakabayashi, H.,** Isolation of atypical *Aeromonas salmonicida* from diseased eel, *Anguilla japonica, Fish Pathol.,* 19, 109, 1984.

19. **Cornick, J. W., Morrison, C. M., Zwicker, B., and Shum, G.,** Atypical *Aeromonas salmonicida* infection in Atlantic cod, *Gadus morhua* L., *J. Fish Dis.,* 7, 495, 1984.

20. **Fijan, N.,** Experimental transmission of infectious dropsy of carp, *Bull. Off. Int. Epizool.,* 65, 731, 1966.

21. **Pol, J. M. A., Bootsma, R., and Berg-Blommaert, J. M.,** Pathogenesis of carp erythrodermatitis (CE): role of bacterial endo- and exotoxin, in *Fish Diseases Third COPRAQ-Session,* Ahne, W., Ed., Springer-Verlag, Heidelberg, 1980, 120.

22. **Evenberg, D., DeGraaff, P., Lugtenberg, B., and Van Muiswinkel, W. B.,** Vaccine-induced protective immunity against *Aeromonas salmonicida* tested in experimental carp erythrodermatitis, *J. Fish Dis.,* 11, 337, 1988.

23. **Wichardt, U.-P.,** Atypical *Aeromonas salmonicida-* infection in sea-trout (*Salmo trutta,* L.). I. Epizootiological studies, clinical signs and bacteriology, *Lasforskningsinstitutet Meddelande,* (Salmon Research Institute Report), 6, 1983.

24. **Paterson, W. D., Douey, D., and Desautels, D.,** Isolation and identification of an atypical *Aeromonas salmonicida* strain causing epizootic losses among Atlantic salmon *(Salmo salar)* reared in a Nova Scotian Hatchery, *Can. J. Fish. Aquat. Sci.,* 37, 2236, 1980.

25. **Håstein, T., Satueit, S. J., and Roberts, R. J.,** Mass mortality among minnows *Phoxinus phoxinus* (L) in Lake Tveitevatn, Norway, due to an aberrant strain of *Aeromonas salmonicida, J. Fish Dis.,* 1, 241, 1978.

26. **Munroe, A. L. S. and Hastings, T. S.,** Furunculosis, in *Bacterial Diseases of Fish,* Inglis, V., Roberts, R. J., and Bromage, N. R., Eds., Blackwell Scientific, Oxford, 1993, 122.

27. **Kimura, T.,** A new subspecies of *Aeromonas salmonicida* as an etiological agent of furunculosis on "Sakuramasu" *(Oncorhynchus masou)* and pink salmon *(O. gorbuscha)* rearing for maturity. I. On the morphological and physiological properties, *Fish Pathol.,* 3, 34, 1969.

Chapter XIV

Eels (Anguillidae)

Generally, eels contract the same types of bacterial diseases as do other warmwater fishes; however, two infectious diseases, "edwardsiellosis", also known as "hepatonephritis", and "red spot disease" seem to have a greater affinity for cultured eels than for other species of fish.

A. EDWARDSIELLOSIS

Edwardsiellosis is a subacute to chronic disease of a variety of fish species, particularly cultured eels[1] in Asia and to a lesser degree of channel catfish[2] in the U.S. The disease is known as hepatonephritis in eels and in catfish it is called emphysematous putrifactive disease of catfish, or fish gangrene, because of the foul odor that is emitted from gas-filled pockets in necrotic muscle tissue. The causative agent of edwardsiellosis is *Edwardsiella tarda* and is the same organism that was called *Paracolobacterum anguillimortiferum* by Wakabayashi and Egusa.[3]

1. GEOGRAPHICAL RANGE AND SPECIES SUSCEPTIBILITY

Edwardsiella tarda is widespread and is found primarily in cultured freshwater and, to some extent, in marine environments. It has been reported from 25 countries in North and Central America, Europe, Asia, Australia, Africa, and the Middle East.[4]

Although the most prominent species of fish infected by *E. tarda* are eels (*Anguilla japonica, A. rostrata,* and *A. anguilla*) and catfishes *(Ictalurus punctatus)*, the organism has been isolated from over 20 species of fish from freshwater and marine environments and include carp *(Cyprinus carpio)*, largemouth bass *(Micropterus salmoides)*, striped bass *(Morone saxatilis)*, red sea bream *(Chrysophrys major)*, flounder *(Paralichthys olivaceous)*, tilapia (*Tilapia* spp.), and yellowtail *(Serila gaingu)*, to name a few.[1-2,4] It has also been isolated from the intestinal content of apparently healthy fish during routine bacteriological surveys.[5-7] Although generally considered a warmwater inhabitant, *E. tarda* has, on at least two occasions, been implicated in infections of salmonids. One was from Rogue River chinook salmon *(Oncorhynchus tshawytscha)* in Oregon when water temperatures were above normal[8] and the second in Nova Scotia, Canada where it was isolated from a small number of migrating adult Atlantic salmon *(Salmo salar).*[9]

E. tarda infections are not limited to fish, but often exist as part of the normal intestinal microflora, especially in fish-eating birds and waterbound animals.[10] Reptiles and amphibians, cattle and swine, marine mammals,[11-13] and other warmblooded animals including humans have also been known to be infected with *E. tarda*.

2. CLINICAL SIGNS AND FINDINGS

Clinical signs of *E. tarda* infections differ slightly between locales and among different fish species. Infected eels exhibit lethargic swimming and tend to float at the surface; their fins become congested and hyperemic; petechial hemorrhages develop on the underside; the anal region is swollen and hyperemic; and pockets of gas may develop between the dermis and muscle (Figure 1). Internally, the liver, kidney, and spleen have a whitish appearance, and may develop abscesses.[14]

An *E. tarda* infection in channel catfish is characterized by lethargic swimming and the presence of small, 3- to 5-mm cutaneous lesions located dorsolaterally on the body.[2] Within the flank muscles or caudal peduncle, these small lesions progress into larger abscesses and develop obvious convex, swollen areas. The skin loses its pigmentation and incised lesions emit a foul smelling gas. As the infection progresses, mobility of the posterior portion of the body is lost. Internally, there is a general hyperemia characteristic of a septicemia and the body cavity has a putrid odor. The kidney, in particular, is enlarged and the liver mottled or abscessed.

A variety of clinical signs occur in other species of fish and include exophthalmia and opaqueness of the eye, swollen and necrotic skin and muscle lesions (Figure 1), rectal protrusion, or abscesses in swollen internal organs.

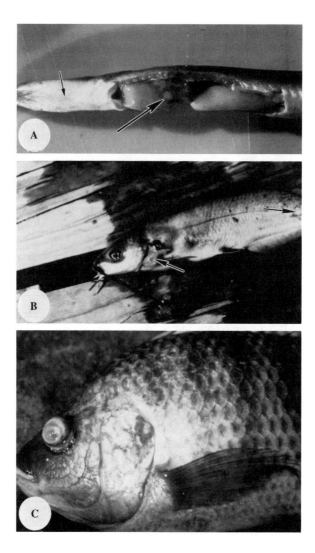

Figure 1   (A) Japanese eel infected with *Edwardsiella tarda;* note the pale, mottled liver with an abscess (large arrow) and slight petechia on the throat (small arrow); (B) channel catfish infected with *E. tarda* and showing depigmented epithelium and petechial hemorrhages (arrows); (C) *Tilapia* sp. with *E. tarda* infection in the eye. (Photograph A by E. B. Shotts; photograph B by F. P. Meyer.)

3. DIAGNOSIS AND BACTERIAL CHARACTERISTICS

E. tarda is easily isolated from internal organs and muscle lesions of clinically infected fish and then identified by conventional bacteriological or serological methods. When working with eels, it is necessary to isolate bacteria from the diseased fish for diagnostic purposes because clinical signs of *E. tarda* and other bacterial septicemias cannot otherwise be differentiated.[15] *E. tarda* is not fastidious, therefore isolation is easily made on brain heart infusion (BHI) agar, tryptic soy agar (TSA), or any other general purpose medium.[2,8] Incubation at 26 to 30°C for 24 to 48 h yields small, circular, convex, transparent colonies approximately 0.5 mm in diameter. The incidence of *E. tarda* isolation from chinook salmon brain was improved from 2 to 19% by first inoculating thioglycolate media and transferring the inoculum to BHI agar rather than inoculating directly onto BHI agar.[8] *E. tarda* will form a small green colony with a black center on EIM (Chapter XII, Figure 6).[16]

Key presumptive characteristics of *E. tarda* for diagnostic purposes are its motility, indole production in tryptone broth, H_2S production on triple sugar iron (TSI) slants, and its tolerance of up to 4% salt.

It produces gas during glucose fermentation and grows at 40°C. Positive identification can be made by specific serum agglutination or fluorescent antibody tests because no serological cross-reactivity occurs between *E. tarda* and *E. ictaluri* or other enteric bacteria.[17] *E. tarda* is phenotypically a homogeneous organism. Waltman et al.[18] have found only slight variation in the biochemical and biophysical characteristics of 116 isolates of *E. tarda* from the U.S. and Taiwan.

The genus *Edwardsiella* is a member of the family Enterobacteriaceae. Members of the genus are small, straight rods about 1 μm in diameter and 2 to 3 μm in length.[19] *E. tarda* is a Gram-negative, motile rod with peritrichous flagella and is facultatively anaerobic. *E. tarda* is catalase positive, cytochrome oxidase negative, ferments glucose, and reduces nitrate to nitrite (Chapter XII, Table 2). It is also lactose negative, indole positive, and produces an alkaline slant and acid butt on TSI agar, but the acid production is often obscured by the H_2S.

Farmer and McWhorter[19] identified a wild type of *E. tarda* and a second Biogroup 1, both of which are easily distinguished from *E. ictaluri* and *E. hoshinae* (the isolates from reptiles and amphibians).[20] Park et al.[21] identified 4 *E. tarda* serotypes (A, B, C, and D) among 445 isolates collected from fish and environmental sources — 72% of the isolates from fish were serotype A, indicating that it may be the predominant disease-causing type.

4. EPIZOOTIOLOGY

Edwardsiellosis of catfish in the U.S. occurs most often in the summer when water temperatures are over 30°C, however it is not restricted to these warm temperatures. According to Egusa,[14] *E. tarda* infections of Japanese eels were more prevalent in summer when water temperatures were highest, a trend also reported by Kuo et al.[22] who found that the optimum water temperature for edwardsiellosis of eels was 30°C. However, Liu and Tsai[23] found that eel infections in Taiwan were most common during January to April, during fluctuating water temperatures. In the U.S. Amandi et al.[8] isolated *E. tarda* from 19% of dead or dying Rogue River chinook salmon in Oregon when water temperatures rose from 17 to 20°C. The range of temperatures at which *E. tarda* can cause disease may reflect an opportunistic capability that is more closely associated with increased species susceptibility under specific stress conditions (fluctuating or high temperature) rather than to the pathogenic nature of the organism.

The source of *E. tarda* is presumably the intestinal contents of carrier aquatic animals, primarily catfish and eels; however, amphibians and reptiles are also considered probable sources of the organism. *E. tarda* is shed from the infected host and probably transmitted through the water column from an infected source (e.g., carrier animal, water, or mud). It has also been reported that the *E. tarda* cell count in eel pond water is higher when clinical disease is present than when there is no disease.

E. tarda is a hardy bacterium that can survive for extended periods in waters at temperatures of 15 to 45°C (optimum 30 to 37°C), pH 4.0 to 10.0 (optimum 7.5 to 8.0), and in salt solutions of 0 to 4% (optimum 0.5 to 1.0% NaCl),[24] therefore it is capable of withstanding diverse environmental conditions. In 20°C pond water *E. tarda* was isolated for over 76 d, indicating that it can be a source of infection for an extended period of time. The organism survives in both fresh and salt water, mud, and on fouling material on nets.[7,24-25]

Apparently *E. tarda* is a common inhabitant of the catfish pond microflora and its presence poses a constant threat of disease. The organism was isolated from 75% of catfish pond water samples, 64% of catfish pond mud samples, and 100% of apparently healthy frogs, turtles, and crayfish from *E. tarda*-positive ponds.[7] Also, *E. tarda* was isolated from the flesh of as many as 88% of dressed domestic catfish but was found in only 30% of imported dressed fish.

Environmental stressors may not be essential precursors to an *E. tarda* infection in fish, but fluctuating and high water temperatures,[10,26] high organic content, generally poor water quality,[2] and crowded conditions can contribute to the onset and severity of the disease. When exposed to environmental stressors (e.g., low dissolved oxygen, high ammonia, and high carbon dioxide) 25 to 50% of juvenile channel catfish that were experimentally infected with *A. hydrophila* developed an *E. tarda* infection.[26] Only 4 to 13% of nonstressed, non-*A. hydrophila*-infected fish developed *E. tarda* infections. This study would indicate that environmental stress, in conjunction with other bacterial infections, can predispose channel catfish to disease from naturally present *E. tarda*. Also, sublethal concentrations of copper (100 to 250 μg/l) in the water reduced the resistance of Japanese eels to *E. tarda* infection.[27]

E. tarda disease is usually more prevalent in larger fish: approximately 0.4 kg or larger channel catfish and in marketable-sized eels; however, elvers and subadult channel catfish are not immune to the organism. Mortality of pond-held channel catfish infected with *E. tarda* seldom exceeds 5%, but if these fish are moved into confined holding tanks the rate of infection quickly increases to 50% with parallel morbidity and mortality.[2] In experimental infectivity studies where 100 g eels were immersed in *E. tarda* 60% mortality occurred;[28] however, Kodoma et al.[29] reported that a naturally *E. tarda*-infected eel population suffered an 80% loss. Aquaculturists in Taiwan feel that if *E. tarda* could be controlled, overall eel production would increase by 30%.

E. tarda can pose a health threat to humans, usually manifesting itself as gastroenteritis and diarrhea; however, extraintestinal infections can produce a typhoid-like illness, meningitis, peritonitis with sepsis, cellulitis, and hepatic abscess.[30-32] The organism has also been associated with infections in wounds precipitated by fish hooks and catfish spines,[33] diarrhea associated with intestinal infection of *Entamoeba histolytica* and other tropical diarrheas, and consumption of contaminated freshwater fish.[34] These infections in humans are usually successfully treated, but death may occasionally occur.

5. PATHOLOGICAL MANIFESTATIONS

In eels *E. tarda* causes either a nephritic or hepatic histopathology with the nephritic form being more common.[35] In the nephritic form, the kidney is enlarged and has various sized abscesses that are initially formed in the sinusoids of hematopoietic tissue. The hematopoietic tissue becomes swollen and the foci of infection develop into abscesses that progress into cavities filled with dark red, odiferous, purulent matter. These large abscesses are walled off by fibrin and contain neutrophils, some of which phagocytize bacteria. Liquifaction of cells follows, and small foci of abscesses can be found in other organs and in the lateral musculature adjoining the kidney. In the hepatitis form of disease, livers are usually enlarged and contain various sized abscesses, some of which leak fluid into the body cavity. Enlarged abscesses involve hepatocytes and blood vessels which form pus-like emboli and pyemia,[14] with extensive bacterial multiplication. Ullah and Arai[36] found evidence that *E. tarda* does not form endotoxins as do other Gram-negative bacteria, but that they produce two exotoxins which may contribute to pathogenicity.

Histopathology of *E. tarda* in Japanese flounder, sea bass, Japanese eel, tilapia, striped bass, and other fish may vary slightly from that previously described, but there are no published descriptions of histopathology of the organism in infected channel catfish.

6. SIGNIFICANCE

The significance of *E. tarda* depends largely on the fish species affected and the presence of environmental stressors. In Taiwan and Japan it is one of the most serious bacterial diseases of cultured eels. However, in the U.S., unless channel catfish are crowded, the disease is generally of little consequence. The greatest problem to catfish farmers is the presence of a subclinical infection when fish are processed that affects the quality of the fish at slaughter and causes contamination problems during dressing.[2] In other species of fish, *E. tarda* may be responsible for some losses, but usually these are chronic rather than acute. Caution must always be taken when handling *E. tarda*-infected fish because of its potential for human infection.

7. MANAGEMENT

Preventing *E. tarda*-infected fish, turtles, snakes, etc., from coming into contact with aquaculture species is impossible in most pond culture systems. Therefore, maintaining a high-quality environment, keeping organic enrichment to a minimum, and preventing extreme temperature fluctuations or shock are positive approaches to management of the organism.

The application of antibiotic medicated feed is the most commonly used control practice for *E. tarda*. The drug of choice for *E. tarda* infection in eel in Taiwan is naladixic acid because a large percentage of the isolates are resistant to most other antibiotics, including oxytetracycline.[37-38] In the U.S. oxytetracycline (Terramycin®) is the drug of choice, although it is not FDA-approved specifically for *E. tarda*.

Development of a vaccine using whole cells, disrupted cells, and cell extracts as immunogens has been pursued in Japan and Taiwan.[39-40] All of these preparations are immunogenic, especially by injection, but a practical, commercially available vaccine has not yet been developed.

B. RED SPOT DISEASE

Red spot disease of eels is a mild to serious bacterial infection of cultured eels. The disease, known as Sekiten-byo in Japan, is caused by *Pseudomonas anguilliseptica*.[41]

1. GEOGRAPHICAL RANGE AND SPECIES SUSCEPTIBILITY

Red spot disease primarily affects cultured Japanese *(A. japonica)* and European *(A. anguilla)* eels in Japan, but the disease and its etiological agent have been identified in Taiwan,[42] Malaysia,[43] Scotland,[44] and more recently in Finland.[45] Although Japanese eel appear to be the more susceptible species, *P. anguilliseptica* has occurred in European eel, black sea bream *(Acanthopagrus schleyeli)*, giant sea perch *(Lates calcarifer),* and estuarine grouper *(Epinephelus tauvina)* in Malaysia.[43] The organism has also been isolated from Atlantic salmon, brown trout *(S. trutta)*, rainbow trout *(O. mykiss)*, and whitefish *(Coregonus* sp.).[45] Experimental infections of *P. anguilliseptica* have been induced in ayu *(Plecoglossus altivelis).*[46]

2. CLINICAL SIGNS AND FINDINGS

Japanese eels infected with *P. anguilliseptica* develop extensive petechial hemorrhage in the subepidermal layer of jaws, underside of head, and along the ventral body surface. Internally, petechia may occur on the peritoneum. Anatomical changes of visceral organs include swelling of the liver, atrophy of spleen and kidney, and pericarditis.[47] European eels react similarly to *P. anguilliseptica,* but generally the response is less severe than in the Japanese eel.[48]

3. DIAGNOSIS AND BACTERIAL CHARACTERISTICS

Red spot disease of eels is diagnosed by isolation of *P. anguilliseptica* from internal organs. The slow-growing organism is cultured on general purpose media, forming 1-mm-diameter colonies in 3 to 4 d at 25°C. Presumptively, *P. anguilliseptica* is a motile (better at 15 than at 25°C), Gram-negative, cytochrome oxidase positive organism, but is negative for soluble pigments, H_2S, indole, and is nonreactive in glucose oxidation-fermentation tests (Table 1).[41]

P. anguilliseptica measures 0.8 μm in diameter and is 5 to 10 μm in length. Growth takes place at 5 to 30°C, but not at 37°C. The organism is not reactive on many substrates; however, it is positive for catalase, degradation of gelatin, and grows on media with 0 to 4% NaCl. The bacterium has low metabolic reactivity for many sources of carbon.[41,44,49]

There are two antigenic groups present in *P. anguilliseptica,*[50] Type I is thermolabile (121°C for 30 min) while Type II is heat stable. The heat labile type from Japan possesses a K antigen and is solely pathogenic to eels. Based on agglutination and precipitin tests, the *P. anguilliseptica* taken from eels in Taiwan is identical to the Japanese K antigen type.[51]

4. EPIZOOTIOLOGY

In Japan, red spot disease of eels occurs primarily during April and May, becomes less apparent during summer, and reoccurs to a mild degree in October.[52] Epizootics appear to coincide with water temperatures of approximately 20°C and mortalities decline as temperatures rise above 25°C. Experimental infections of Japanese eels confirmed that a relationship between water temperatures and *P. anguilliseptica* epizootics existed because nearly all infected eels held below 20°C died, while few fish held above 27°C succumbed to the disease.[53] In these studies, 100% of Japanese eels and 71% of European eels died at 19 to 21°C. Stewart et al.[44] demonstrated the affinity of *P. anguilliseptica* for young eels when 96% of a population of European eel elvers and only 3.9% of adult eels died. A concentration of copper at 25 to 100 μg/l in water also reduced resistance of Japanese eels to *P. anguilliseptica* by adversely affecting phagocytosis and other cell-mediated defense mechanisms.[54]

P. anguilliseptica infection in giant sea perch and estuarine grouper in Malaysia occurred in off-shore caged fish.[43] Mortality was 20 to 60% in these fish during the cooler monsoon season of November to December and February to March, which also coincided with poor water quality. It was speculated that the *P. anguilliseptica* infection was secondary to stressful environmental conditions and poor nutrition.

P. anguilliseptica infection may be more severe in brackish water ponds than in freshwater ponds because of the organism's ability to withstand up to 4% NaCl. Muroga[55] summarized the epizootiological

Table 1 **Biophysical and biochemical characteristics of
Pseudomonas fluorescens and *Pseudomonas
anguilliseptica*[41,59]**

Characteristic	*Pseudomonas fluorescens*	*Pseudomonas anguilliseptica*
Size (μm)	0.8 × 2.3–2.8	0.8 × 5–10
Gram stain	−	−
Motility	+ (weak)	+ (15°C)
Fluorescent pigment	+	−
Optimum temperature	25–30°C	25°C
Growth at		
30°C	+	+
40°C	−	−
Glucose motility deep	Oxidative	?
Growth 0–4% NaCl	−	+
Catalase	+	+
Cytochrome oxidase	+	+
Hydrolysis of		
Arginine	+	−
Gelatin	+	+
Starch	+	+
Indole	−	−
Nitrate reduction	+	−
β-Galactosidase	−	−
Production of acid from		
Arabinose	+	−
Arginine	+	−
D-Alanine	+	?
Fructose	?	−
Galactose	?	−
Glucose	+	−
Glycerol	?	−
Inositol	+	−
Lactose	−	−
Maltose	+	−
Mannitol	+	−
Mannose	?	−
Raffinose	?	−
Rhamnose	?	−
Salicin	−	−
Sucrose	+	−
Trehalose	+	?
Xylose	+	−

characteristics of red spot as follows: (1) prevailed in brackish water ponds; (2) prevailed when water temperature was below 20°C and ceased when it was above 27°C; and (3) prevailed among Japanese eels.

5. PATHOLOGICAL MANIFESTATIONS

Histopathological studies by Miyazaki and Egusa[56] showed that lesions appeared initially in the dermis, subcutaneous adipose tissue, interstitial tissue of the musculature, vascular walls, bulbous arteriosis, and heart. Bacteria multiplied in the vascular walls, resulting in inflammation with serous exudation and cellular infiltration. Many small hemorrhages occurred in dermal connective tissue (Figure 2). Visceral organs experienced congestive edema and fatty degeneration of liver hepatic cells, serous

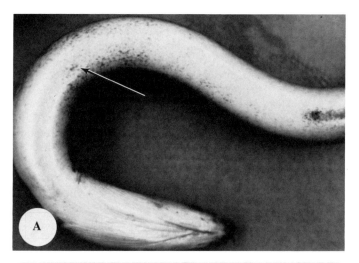

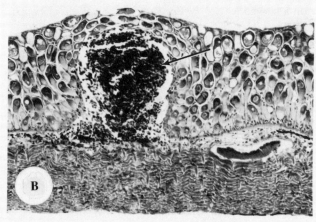

Figure 2 (A) Japanese eel infected with *Pseudomonas anguilliseptica* showing petechia (arrow); (B) histological section of epidermis illustrating the petechial hemorrhage (arrow). (Photographs courtesy of T. Miyazaki.)

exudation and cellular proliferation of the spleen, glomerulitis, activation of the reticuloendothelial cells lining the sinusoids, and atrophy of hematopoietic tissue of the kidney.

Within several days of being injected with *P. anguilliseptica,* eels become moribund and show clinical signs similar to naturally infected fish with mortality occurring in 6 to 10 d.[41] Although no exotoxin production by *P. anguilliseptica* has been demonstrated, it is speculated that pathogenicity is the result of these substances. Nakai et al.[51] traced the progression of a red spot infection in Japanese eels following an intramuscular injection of 10^7 or 10^9 cells per 100 g of fish which were held at 12 and 20°C. Viable cells in the blood decreased for 1 to 12 h postinjection, followed by a stationary period when cells remained constant or increased slightly. Cell counts then increased to 10^8 to 10^{10} cells per gram of tissue or per milliliter of blood, thus establishing a septicemia which persisted until death. The period of bacterial growth from injection to maximum cell concentration took approximately 12 d at 12°C and 5 d at 20°C. When inoculated fish were held at 28°C, bacterial cells disappeared from internal organs within 2 d.

6. SIGNIFICANCE
Red spot disease has become one of the most serious diseases of cultured eels in Japan, and with its appearance in Scotland and Europe it is now of global concern. The ability of *P. anguilliseptica* to infect other culture species only adds to its importance.

7. MANAGEMENT

Red spot disease is a seasonal infection that occurs when water temperatures are 15 to 20°C; therefore, it can be averted by raising water temperatures above 27°C when practical. Muroga[55] proposed the following prophylaxis for cultured Japanese eels: (1) maintain water above 27°C; (2) culture eels in freshwater ponds; and (3) where the disease is endemic, culture European eels which are not as susceptible to *P. anguilliseptica*. Medicated feed containing an effective antibiotic is also applicable to control of the disease. Wicklund and Bylund[45] found that oxytetracycline was not effective in *P. anguilliseptica*-infected salmon, but Romet® was effective. Red spot disease of Japanese eels was controlled by 9 consecutive d of baths in oxolinic acid (2 to 10 mg/l) or naladixic acid (2 to 10 mg/l) in which there was no mortality; this compared to 100% mortality in control fish.[57] Also, oxolinic acid at 5 to 20 mg/kg of fish fed for 3 consecutive d resulted in 100% survival.

Vaccination experiments have shown that formalin-killed bacterins are effective immunogens when injected into eels.[58] Injected vaccines stimulated high agglutinating antibodies and were highly protective; however, immersion apparently had no immunological effect.

REFERENCES

1. **Hoshina, T.,** On a new bacterium, *Paracolobactrum anguillimortiferum* sp., *Bull. Jpn. Soc. Sci. Fish.,* 28, 162, 1962.
2. **Meyer, F. P. and Bullock, G. L.,** *Edwardsiella tarda,* a new pathogen of channel catfish *(Ictalurus punctatus), Appl. Microbiol.,* 25, 155, 1973.
3. **Wakabayashi, H. and Egusa, S.,** *Edwardsiella tarda (Paracolobactrum anguillimortiferum)* associated with pond-cultured eel disease, *Bull. Jpn. Soc. Sci. Fish.,* 39, 931, 1973.
4. **Austin, B. and Austin, D. A.,** *Bacterial Fish Pathogens: Disease in Farmed and Wild Fish,* Ellis Horwood, Chichester, 1987.
5. **Kanai, K., Tawaki, S., and Uchida, Y.,** An ecological study of *Edwardsiella tarda* in flounder farm, *Fish Pathol.,* 23, 41, 1988.
6. **van Damme, L. R. and Vandepitte, J.,** Frequent isolation of *Edwardsiella tarda* and *Plesiomonas shigelloides* from healthy Zairese freshwater fish: a possible source of sporadic diarrhea in the tropics, *Appl. Environ. Microbiol.,* 39, 475, 1980.
7. **Wyatt, L. E., Nickelson, R., II., and Vanderzant, C.,** *Edwardsiella tarda* in freshwater catfish and their environment, *Appl. Environ. Microbiol.,* 38, 710, 1979.
8. **Amandi, A., Hiu, S. F., Rohovec, J. S., and Fryer, J. L.,** Isolation and characterization of *Edwardsiella tarda* from fall chinook salmon *(Oncorhynchus tshawytscha), Appl. Environ. Microbiol.,* 43, 1380, 1982.
9. **Martin, J. D.,** Atlantic salmon and alewife passage through a pool and weir fishway of the Mayaguadaric River, New Brunswick, Canada during 1983, *Can. Manu. Rep. Fish. Aqua. Sci.,* 1983 (Abstr.).
10. **Bauwens, L. D., Meurichy, W., Lemmens, P., and Vandepitte, J.,** Isolation of *Plesiomonas shigelloides* and *Edwardsiella* spp. in the Antwerp Zoo Belgium, *Acta Zool. Pathol. Antverp.,* 0(77), 61, 1983.
11. **White, F. H., Neal, F. C., Simpson, C. F., and Walsh, A. F.,** Isolation of *Edwardsiella tarda* from an ostrich and an Australian skink, *J. Am. Vet. Med. Assoc.,* 155, 1057, 1969.
12. **Ewing, W. H., McWhorter, A. C., Escobar, M. R., and Lubin, A. H.,** *Edwardsiella,* a new genus of Enterbacteriaceae based on a new species, *Edwardsiella tarda, Int. Bull. Bacteriol. Nomencl. Taxon.,* 15, 33, 1965.
13. **Coles, B. M., Stroud, R. K., and Sheggeby, S.,** Isolation of *Edwardsiella* from 3 Oregon sea mammals, *J. Wildl. Dis.,* 14, 339, 1978.
14. **Egusa, S.,** Some bacterial diseases of freshwater fishes in Japan, *Fish Pathol.,* 10, 103, 1976.
15. **Nichibuchi, M., Muroga, K., and Jo, Y.,** Pathogenic *Vibrio* isolated from eels, diagnostic tests for the disease due to present bacterium, *Fish Pathol.,* 14, 124, 1980.
16. **Shotts, E. B. and Teska, J. D.,** Bacterial pathogens of aquatic vertebrates, in *Methods for Microbiological Examination of Fish and Shellfish,* Austin, B. and Austin, D. A., Eds., Ellis Horwood, Chichester, 1989, 164.

17. **Klesius, P., Johnson, K., Durborow, R., and Vinitnantharat, S.,** Development and evaluation of an enzyme-linked immunosorbent assay for catfish serum antibody to *Edwardsiella ictaluri, J. Aquat. Anim. Health,* 3, 94, 1991.

18. **Waltman, W. D., Shotts, E. B., and Hsu, T. C.,** Biochemical and enzymatic characterization of *Edwardsiella tarda* from the United States and Taiwan, *Fish Pathol.,* 21, 1, 1986.

19. **Farmer, J. J. and McWhorter, A. L.,** Genus X *Edwardsiella* Ewing and McWhorter 1965, 37^AL, in *Bergey's Manual of Systematic Bacteriology,* vol. 1, Williams & Wilkins, Baltimore, 1984, 436.

20. **Grimont, P. A. D., Grimont, F., Richard, C., and Sakazaki, R.,** *Edwardsiella hoshinae,* a new species of Enterobacteriaceae, *Curr. Microbiol.,* 4, 347, 1980.

21. **Park, S., Wakabayashi, H., and Watanabe, Y.,** Serotype and virulence of *Edwardsiella tarda* isolated from eel and their environment, *Fish Pathol.,* 18, 85, 1983.

22. **Kuo, S.-C., Chung, H.-Y., and Kou, G.-H.,** *Edwardsiella anguillimortifera* isolated from edwardsiellosis of cultured eel *(Anguilla japonica), JCRR Fish. Ser. No. 29, Fish Dis. Res.,* I, 1, 1977.

23. **Liu, C. I. and Tsai, S. S.,** Edwardsiellosis in pond-cultured eel in Taiwan, *CAPD Fish. Ser. No. 8, Rep. Fish Dis. Res.,* IV, 92, 1982.

24. **Chowdhury, M. B. R. and Wakabayashi, H.,** Survival of four major bacterial fish pathogens in different types of experimental water, *Bangladesh J. Microbiol.,* 7, 47, 1990.

25. **White, F. H., Simpson, C. F., and Williams, L. E., Jr.,** Isolation of *Edwardsiella tarda* from aquatic animal species and surface waters in Florida, *J. Wildl. Dis.,* 9, 204, 1973.

26. **Walters, G. and Plumb, J. A.,** Environmental stress and bacterial infection in channel catfish, *Ictalurus punctatus* Rafinesque, *J. Fish Biol.,* 17, 177, 1980.

27. **Mushiake, K., Muroga, K., and Nakai, T.,** Increased susceptibility of Japanese eel *Anguilla japonica* to *Edwardsiella tarda* and *Pseudomonas anguilliseptica* following exposure to copper, *Bull. Jpn. Soc. Sci. Fish.,* 50, 1797, 1984.

28. **Ishihara, S. and Kusuda, R.,** Experimental infection of elvers and anguillettes with *Edwardsiella tarda, Bull. Jpn. Soc. Sci. Fish.,* 47, 999, 1981.

29. **Kodoma, H., Murai, T., Nakanishi, Y., Yamamoto, F., Mikami, T., and Izawa, H.,** Bacterial infection which produces high mortality in cultured Japanese flounder *Paralichthys olivaceus* in Hokkaido Japan, *Jpn. J. Vet. Res.,* 35, 227, 1987.

30. **Bockemuhl, J., Pan-Rrai, R., and Burkhardt, F.,** *Edwardsiella tarda* associated with human disease, *Pathol. Microbiol.,* 37, 393, 1971.

31. **Clarridge, J. E., Musher, D. M., Fainstein, V., and Wallace, R. J.,** Extraintestinal human infection caused by *Edwardsiella tarda, J. Clin. Microbiol.,* 11, 511, 1980.

32. **Zighelboim, J., Williams, T. W., Bradshaw, M. W., and Harris, R. L.,** Successful medical management of a patient with multiple hepatic abscesses due to *Edwardsiella tarda, Clin. Infect. Dis.,* 14, 117, 1992.

33. **Hargreaves, J. E. and Lucey, D. R.,** *Edwardsiella tarda* soft tissue infection associated with catfish puncture wound, *J. Infect. Dis.,* 162, 1416, 1990.

34. **Gilman, R. H., Madasamy, M., Gan, E., MaRiappan, M., Davis, E., and Kyser, K. A.,** *Edwardsiella tarda* in jungle diarrhea and a possible association with *Entamoeba histolytica, Southeast Asian J. Trop. Med. Public Health,* 2, 186, 1971.

35. **Miyazaki, T. and Kaige, N.,** Comparative histopathology of edwardsiellosis in fishes, *Fish Pathol.,* 20, 219, 1985.

36. **Ullah, A. and Arai, T.,** Pathological activities of the naturally occurring strains of *Edwardsiella tarda, Fish Pathol.,* 18, 65, 1983.

37. **Chen, S. C., Tung, M. C., Lu, C. F., and Huang, S. T.,** Sensitivity *in vitro* of various chemotherapeutic agents to *Edwardsiella tarda* of pond cultured eels, *CAPD Fish. Ser. No. 10,* 100, 1984.

38. **Liu, C. I. and Wang, J. H.,** Drug resistance of fish-pathogenic bacteria. II. Resistance of *Edwardsiella tarda* in aquaculture environment, *COA Fish. Ser. No. 8, Fish Dis. Res.,* 8, 56, 1986.

39. **Song, Y. L., Kou G. H., and Chen, K. Y.,** Vaccination conditions for the eel *(Anguilla japonica)* with *Edwardsiella anguillimortifera* bacterin, *J. Fish. Soc. Taiwan,* 4, 18, 1982.

40. **Salati, F., Kawai, K., and Kusuda, R.,** Immunoresponse of eel against *Edwardsiella tarda* antigens, *Fish Pathol.,* 18, 135, 1983.

41. **Wakabayashi, H. and Egusa S.,** Characteristics of a *Pseudomonas* sp. from an epizootic of pond-cultured eels *(Anguilla japonica), Bull. Jpn. Soc. Sci. Fish.,* 38, 577, 1972.

42. **Kuo, S. C. and Kou, G. H.,** *Pseudomonas anguilliseptica* isolated from red spot disease of pond-cultured eel, *Anguilla japonica, Rep. Inst. Fish. Biol. Minis. Econ. Affairs Nat. Taiwan Univ.,* 3, 19, 1978.

43. **Nash, G., Anderson, I. G., Schariff, M., and Shamsudin, M. N.,** Bacteriosis associated with epizootic in the giant sea perch, *Lates calcarifer,* and the estuarine grouper, *Epinephelus tauvine,* cage cultured in Malaysia, *Aquaculture,* 67, 105, 1987.

44. **Stewart, D. J., Woldemariam, K., Dear, G., and Mochaba, F. M.,** An outbreak of "Sekiten-byo" among cultured European eels, *Anguilla anguilla* L., in Scotland, *J. Fish Dis.,* 6, 75, 1983.

45. **Wicklund, T. and Bylund, G.,** *Pseudomonas anguilliseptica* as a pathogen of salmonid fish in Finland, *Dis. Aquat. Org.,* 8, 13, 1990.

46. **Nakai, T., Hanada, H., and Muroga, K.,** First records of *Pseudomonas anguilliseptica* infection in cultured ayu *Plecoglossus altivelis, Fish Pathol.,* 20, 481, 1985.

47. **Jo, Y., Muroga, K., and Onishi, K.,** Studies on red spot disease of pond cultured eels. III. A case of the disease in the European eels *(Anguilla anguilla)* cultured in Tokushima Prefecture, *Fish Pathol.,* 9, 115, 1975.

48. **Ellis, A. E., Dear, G., and Steward, D. J.,** Histopathology of Sekiten-byo caused by *Pseudomonas anguilliseptica* in the European eel, *Anguilla anguilla* L., in Scotland, *J. Fish Dis.,* 6, 77, 1983.

49. **Muroga, K., Nakai, T., and Sawada, T.,** Studies on red spot disease of pond-cultured eels. IV. Physiological characteristics of the causative bacterium, *Pseudomonas anguilliseptica, Fish Pathol.,* 12, 33, 1977.

50. **Nakai, T., Muroga, K., and Wakabayashi, H.,** Serological properties of *Pseudomonas anguilliseptica* in agglutination, *Bull. Jpn. Soc. Sci. Fish.,* 47, 699, 1981.

51. **Nakai, T., Muroga, K., Chung, J.-Y., and Kou, G.-H.,** A serological study on *Pseudomonas anguilliseptica* isolated from diseased eels in Taiwan, *Fish Pathol.,* 19, 259, 1985.

52. **Muroga, K., Jo, Y., and Yano, M.,** Studies on red spot disease of pond-cultured eels. I. The occurrence of the disease in eel culture ponds in Tokushima Prefecture in 1972, *Fish Pathol.,* 8, 1, 1973.

53. **Muroga, K., Jo, Y., and Sawada, T.,** Studies on red spot disease of pond-cultured eels. II. Pathogenicity of the causative bacterium, *Pseudomonas anguilliseptica, Fish Pathol.,* 9, 107, 1975.

54. **Mushiake, K., Muroga, K., and Nakai, T.,** Increased susceptibility of Japanese eel *Anguilla japonica* to *Edwardsiella tarda* and *Pseudomonas anguilliseptica* following exposure to copper, *Bull. Jpn. Soc. Sci. Fish.,* 50, 1797, 1984.

55. **Muroga, K.,** Red spot disease of eels, *Fish Pathol.,* 13, 35, 1978.

56. **Miyazaki, T. and Egusa, S.,** Histopathological studies of red spot disease of the Japanese eel *(Anguilla japonica).* I. Natural infection, *Fish Pathol.,* 12, 39, 1977.

57. **Jo, Y.,** Therapeutic experiments on red spot disease, *Fish Pathol.,* 13, 41, 1978.

58. **Nakai, T., Muroga, K., Ohnishi, K., Jo, Y., and Tanimoto, H.,** Studies on red spot disease of pond-cultured eels. IX. A field vaccination trial, *Aquaculture,* 3, 131, 1982.

59. **Palleroni, N. J.,** Family I. Pseudomonadaceae Winslow, Broadhurst, Buchanan, Krumwiede, Rogers and Smith 1917, 555, in *Bergey's Manual of Systematic Bacteriology,* Vol. 1, Krieg, N. R. and Holt, J. G., Eds., Williams & Wilkins, Baltimore, 1984, 141.

Chapter XV

Salmonids (Salmonidae)

Bacterial infections of salmonids occur worldwide wherever they are cultured. There is a certain mystique associated with salmonids that no other group of fish enjoys, and this is in part responsible for the extensive attention they receive from aquaculturists and pathologists. However, the high economic value of wild and cultured salmonids demands attention to their health. Many diseases that affect salmonids have a significant impact on cultured trout and salmon and, under certain conditions, on wild populations as well.

A. FURUNCULOSIS

Furunculosis is one of the oldest and best known bacterial diseases of fish. It is generally considered to be a disease of salmonids but is increasingly associated with other cool and occasionally warmwater fishes.[1] Its etiological agent is *Aeromonas salmonicida* subsp. *salmonicida,* which is also known as "typical *A. salmonicida*".[2-5]

1. GEOGRAPHICAL RANGE AND SPECIES SUSCEPTIBILITY

Furunculosis of salmonids is found in most regions of the world where trout occur, including North America, Great Britain, Europe, Asia (Japan and Korea), South Africa, and Australia.[5-7] *A. salmonicida* is endemic in many trout hatcheries in North America and probably no waters with resident trout populations should be considered free of the disease. Brook trout *(Salvelinus fontinalis)* is generally considered to be the most susceptible species, however, there are resistant strains. All salmonids, especially brown trout *(Salmo trutta),* rainbow trout *(Oncorhynchus mykiss),* lake trout *(S. nemaicush),* and the anadromous salmonids are susceptible to the disease. Wichardt et al.[8] indicated that in Sweden, acute, typical *A. salmonicida* infections occurred in Atlantic salmon *(Salmo salar)* and arctic char *(Salvelinus alpinus).* Nomura et al.[9] noted that in Japan, rainbow trout are not severely affected by *A. salmonicida* but cultured amago *(O. rhodurus)* and masou salmon *(O. masou)* are, and as the culture of these latter species increases *A. salmonicida* is likely to have an increasing impact on them. However, in experimental infections of *A. salmonicida* in different populations of pink salmon *(O. gorbuscha),* no difference in susceptibility was evident.[10] The disease may occur in species of cyprinids, pikes, perches, bullheads, and some nonsalmonid marine fish species.[11-12] When nonsalmonids are affected, they have often been in close proximity to *A. salmonicida*-infected trout.

2. CLINICAL SIGNS AND FINDINGS

Furunculosis is classified into four categories depending upon severity: (1) acute, (2) subacute, (3) chronic, or (4) latent.[7] Generally, fish infected with *A. salmonicida* become listless and float downstream or lie on the bottom of ponds or tanks where they respire weakly.[7,13] Few external clinical signs accompany acute furunculosis but fish exhibit darkened pigmentation of the skin, lethargic swimming, and inappetence. Discrete petechiae may occur at the base of fins. Internally the spleen may be enlarged, and septicemia develops with high mortalities occurring within 2 to 3 d.

Subacute infections develop more gradually but are associated with high morbidity, which along with mortality, slowly and consistently increases. Various clinical signs are apparent including dark pigmentation, "furuncles" or boil-like lesions on the body surface, and hemorrhage at the base of frayed fins (Figure 1). Internally the peritoneal cavity may contain a bloody, cloudy fluid and the intestine is flaccid, inflamed, and filled with a bloody fluid that exudes from a red, prolapsed anus. The kidney is generally edematous and hemorrhagic and the spleen is usually enlarged and dark red.

Chronic furunculosis results in a low-grade but consistent mortality over an extended period of time. Clinical findings described for subacute infections are also present in chronically infected moribund fish, but an inflamed intestine is often the only internal pathological manifestation.

The latent form of furunculosis occurs when the organism is dormant in a fish population, during which time fish show no clinical or internal pathological signs of disease.

Figure 1   Brown trout with furuncle (arrow), caused by *Aeromonas salmonicida,* in the muscle under the skin. These furuncles will ulcerate, producing an open lesion exposing the muscle. (Photograph courtesy of the U.S. Fish and Wildlife Service.)

3. DIAGNOSIS AND BACTERIAL CHARACTERISTICS

When overt furunculosis occurs diagnosis is relatively straight forward. Basic external and internal findings may be helpful in diagnosis, but isolation of the bacterium from diseased fish is essential for positive identification. These isolates should be taken from muscle lesions, kidney, spleen, or liver; however, Cipriano et al.[14] showed that *A. salmonicida* could be isolated from the skin mucus (a nonlethal sampling procedure) of 56% of infected lake trout compared to isolation from only 6% of the kidneys. A similar ratio was seen in mucus vs. kidney of Atlantic salmon. *A. salmonicida* will grow on tryptic soy agar (TSA), or brain heart infusion (BHI) agar, or most other general bacteriological media when incubated at less than 25°C but preferentially at 20°C. It grows poorly at 30°C and not at all at 37°C. Upon initial isolation, most colonies of *A. salmonicida* are hard and friable with some being smooth, soft, and glistening. Subcultivation of hard colonies (rough) will autoagglutinate in broth and are virulent, while the smooth colonies do not autoagglutinate and are avirulent.[15] As discussed later, the autoagglutination and virulence is associated with the presence of an additional cell surface protein called the "A-layer".[16]

Typical *A. salmonicida* grown on media with tyrosine produces a brown water-soluble pigment that diffuses throughout the medium in 2 to 4 d at incubation temperatures below 25°C. Presumptive characteristics of typical *A. salmonicida* are brown water-soluble pigment, small (0.9 × 1.3 μm) Gram-negative, nonmotile rods that ferment and oxidize glucose, and are catalase and cytochrome oxidase positive.[4] Definitive identification is achieved by additional biochemical reactions (Chapter XIII, Table 2), or serologically. Typical *A. salmonicida* is a relatively morphologically and biochemically homogeneous species with several exceptions, the most notable of which are variation in pigment production, results of the Voges-Proskauer test, and the ability to ferment selected sugars.[1,17-18] Also, some isolates of *A. salmonicida* lose their capacity to produce pigment after numerous transfers on bacteriological media while some motile *A. hydrophila* isolates will acquire the capacity to do so after numerous passages on media.

Serological procedures offer a more rapid identification than does the culture method. *A. salmonicida* can be identified serologically by fluorescent antibody, ELISA[19] using either infected tissue or the cultured bacterium, or by the serum agglutination test.[20]

It appears that there is only one serological strain of *A. salmonicida salmonicida.* Two subspecies of *A. salmonicida*, *A. salmonicida achromogens* (atypical) and *A. salmonicida masoucida,* do not produce pigment and diseases associated with these subspecies will be discussed separately. McCarthy and Rawle[21] showed that a strong serological homogeneity of typical *A. salmonicida* strains exist but that these strains will not cross-react to other aeromonads. Rockey et al.[22] used monoclonal antibody against *A. salmonicida salmonicida* lipopolysaccharide (LPS) to demonstrate the homogeneity of this antigenic property.

4. EPIZOOTIOLOGY

Since early descriptions of furunculosis, *A. salmonicida* has been considered to be an obligate pathogen[18] with a short life span outside of the host, and that was passed from fish to fish or generation to generation through carrier fish. Michel and Dubois-Darnaudpeys[23] demonstrated that the organism will survive for 6 to 9 months in pond bottom mud where it retains its pathogenicity. Allen-Austin et al.[24] reported that the organism will remain viable in fresh water at undetectable levels for over 1 month, but Rose et al.[25] showed that *A. salmonicida* survived for less than 10 d in sea water. Nomura et al.[9] maintained viable *A. salmonicida* for 60 d in sterilized fresh water but for only 4 d in nonsterile water, concluding that the organism could survive in water long enough to infect other fish. In spite of data supporting its extended survival outside of the host, *A. salmonicida* is still considered to be an obligate pathogen.

McCarthy[26] reviewed the literature on transmission of furunculosis and concluded that infection occurred by fish to fish contact, either through the skin or by ingestion. Lesions of the skin caused by injury or parasites enhance invasion of the bacterium. Cipriano[27] demonstrated infection of the disease in brook trout by contact exposure. Carrier fish which show no overt signs of disease, and from which the bacterium can not be easily isolated, may also be implicated in the transmission cycle. Nonsalmonid fish that are infected with *A. salmonicida* can serve as a reservoir of the pathogen; Ostland et al.[12] demonstrated transmission of furunculosis from *A. salmonicida*-infected common shiner *(Notropis cornutus)* to coho *(O. kisutch)* salmon and brook trout. Contaminated water and/or fish farm equipment can serve as a source and means of transmission of *A. salmonicida*. Wichardt et al.[8] pointed out that in Sweden it had been documented that fish at several trout and salmon farms became infected after release of *A. salmonicida*-infected fish in upstream water supplies, thus substantiating horizontal transmission through water.

Bullock and Stuckey[28] demonstrated that rainbow trout could carry *A. salmonicida* for up to 2 years after initial infection without reexposure. In the same study, they showed that if rainbow trout were stressed by a water temperature of 18°C and injected with an immunosuppressant (Kenolog 40 — triamcinolone acetonide), that latently infected fish could become active shedders. Following immunosuppression with Kenolog 40, 33% of the trout exhibited overt disease compared to 4% of fish stressed only with heat. There was also 73% mortality in chemically immunosuppressed fish compared to 33% mortality in solely heat-stressed fish. More recently, Nomura et al.[9] isolated *A. salmonicida* from the coelomic fluid and kidney of chum *(O. keta)* and pink and masou salmon, which showed no frank signs of disease after they entered fresh water.

Once an infected population of trout has overcome the disease, because of natural defenses or chemotherapy, at least some of the survivors will become carriers.[26] In such populations furunculosis may appear without any apparent environmental insult to the fish, or more often, it can be a stress-mediated disease. If the fish are stressed as a result of handling, elevated water temperatures, low oxygen, or other adverse environmental conditions that reduce their resistance, the carrier state can become a chronic, subacute, or acute infection. Nomura et al.[9] found that 12.4% of chum salmon held at high density ($14.7/m^2$) had an *A. salmonicida* infection and that lightly stocked fish ($4.9/m^2$) were disease-free. They also found that prevalence of *A. salmonicida* was significantly higher in fish held in water with low dissolved oxygen (6 to 7 mg/l) than those in water with 10 mg/l O_2. Survival in the high density-low oxygen water was approximately 40% less than in low density-high oxygen conditions. Kingsbury[29] also correlated outbreaks of furunculosis with depressed oxygen levels. When oxygen concentrations dropped to less than 5 mg/l during the night, losses due to *A. salmonicida* increased. Water temperatures also play an important role in furunculosis outbreaks. The optimum water temperature for development in trout is 15 to 20°C. When infected fish are held in water at 20°C the presence of *A. salmonicida* increases dramatically in the blood in 48 h, but at 10°C, 168 h are required for the bacteria to reach the same concentration. Mortality may reach 5 to 6% of a population per week and as high as 85% total mortality has been reported in untreated populations.

Vertical transmission of *A. salmonicida* has been thought to be important in the epizootiology of furunculosis. Transmission of the disease during spawning has always been considered because the organism can be readily isolated from ovaries, eggs, and ovarian fluid at spawning time.[26] Bullock and Stucky[28] were unable to effect vertical transmission in four attempts and concluded that this means of *A. salmonicida* transmission does not occur. In a recent report by Nomura et al.,[9] it was shown that *A. salmonicida* in the presence of chum salmon eggs was reduced by a log of about 10^6 colony-forming units in about 48 h. They also showed that the organism could not be isolated from eggs 5 d after beginning incubation, and therefore agreed that vertical transmission of *A. salmonicida* was unlikely.

The disease is most prevalent during late spring and summer when subacute epizootics occur, however, the disease can occur at any time of the year. Piper et al.[30] noted that twice as many furunculosis cases occurred in July than in any other month.

5. PATHOLOGICAL MANIFESTATIONS

Death of *A. salmonicida*-infected trout is generally attributed to a massive septicemia that interferes with the host's blood supply and results in massive tissue necrosis. Skin lesions are characterized by loss of scales, necrosis of epithelium and muscle, capillary dilation, and hemorrhage at the periphery of the lesion.[31] Klontz et al.[32] experimentally infected rainbow trout with *A. salmonicida* and found that the most consistent pathological aberration was an enlarged spleen that appeared 16 h after injection and persisted throughout the infection. Sinusoids of the spleen became congested and engorged with erythrocytes. Bacteria could be isolated at 8 h postinjection, but not during the ensuing 48 h. Following this latent period a systemic infection developed. At the site of injection extensive inflammation and severe muscle necrosis developed.

Hematopoietic tissue of the kidney was most severely affected by an increase in lymphoid hemoblasts, macrophages, neutrophils, and lymphocytes until, finally, hematopoietic activity of the kidney ceased. Renal tissues of the kidney are largely spared from injury. These histopathological changes were similar to those observed in naturally infected trout, especially in hematopoietic organs.

Pathogenicity of *A. salmonicida* was initially associated with colonial morphology on solid bacteriological media, where organisms from "smooth" colonies were considered pathogenic and organisms from "rough" colonies were not.[33] However, the rough colonies have been clarified. Udey and Fryer[16] found that virulence of *A. salmonicida* was associated with the presence of an A-layer (additional-layer) which was absent on nonvirulent strains. They also observed that virulent cells which possessed the A-layer autoagglutinated, and that avirulent cells lacking this protein did not. This aspect of *A. salmonicida* has also been studied by numerous investigators and they concur that the A-layer (A-protein monomer) is correlated to virulence.[34-36] The surface protein seems to protect *A. salmonicida* from the fish's natural defense mechanisms, thus giving the bacterium a pathogenic advantage.

Extracellular products (ECP), in the form of toxins and enzymes, are produced by *A. salmonicida*. The ECPs include several proteases, hemolysins, and a leukocidin.[37-39] *A. salmonicida* ECPs are capable of producing a syndrome similar to chronic furunculosis that was manifested by muscle necrosis and edema at the site of injection.[40]

It would appear that the pathological manifestation of *A. salmonicida* can be attributed to a combination of cellular and extracellular components, with the precise mechanisms yet to be fully understood.

6. SIGNIFICANCE

The severity, host susceptibility, and wide geographic range and history of furunculosis makes it a very significant disease in coldwater aquaculture. In some geographical areas where trout and other salmonids are intensively cultured, it may be the "most" serious disease of cultured salmonids, particularly in salmon farming in the U.K. and Scandanavia.[15,41] Also, the fact that *A. salmonicida* is capable of causing infections and mortality in nonsalmonid fishes contributes to its significance. The current practice of increased stocking rates in culture units, which leads to a general degradation of environmental conditions, has contributed to the impact and severity of furunculosis on salmonid populations.

7. MANAGEMENT

Although furunculosis is caused by a bacterium that is generally considered to be an obligate pathogen, disease outbreaks are often associated with unfavorable environmental conditions and questionable fish culture practices; therefore, environmental management is critical in controlling and preventing the disease.[42] Maintaining water temperatures for trout at 15°C or below is desirable, but moderate stocking densities, use of aeration to provide near saturation of oxygen, and ensuring an adequate water flow to remove metabolites, uneaten feed, and solid wastes are essential. Eggs, fry, and fingerlings should be cultured in water from a fish-free source. Management of hatcheries and farms where *A. salmonicida* is enzootic should include utilization of strains of fish, especially brook trout, that are more resistant to furunculosis.[43] Destruction of *A. salmonicida*-infected fish, sanitation of the facility, and restrictions on the use of eggs from infected brood stock have been used successfully to keep the disease limited to only 2% of the "health-controlled" farms in Sweden.[8]

A high standard of sanitation is another important approach to prevention and/or control of furunculosis and should be an intricate part of culture management, particularly in hatcheries where *A. salmonicida* is not present. Newly arrived eggs should be disinfected to kill any surface-associated bacteria. The most reliable disinfectants for this purpose are iodophores, with Betadine® and Wescodyne® being the disinfectants of choice in the U.S.[44-45] Eyed eggs are immersed in 25 to 200 mg/l of either compound for 10 to 15 min at a temperature of 10 to 15°C. Rinsing eggs after treatment or immediately putting them into flowing water incubators is recommended.

Where *A. salmonicida* is enzootic, the disease will likely manifest itself sooner or later, regardless of preventive measures. When it does appear, chemotherapy is necessary to reduce mortalities and to help fish overcome the infection. Sulfamerazine® had previously been used to treat furunculosis[42] but is no longer available because of the manufacturer's withdrawal of its label for fish application. Although oxytetracycline (Terramycin®) is an FDA-approved drug for furunculosis, many of the presently isolated *A. salmonicida* are resistant to it. When Terramycin®-sensitive isolates are encountered, it is recommended that this drug be incorporated into the diet so that fish receive 50 to 80 mg/kg of body weight daily. Terramycin® is fed 10 to 14 d and requires a 21-d withdrawal period before slaughter. The potentiated sulfonamide Romet-30®, applied at 50 mg/kg of fish daily for 5 d, is also FDA-approved for furunculosis with a 42-d withdrawal period.[46] Oxolinic acid is used extensively in European salmonid culture at 10 mg/kg body weight daily. Before chemotherapy is applied, it is strongly recommended that antibiotic sensitivity be determined for a particular isolate of *A. salmonicida* because of the potential for antibiotic resistance. Antibiotics should not be used indiscriminately for long periods of time or at less than recommended concentrations, because these practices will lead to increased drug resistance by the pathogen. To illustrate the importance of this problem, Inglis and Richards[41] reported that between 1989 and 1991 53% of the *A. salmonicida* isolates from salmon were resistant to oxytetracycline and 43% were resistant to oxolinic acid.

Hastings[47] reviewed vaccination of salmonids against furunculosis. Generally, injectable vaccines against furunculosis are effective, however application on a production scale is usually cost prohibitive. However, there are injectable *A. salmonicida* vaccines that are being used. Rodgers[48] demonstrated the effectiveness of vaccinating juvenile rainbow trout against *A. salmonicida* by immersion. A vaccine composed of whole cells and extracellular products significantly enhanced protection from a natural challenge of *A. salmonicida*. Mortality of vaccinated fish was about 11% compared to 37% for non-vaccinated controls. Vaccinated fish also grew more rapidly than did the nonvaccinates. Oral vaccination of salmonids against *A. salmonicida* has been tried by several researchers. Paterson et al.[49] described protection of Atlantic salmon using a pelletized diet with a dried and coated culture preparation in laboratory and field trials.

B. VIBRIOSIS

It has been speculated that fish, particularly of marine origin, have been infected by members of the bacterial genus *Vibrio* since the middle of the 18th century.[50] Austin and Austin[1] listed seven members of the genus that have been reported to cause infections in fish: *V. alginolyticus*, *V. anguillarum*, *V. carchariae*, *V. cholerae*, *V. damsela*, *V. ordalii*, and *V. vulnificus*; *V. parahaemolyticus* was included by Colwell and Grimes[50] and Egidius et al.[51] added *V. salmonicida*, a bacterium that causes a distinct disease in salmonids. Some *Vibrio* spp. isolates that are dissimilar to the known species also produce diseases in fish, but they are of minor importance in aquaculture and not discussed here.[52] The ensuing discussion will combine diseases that are caused by *V. anguillarum* and *V. ordalii* and which collectively are referred to as "vibriosis". Both of these organisms produce hemorrhagic septicemias but the diseases are slightly different. Cold-water vibriosis (i.e., Hitra disease) caused by *V. salmonicida* will be discussed separately. *Vibrio* spp. infections have a variety of common names in addition to vibriosis, including red pest, saltwater furunculosis, boil disease, and ulcer disease.

1. GEOGRAPHICAL RANGE AND SPECIES SUSCEPTIBILITY

Although vibriosis is considered to be a saltwater fish disease, it does occasionally occur in freshwater fish and reports of its appearance in this environment have increased over the years.[53-56] *V. anguillarum* is found worldwide and is particularly significant along the Pacific, Atlantic, and Gulf of Mexico coasts of North America; the North Sea, Atlantic and Mediterranean coasts of Europe and North Africa, and throughout Asia. *V. ordalii* occurs along the Northwest Pacific Coast of North America and in Japan.[57-59]

A variety of species of fish are susceptible to vibriosis. Anderson and Conroy[60] listed approximately 50 species of fish, mostly marine, from which *V. anguillarum* has been isolated; since that time numerous other species have been added to the list. Colwell and Grimes[50] listed 12 groups or families and 42 species of marine and freshwater fishes that are affected by *Vibrio* spp., mostly *V. anguillarum*. Nevertheless, salmon are the most notable of the susceptible species of fish in which *V. anguillarum* occurs in wild populations but more significant infections occur in cultured salmon especially coho, chinook (*O. tshawytscha*), and sockeye (*O. nerka*) salmon.[61] Essentially all salmonids can become infected, particularly those anadromous species cultured in saltwater pens or reared to smolts in fresh water and released into salt water. Some of the most severely affected nonsalmonid cultured species are striped bass (*Morone saxatilis*), European eel (*A. anguilla*) and the Japanese eel (*A. japonica*), and milkfish (*Charos charos*) grown in brackish water ponds, and ayu (*Plecoglossus altivelis*) grown in freshwater ponds. Other commonly affected families include cod, flounder, sole and turbot, mullets, sea basses, and groupers. It is unlikely that any group of marine or brackish water fish is resistant to *V. anguillarum* infections, especially in mariculture. *V. ordalii* is specific for anadromous salmon and freshwater trout.[57]

2. CLINICAL SIGNS AND FINDINGS

Clinical signs of vibriosis are similar to those of motile *Aeromonas* septicemia because both cause the same type of syndrome — a hemorrhagic septicemia (Figure 2).[1] Most vibrio-infected salmonids, as well as other species of cultured fish, exhibit diminished feeding activity and swim lethargically around the edges of ponds, raceways, or pens accompanied by intermittent erratic or spinning patterns.[62] Infected fish also display pale gills indicative of anemia, discoloration of skin, and have red (hemorrhagic), ulcerative necrotic skin and muscle lesions. These lesions may also develop on the head, within the mouth, around the vent, and at the base of fins.[61] The vent is red and swollen and exudes a bloody discharge. Abdominal distension and exophthalmia occur in many fish.

Internal gross pathological changes may or may not be striking. The body cavity often contains bloody ascitic fluid, intestines are inflamed and flaccid, and petechiae are present throughout the viscera including the liver, kidney, and adipose tissue. Evelyn[61] reported that death in Pacific salmon usually came before large necrotic muscle lesions could develop, or before massive changes occurred within the internal organs. However, Novotny and Harrell[62] noted that the spleen may sometimes be two to three times larger than normal.

3. DIAGNOSIS AND BACTERIAL CHARACTERISTICS

Diagnosis of vibriosis is accomplished by isolation of the causative organism on general laboratory media (BHI or TSA, etc.). Isolation can be enhanced by the addition of 0.5 to 3.5% NaCl to the media.[63-64] Inoculated media is incubated at 25 to 30°C for 24 to 48 h at which time round, raised, entire and shiny cream-colored colonies will be obvious.[1,50] Isolates of *V. anguillarum* can be obtained from any internal organ, especially those that have a copious blood supply because the blood is heavily laden with bacteria.[64-65] *V. anguillarum* is a Gram-negative, slightly curved rod which measures about 0.5 × 1.5 μm, is motile, cytochrome oxidase positive, ferments carbohydrates without gas production, and is sensitive to novobiocin and the vibriostat 0/129 (2,4-diamino-6,7-disopropyl pteridine phosphate) (Table 1).

Biochemically *V. anguillarum* is a heterogeneous organism. At least five different biotypes of *V. anguillarum* have been described.[1,66-67] The species is further fractionated by serological diversity. Initially, three serotypes were identified from salmonids in northwestern U.S. and Europe[68] and other researchers extended the serological groupings to as many as six serotypes. Clearly, *V. anguillarum* is a diverse species that may reflect its wide geographical distribution and the multiple species susceptibility. However, for our discussion two biogroups (Biogroup I and II) as proposed by Hastein and Smith[69] will suffice, because most infections in salmon are caused by members of Biogroup I. A third serologically and biochemically distinct biogroup (III) is considered to include *V. ordalii*.

Isolation of *V. ordalii* is enhanced greatly by utilization of seawater agar with up to 3% NaCl and incubated at 15 to 25°C for 7 d. *Vibrio ordalii* is not morphologically dissimilar from *V. anguillarum*.[58] Colonies are off-white, circular, and convex and measure about 1 to 2 mm in diameter. Although *V. ordalii* and *V. anguillarum* share numerous biophysical and biochemical characteristics, as well as some serological cross-reactivity,[70] there are distinct taxonomical differences between these two pathogens and *V. salmonicida* (Table 1).

4. EPIZOOTIOLOGY

According to West and Lee,[71] *V. anguillarum* is a normal marine microbe that may be found free-living or in fish. Isolates from fish may possess low to high virulence, and the species is quite varied in its habitat requirements. It is not known if different strains are obligate pathogens while others are

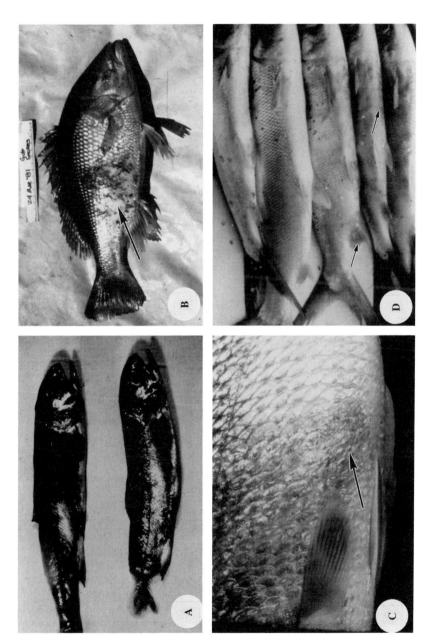

Figure 2 Clinical signs of Vibriosis. (A) Pacific salmon with epidermal lesion; (B) red grouper with hemorrhagic lesion (arrow) on the skin; (C) red drum with diffused hemorrhage (arrow) in epithelium; (D) milkfish with hemorrhagic lesions (arrow) in the scale pockets and epidermis. (Photograph A courtesy of J. Rohovec; photographs B and C courtesy of J. Hawke; source of photograph D unknown.)

Table 1   **Biochemical comparison of *Vibrio anguillarum*,
Vibrio ordalii, and *Vibrio salmonicida*[50,58,95,102]**

| | Reaction | | |
Test	*Vibrio anguillarum*	*Vibrio ordalii*	*Vibrio salmonicida*
Growth on			
0% NaCl	±	−	−
3% NaCl	+	+	+
Growth at 37°C	+	−	−
Voges-Proskauer	+	−	−
Arginine	+	−	−
Christiensen citrate	+	−	−
Starch hydrolysis	+	−	−
Gelatin hydrolysis	+	+	−
ONPG	+	−	−
Lipase	+	−	−
Indole	+	−	−
Glycogen	+	−	−
Nitrate reduction	+	(+)	−
Acid from			
Cellobiose	+	−	(+)
Fructose	+	−	+
Galactose	+	−	−
Glycerol	+	−	(+)
Glucose	+	+	+
Sorbitol	+	−	−
Trehalose	+	−	+
Maltose	+	+	−
Sucrose	+	+	−
Gentiobiose	−	−	+
Gluconate	+	−	+
D-Mannose	+	−	+

Note: + = positive, (+) = weakly positive, − = negative.

facultative, but Larsen[72] showed that environmental strains of *V. anguillarum* were less pathogenic than reference strains from fish. *V. anguillarum* has its greatest impact on cultured fish, often during periods of environmentally induced stress. The disease is more prevalent during the summer when water temperatures are high and dissolved oxygen is low. Overcrowding or poor management hygiene will contribute to epizootics. However, some highly virulent isolates can induce disease without an exogenous stressor when fish are exposed to a low number of bacteria.

V. ordalii is not as widespread as the ubiquitous *V. anguillarum*. The former has only been isolated from fish and has not been found in water or bottom sediments. This organism also preferentially infects skeletal muscle, heart, gills, and intestinal tract of salmonids, and does not produce as extensive a septicemia as *V. anguillarum*.[57]

V. anguillarum and *V. ordalii* are capable of infecting and killing large numbers of cultured fishes. Up to 100% mortality has been reported in Atlantic salmon[73] after experimental exposure to *V. anguillarum*. In contrast, Thorburn[74] noted mortalities ranging from 0 to 17% in rainbow trout with natural infections on various Swedish farms. These mortalities were correlated to transport of fish, size of farm, and the number of years the farm had been in production. Less severe effects occurred at older and larger farms, thus suggesting that experience in management had a positive influence on the outcome of the infection.

Increased water temperature has a positive effect on mortality of salmonids infected with *V. anguillarum*. For example, coho salmon suffered 58 to 60% mortality when held at 18 to 20°C, 40% at 15°C, 28% at 12°C, and only 4% at 6°C.[75]

While salmon smolts are in fresh water they have little if any exposure to *V. anguillarum* and when they reach salt water they are highly vulnerable to infection. In a vaccination experiment, Harrell[76] found that over 90% of unvaccinated 0-age sockeye salmon were killed by *V. anguillarum* during their first 50 d in saltwater pens, and mortalities of 60 to 90% are common among young salmon when stocked into saltwater net pens. In Norway, mortality of Atlantic salmon that were transferred from fresh water to salt water was influenced by the time of transfer; fish moved late in the growing season suffered higher mortality than those moved early.[74]

Epizootilogically, salmonids have been the focus of the vast majority of research on vibriosis. However, wild and cultured flounder, eels, and milkfish are also highly susceptible to the pathogen and may develop acute or chronic vibriosis. In winter flounder *(Pseudopleuronectus americanus)*, the acute form occurs year-round with onset of clinical signs appearing 12 to 24 h after inoculation and a disease duration of 2 to 4 d.[77] Gross clinical signs are not as evident as in the chronic form of disease which occurred in the summer. Onset of chronic vibriosis appears 1 to 4 d after exposure and has a duration of 2 to 6 weeks, after which time some fish recover. Gross clinical signs are more evident than in the acute form and consist of petechiae, ecchymosis, hematomas, ulcerations, and erosion of the fins.

In Japan, juvenile ayu are susceptible to *V. anguillarum* where the problem is more severe in sea run fish. Incidence of infection can be as high as 17% in ayu from the sea compared to less than 1% after stocking into fresh water.[78] In addition, ayu from saltwater environs possess higher numbers of bacteria in their tissues. A similar pattern is true for cultured eels in Japan, where *V. anguillarum* is a greater problem in brackish water ponds than strictly freshwater ponds. In both instances, it is concluded that *V. anguillarum* is introduced into the freshwater environs with fish that are captured in salt water and transferred to fresh water.

5. PATHOLOGICAL MANIFESTATIONS

Pathological manifestations of vibriosis in salmonids is similar whether fish are infected by injection or naturally exposed.[65] A bacteremia occurs in the early stages of *V. anguillarum* infection with bacterial cells most abundant in the blood, which otherwise are uniformly dispersed throughout affected tissue. Shortly after infection the pathogen is sequestered in the spleen, where bacterial numbers increase and then proliferate to the kidney. Death of the fish results from a septicemia accompanied by extensive necrosis in the kidney, spleen, posterior intestine, and liver. The gills become congested with necrotic epithelium. Harbell et al.[79] showed that *V. anguillarum*-infected salmon suffered a loss of electrolytes and low plasma osmolarity, which is indicative of kidney dysfunction. They also found anemia, leukopenia, and cell necrosis. Miyazaki and Kubota[80] found similar histopathological responses in eels injected with *V. anguillarum*.

It has been proposed that *V. ordalii* infection is initiated in the rectum and posterior gastrointestinal tract or possibly by invasion of the integument.[81] The pathology of *V. ordalii* differs to some degree from that of *V. anguillarum*. *V. ordalii* is not dispersed evenly in the various organs and tissues but produces microcolonies in the skeletal and heart muscle, gill tissue, and throughout the gastrointestinal tract.[65] There is a lower number of *V. ordalii* cells in the blood than is the case with *V. anguillarum* infections, possibly because of the late development of a bacteremia in the disease cycle.

Pathogenesis of *V. anguillarum* was reported to be facilitated by an extracellular toxin that is heat stable at 100°C for 15 min.[82] Live and heat-killed bacteria and heat-treated cell-free culture fluid killed 53 to 80% of injected goldfish. Conversely, however, no evidence was found by Harbell et al.[79] that endotoxin, culture supernatant, or cell lysate caused pathology or death when injected into coho salmon.

6. SIGNIFICANCE

The fact that vibriosis occurs worldwide and affects a wide variety of fish species makes it a significant aquaculture disease, especially in marine and brackish waters. Vibriosis has been an important disease of eels in Europe and Asia for decades and continues to be today. In general, vibriosis is the major bacterial disease of marine species and Novotny and Harrell[62] stated that vibriosis is the most common, and probably the most important disease of cultured marine fishes, particularly salmonids, in the U.S. Likewise, Muroga[54] and Kitao et al.[55] stated that vibriosis was the most important infectious disease of cultured ayu in Japan. Salmon smolts are particularly vulnerable as they move from a freshwater to a marine environment, often incurring losses up to 90%. Any fish species cultured in salt or brackish waters is highly vulnerable to vibriosis.

7. MANAGEMENT

Vibriosis is often associated with environmental stress brought on by crowding, increased water temperatures, depressed oxygen concentrations, elevated metabolites or metallic ions, rapid acclimation of fish from fresh water to sea water, or other unidentified factors. In view of this, good fish husbandry practices and proper hygiene are very important in managing the disease.

The treatment of clinical vibriosis with antimicrobials incorporated into feed has been successful as long as the treatment is begun early in an epizootic while affected fish are still feeding.[1,83] The most commonly used antibiotics for treatment of vibriosis are oxytetracycline (Terramycin®) at 50 to 75 mg/kg of body weight daily for 10 to 14 d, sulfamerazine at 200 mg/kg of body weight daily for 10 d, and Romet® (potentiated sulfonamide) at 50 mg/kg body weight daily for 5 d. A 21-d withdrawal period is required for Terramycin® and sulfamerazine, while Romet® requires a 42-d withdrawal. These withdrawal periods may vary among countries where the drugs are used. Additional drugs used to treat vibriosis in parts of the world other than the U.S. include chloramphenicol, nitrofuran compounds, Halquinol, and oxolinic acid.

Vaccination against vibriosis, particularly *V. anguillarum* and *V. ordalii,* has become a widely accepted practice for cultured fish. Initial vaccination was accomplished by injecting a formalin-killed bacterin into salmon smolts prior to their transfer to sea water.[84] However, vaccination by immersion and/or spraying has proved to be a more efficient and effective method of immunization on a large scale.[85] Bivalent vibrio vaccines containing *V. anguillarum* and *V. ordalii* are commercially available. Numerous studies have demonstrated the positive effects of vaccinating salmon and other species such as ayu, eels, milkfish, and striped bass.[86-89] By vaccinating salmon at appropriate times survival can be improved as much as 90%. Coho salmon mortalities due to vibriosis were reduced from 52% in unvaccinated controls to 4% in fish vaccinated by 20 to 30 s immersion and to 1% in spray-vaccinated fish.[85] Horne et al.[90] reduced mortality from 100% in unvaccinated rainbow trout to 53% by immersion vaccination. In fish that are to be transferred to salt water, vaccination should be carried out 2 to 4 weeks prior to release. Salmonids can be vaccinated at any size over 2 g but most efficiently up to 10 g.

Evidence exists that vaccination of salmonids not only significantly reduces mortality due to *V. anguillarum* and *V. ordalii,* but the vaccinated fish often have an increased growth rate and a lower feed conversion.[91] It was found by Thornburn et al.[92] that in Swedish marine net-pen farms the decision to vaccinate or not vaccinate depended upon the size of the farm and the anticipated risk of vibriosis, but overall, one might conclude that when fish are to be exposed to an environment where *V. anguillarum* or *V. ordalii* are indigenous, vaccination is economical and management-wise.

C. COLD-WATER VIBRIOSIS

Cold-water vibriosis (CV), also known as "Hitra disease" and hemorrhagic syndrome, is a disease primarily of the Norwegian Atlantic salmon farming industry.[93] The disease was first described by Egidius et al.[94] on the island of Hitra, and since that time has been reported annually throughout Norway and elsewhere. The causative agent of CV was described by Holm et al.[95] and Egidius et al.[96] who proposed a new species, *Vibrio salmonicida.* The etiological agent of Hitra disease was disputed by Fjølstad and Heyeroaas[97] and Poppe et al.,[98-99] who hypothesized that it is a multifactorial disease that could include a nutritional deficiency (vitamin E), and environmental stress. However, *V. salmonicida* is generally accepted as the etiological agent of Hitra disease (cold-water vibriosis).[100-101]

1. GEOGRAPHICAL RANGE AND SPECIES SUSCEPTIBILITY

Cold-water vibriosis has been reported from along the coast of Norway and in cultured Atlantic salmon on the Shetland Islands of northern Scotland, and the Faroe Islands.[101-102] To date, the disease has not been reported from other parts of Europe, but cold-water vibriosis was detected by Mitchel[103] in eastern Canada and the northeastern U.S. in 1989 and again in 1993.

Atlantic salmon is the most susceptible species of fish to CV, while rainbow trout can be infected but are far less susceptible.[104] The disease is found primarily in saltwater or brackish water net-pen populations in Norway and *V. salmonicida* occurred under similar circumstances in Atlantic salmon in the U.S. *V. salmonicida* was also isolated from highly stressed juvenile cod *(Gadus morhua)* that had been captured in the wild and stocked into net-pens in close proximity to infected Atlantic salmon.[105] Schroder et al.[106] showed that this incidence was unusual and cod should not be considered a likely host.

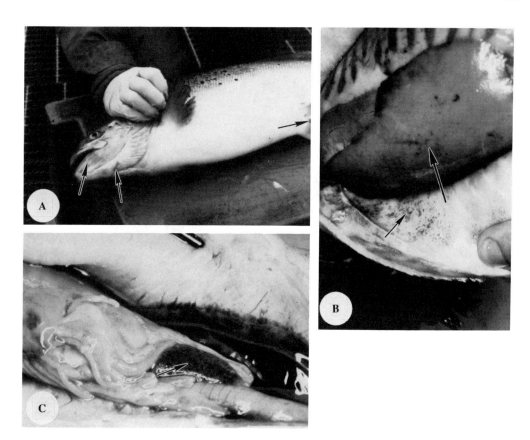

Figure 3   Atlantic salmon infected with *Vibrio salmonicida*. (A) Hemorrhage (arrow) on the gill and skin; (B) petechia (large arrow) on pale liver and peritoneum (small arrow); (C) granular-appearing spleen. (Photograph A courtesy of H. Mitchell; photographs B and C courtesy of B. Hjeltness.)

2. CLINICAL SIGNS AND FINDINGS

Initial clinical signs of CV are fish swimming on their sides near the surface of pens.[94,100] Infected fish, which generally appear well nourished, may or may not exhibit extensive hemorrhage on the skin depending upon the stage of the disease. When hemorrhages do appear, they are at the base of fins and on the abdominal region (Figure 3). Gills are generally pale, but the gill cavity may be hemorrhaged. The anus may be reddish and prolapsed. Internally, a bloody fluid accumulates in the peritoneal cavity with hemorrhages on tissue surfaces ranging from petechiae to ecchymosis (Figure 3). The spleen is grayish; the liver is usually yellowish and anemic; and hemorrhages occur on the air bladder, in fatty tissue around the pyloric cecae, and throughout the visceral cavity. The gut, particularly the posterior region, is hemorrhagic and the lumen contains a watery, bloody fluid. Occasionally, hemorrhages are found in the muscle.

3. DIAGNOSIS AND BACTERIAL CHARACTERISTICS

V. salmonicida is not easily isolated on culture media but it can be recovered from internal organs of infected fish where high numbers of bacteria are present in a septicemia condition (Figure 4). Initially, *V. salmonicida* was isolated on nutrient agar with 5% human blood and 1.5 to 2% NaCl,[104] but either TSA or BHI are suitable media for isolation as long as they contain salt. The bacterium prefers 0.5 to 4% NaCl, grows optimally at 15 to 17°C, and will grow at 1 to 22°C but not at 26°C. At 15°C slight growth will appear in 24 h with more pronounced growth at 72 h. Colonies are smooth, grayish, opaque, slightly raised with entire margins, and range in size from punctate to 1 or 2 mm in diameter. The organism is not hemolytic to human erythrocytes but is Gram-negative, cytochrome oxidase positive, motile, and sensitive to vibriostat O/129. *V. salmonicida* is a slightly curved rod that measures 0.5 to

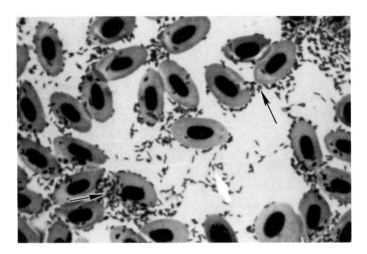

Figure 4 Blood smear of Atlantic salmon infected with cold-water vibriosis and a high number of *V. salmonicida* (arrows) in the blood. (Photograph courtesy of B. Hjeltness.)

2 or 3 μm and may be pleomorphic upon initial isolation.[95-96] Biophysical and biochemical comparisons of *V. anguillarum*, *V. ordalii*, and *V. salmonicida* are included in Table 1.

Espelid et al.[107] utilized a monoclonal antibody against the surface antigen of *V. salmonicida* to identify the organism by ELISA. All isolates possess a common outer membrane antigen designated as VS-P1, which is a protein-lipopolysaccharide complex. Immunohistochemistry, based on the avidin-biotin complex, was used by Evensen et al.[108] to identify *V. salmonicida* in formalin-fixed and histologically prepared tissues from infected Atlantic salmon. This method was also used to successfully identify the organism in tissue that had been preserved from the original Hitra disease episode in 1977.

Holm et al.[95] determined that *V. salmonicida* is a serologically uniform species that is distinct from *V. anguillarum* and *V. ordalii*. Schroder et al.[106] found some serological diversity among *V. salmonicida* isolates from Atlantic salmon and Atlantic cod, but Jørgensen et al.[105] determined by DNA analysis that the isolates from both species of fish were identical. Also, Wiik et al.[109] found 4 different plasmid profiles among 32 isolates which provide evidence of some possible species diversity. In spite of the possible serological and plasmid profile differences, all strains are biochemically similar.

4. EPIZOOTIOLOGY

Although there has been some disagreement as to the etiology of Hitra disease, it has been shown that this disease is synonymous to cold-water vibriosis and *V. salmonicida* infection. Fjølstad and Heyeroaas[97] and Poppe et al.[98-99] viewed the pathology of Hitra disease as similar to the pathology of vitamin E deficiency in higher vertebrates and, coupled with their inability to consistently isolate a pathogenic organism, proposed a nutritional etiology. Observations by Poppe et al.[98] cannot be totally ignored but most scientists working with *V. salmonicida* concur that cold-water vibriosis and Hitra disease are one and the same. Hjeltnes et al.[110] and Hjeltnes and Julshamn[111] clearly correlated Hitra disease and cold-water vibriosis to *Vibrio salmonicida* infection; however, they recognized that environmental conditions and nutrition may play an important role in the disease.

Carrier Atlantic salmon are the most likely source of infection for farm-raised fish, however, the organism survives in sea water and sediments for over a year. Cold-water vibriosis can be transmitted to noninfected Atlantic salmon and, to a mild degree, to rainbow trout. Hjeltnes et al.[110] fulfilled Koch's postulates with *V. salmonicida* and showed that it can be transmitted through the water from infected fish. Superficial injury enhances infectivity, but the pathogen primarily enters the fish through the gills. Atlantic salmon infected via the gills began showing behavioral and clinical signs after 1 week and started to die 2 weeks after exposure.

Mortality of *V. salmonicida*-infected fish can be high. In natural infections acute mortality rates of 5%/d have been noted. Injured fish that were exposed to waterborne bacteria suffered 80 to nearly 100% mortality in about 20 d. Fish challenged more naturally by cohabitation suffered from 24 to 46%

mortality in 24 and 35 d, respectively, and experimental waterborne exposure of the more susceptible Atlantic salmon resulted in 90% mortality over a 45-d period compared to 40% mortality in similarly infected cod.[106] Rainbow trout, infected by injection with very high numbers of bacteria, showed first signs of disease 9 d postinfection and suffered 80 to 100% mortality after 14 d when held at 9°C.[94] These results may be misleading because of the high number of bacteria injected. Generally, cod are not susceptible to *V. salmonicida,* but highly stressed juvenile cod that were naturally infected with *V. salmonicida* suffered 10 to 90% (50% average) mortality, apparently under unusual circumstances, among 15 farms.[105] The disease affects fish that range from yearlings to market size.

The majority of *V. salmonicida* infections have occurred during autumn and winter when water temperatures range between 4 and 9°C,[94] however Mitchell[103] noted 0.5% mortality per day in some cages in Maine when the water temperature was 1°C. Between the years of 1977 and 1981, the maximum number of cases of CV diagnosed in 1 year at the National Veterinary Institute, Oslo, was 11, but in 1982 the number increased to over 70% of their cases and during the ensuing years caused increasing losses. In recent years more cases have been reported during the warmer months.[112] This pattern has also been noted in North American epizootics. CV severity increased during May and June when water temperature rose to 5 to 6°C.

5. PATHOLOGICAL MANIFESTATIONS

Bruno et al.[100] found the histopathology of CV in experimentally infected Atlantic salmon to be similar to that of naturally infected fish. The most significant pathology occurred in the pyloric cecae and mid- and hind-gut where necrotic tissue was sloughed into the lumen. Blood vessels were vasodilated and congested, hemorrhage was present in the lamina propria, and in some cases the entire mucosal epithelium was necrotic. Focal necrosis occurred in the hematopoietic tissue of the kidney where nuclei of glomerular cells were often swollen and occasionally necrotic. The ellipsoid system of the spleen was engorged with macrophage-like cells; focal necrosis was seen in the reticuloendothelial cells; and gills exhibited epithelial necrosis and sloughing, but the liver and pancreas appeared normal.

Totland et al.[113] found that *V. salmonicida* caused the most severe pathological damage in organs and tissues with rich blood supplies and the blood heavily laden with bacteria (Figure 4). Initial histopathology was detected in the cell membrane on the luminal side of endothelial cells of the capillaries and intracellular bacteria were later seen within the endothelial cells. Leukocytes contained bacteria that caused severe damage to the host cell. The histopathology substantiates the causative relationship of *V. salmonicida* in cold-water vibriosis.

The pathogenesis of cold-water vibriosis is not fully understood because knowledge of the factors responsible for its pathogenicity is limited, but pathogenicity or virulence does not appear to be a plasmid-mediated iron sequestering system nor due to specific plasmid profiles.[109]

6. SIGNIFICANCE

On Norwegian fish farms, 80% of lost revenue from disease could be attributed to Hitra disease during the 1980s.[98] CV-related losses of farmed Atlantic salmon in Norway from 1980 to 1987 have been huge and economically devastating; however, since 1989, mortalities due to Hitra disease have decreased.[112] The fact that the disease is no longer confined to Norway and that it not only affects Atlantic salmon, but also to a much lesser extent rainbow trout and cod, adds to its significance.

7. MANAGEMENT

Because of differing opinions concerning the etiological agent of cold-water vibriosis, it is prudent for aquaculturists to practice good fish culture including maintenance of a healthy environment, avoidance of stress on the fish, and a high degree of hygiene. Being a multifactorial disease, severity of CV depends to a large extent on fish culture practices and environmental conditions.[110]

Practical steps to minimize the effects of cold-water vibriosis were suggested by Mitchell.[103] Early detection of the disease is essential, and once its presence is confirmed, handling of the fish should be avoided. After fish that have survived an epizootic are removed from pens, the nets should be cleaned and sanitized by drying in sunlight. Sick, moribund, and dead fish should be promptly removed and disposed of by burial, incineration, or composting. Personal clothing and equipment coming in contact with diseased fish should be disinfected, and effluent from slaughter areas should be disinfected so that it will not introduce large numbers of bacteria back into the culture pen areas.

When cold-water vibriosis occurs, Tribrissen, Terramycin®, or furazolidone have been used in the feed at 75 to 100 mg/kg of body weight daily for 10 d. Poppe et al.[98] reported that furazolidone was the most effective antibiotic against CV, however, Bruno et al.[100] held cumulative mortality to 3% using Terramycin®. Mortality due to *V. salmonicida*-infected Atlantic salmon may be reduced in 3 to 4 d by feeding oxolinic acid, the treatment of choice, at 75 to 90 mg/kg of body weight daily for 10 d. Because of the variety of antibiotics that have been used to control CV, multiple antibiotic resistance of *V. salmonicida* to oxytetracycline, sulfonamides, and oxolinic acid has been detected.[110]

Vaccination of Atlantic salmon in formalin-killed whole-cell bacterins has shown promise and has become the primary preventive measure for cold-water vibriosis.[106,114-116] Immersion vaccination of Atlantic salmon prior to stocking into net-pens in the sea reduced mortality due to *V. salmonicida* from 7.8 to 0.4%.[114] Lillehaug et al.[116] vaccinated Atlantic salmon on Norwegian fish farms and reduced the mortality due to CV from 24.9% in the nonvaccinated fish to 1.87% in the vaccinated groups. Hjeltnes et al.[117] showed that vaccination by injection provided the most dependable protection against *V. salmonicida* but double immersion is probably the most practical and economical method. Schroder et al.[106] pointed out that vaccination of Atlantic salmon against *V. salmonicida* is a practical and beneficial management tool, however, protective immunization breaks down in 1.5 to 2 years.[118]

D. ENTERIC REDMOUTH

Enteric redmouth (ERM) is a systemic bacterial disease of trout that is caused by *Yersinia ruckeri*,[119] a member of the family Enterobacteriaceae. ERM is a chronic to acute infection that primarily affects farm-raised trout. The disease is also known as Hagerman redmouth disease, redmouth, salmonid blood spot, and "yersiniosis" in Norway.

1. GEOGRAPHICAL RANGE AND SPECIES SUSCEPTIBILITY

ERM was initially diagnosed in cultured rainbow trout in the Hagerman Valley of Idaho, in the early 1950s and has since been found in most states where trout are grown. Isolates that were obtained from rainbow trout and brook trout in West Virginia in 1952, and from rainbow trout in Australia in 1963, appeared to be identical to *Y. ruckeri*.[120] In view of this, it is probable that the detection of *Y. ruckeri* in various areas during the past 30 years may not have been totally because of its dissemination from Idaho but was endemic in at least some of those areas. Confirmation of the presence of *Y. ruckeri* has come forth from many other countries including Australia, Bulgaria, Canada, Denmark, Finland, France, Germany, Great Britain, Italy, Norway, Scotland, South Africa, and Switzerland.[1,121-122] However, to date it has not been reported in Japan's extensive trout and salmon farms.

Rainbow trout is the species most severely affected by ERM, however, all salmonids can be affected. Atlantic salmon in Norwegian farms experience serious mortalities.[122,124] Eight species of salmonids and seven nonsalmonids were identified by Stevenson et al.[122] in which *Y. ruckeri* has caused disease. Nonsalmonid species include emerald shiner *(Notemigonus atherinoides)*, fathead minnow, three species of the genus *Coregonus* (ciscos and whitefish), sturgeon, and turbot. Additionally, the bacterium has been isolated from numerous apparently healthy fish species, invertebrates, mammals (muskrat), sea gulls, a human clinical specimen, sewage, and river water.[122,125-127] It is quite possible that *Y. ruckeri* is a more common bacteria in the environment than once thought.

2. CLINICAL SIGNS AND FINDINGS

ERM is categorized into acute, chronic, and latent phases and clinical signs vary within each phase. In the acute phase, fish often exhibit red mouths, heads, and jaws as a result of subcutaneous hemorrhaging.[119,125,128] Acutely affected fish are dark, have inflammation and/or hemorrhaging at the base of fins and around the vent, and exhibit unilateral or bilateral exophthalmia with orbital hemorrhage; gills will appear hemorrhaged near tips of the filaments and fish are anorexic and sluggish with a tendency to accumulate in downstream regions of ponds or raceways. However, Frerichs et al.[129] described *Y. ruckeri* infections in Atlantic salmon in which there was no reddening of the mouth and opercula. In view of this, *Y. ruckeri* should not be ruled out if the classic red mouth is absent.[130]

Fuhrmann et al.[131] indicated that *Y. ruckeri* often produced internal pathological conditions that resembled "furunculosis". Petechiae may be present in muscles, visceral fat, intestines, surface of the liver, pancreas, pyloric cecae, and swim bladder[132] (Figure 5). The lower intestine is flaccid, hemorrhaged, and inflamed and contains a thick, opaque, purulent material. The kidney and spleen may be enlarged.[119]

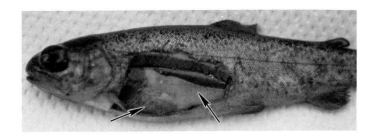

Figure 5 Rainbow trout infected with *Yersinia ruckeri*. The fish has exophthalmia, and petechial hemorrhage (arrows) in the pyloric cecae area and in visceral adipose tissue. (Source of photograph unknown.)

As the disease progresses into the chronic phase, exophthalmia increases (bilaterally or unilaterally), often rupturing the eye. Fins become frayed and eroded, gills are pale, and the abdomen is distended with an accumulation of bloody, ascitic fluid in the body cavity. Fish darken and become lethargic and emaciated before death. At the disappearance of these clinical signs, mortality abates and survivors move into a latent phase of infection.[128]

3. DIAGNOSIS AND BACTERIAL CHARACTERISTICS

Enteric redmouth is diagnosed by isolation of *Y. ruckeri* from internal organs on general purpose bacteriological media (BHI or TSA).[119] The organism grows aerobically and anaerobically at temperatures of 9 to 37°C, with an optimum range of 22 to 25°C at which colonies form in 24 to 48 h. Isolates that grow at 37°C are generally avirulent to trout. The translucent, white to cream colored colonies are 1 to 2 mm in diameter, smooth, slightly convex, raised, round, and with entire edges. Waltman and Shotts[133] described a selective media (Shotts-Waltman: SW) for *Y. ruckeri* on which it forms a green colony surrounded by a zone of hydrolysis. Hastings and Bruno[134] evaluated the SW selective media and stated that it alone was not reliable in isolating and identifying *Y. ruckeri* because it was not sufficiently specific. Rodgers[135] also described a selective *Y. ruckeri* media (ROD) which differentiates the organism by its yellow color. These selective media are particularly helpful in isolation of *Y. ruckeri* from carrier fish, especially when using fecal or kidney material.

Once *Y. ruckeri* is isolated it can be identified by conventional biochemical characteristics.[119,136-137] It is a motile (peritrichous flagellation), Gram-negative, cytochrome oxidase negative rod that measures 1.0×1.0 to 3 μm. The following biochemical reactions, which are fairly homogeneous for the species, are the most informative: positive for fermentative metabolism, production of catalase and β-galactosidase, lysine and ornithine decarboxylation, methyl red test, nitrite reduction, and degradation of gelatin (Table 2). It will grow on media containing up to 3% NaCl and utilizes citrate. *Y. ruckeri*'s inability to produce H_2S, indole, oxidase, and phenylalanine deaminase, and its negative reaction for Voges-Proskauer are significant features. Some strains can vary in the methyl red, Voges-Proskauer, lysine decarboxylase, arginine dihydrolase, and lactose fermentation tests.[119,125,132] Fermentation of sorbitol has received some attention in discriminating between pathogenic strains (Serotype I) and nonpathogenic strains (Serotype II). Serotype I does not ferment sorbitol but Serotype II does, consequently Cipriano and Pyle[138] developed a sorbitol-based medium that can be used to distinguish between them. However, Valtonen et al.[139] questioned the sorbitol reaction as being indicative of pathogenicity for Norwegian isolates. Austin and Austin[1] stated that *Y. ruckeri* could be confused with *Hafnia alvei* if the API 20 system is used for identification; however, de Grandis et al.[140] showed that *H. alvei* is L-arabinose and L-rhamnose positive while *Y. ruckeri* is uniformly negative for these characteristics.

Y. ruckeri can be positively identified by IFAT[121] but one must be aware of the potential cross-reactivity with other enteric bacteria such as *H. alvei*.[141] Serological cross-reaction does not occur with other fish pathogens (i.e., *A. salmonicida, A. hydrophila, V. anguillarum,* or *R. salmoninarum*) that one might encounter in salmonids.[142]

Based on agglutination patterns *Y. ruckeri* was originally separated into two serotypes; Serotype I, the Hagerman strain which is pathogenic and is most commonly encountered in clinical diagnosis of diseased fish,[119,125] and Serotype II, the Oregon strain which is normally not highly pathogenic;[120] however, Cipriano et al.[143] found this serotype to be pathogenic to fall chinook salmon. Currently, six

Table 2   **Biochemical characteristics of *Yersinia ruckeri*[121]**

Characteristic	% Positive	Characteristic	% Positive
Motility	82	Growth on MacConkey	99
Fermentation	100	Nitrite reduction	99
Oxidase	0	Gas from glucose	8
Catalase	100	Acid from	
β-Galactosidase	100	Amygdalin	0
Ornithine decarboxylase	100	Arabinose	0
Lysine	100	Fructose	100
Arginine dihydrolase	0	Galactose	99
Tryptophan deaminase	0	Glucose	100
Urease	0	Inositol	0
H_2S	0	Lactose	0
Indole	0	Maltose	99
Citrate utilization	99	Mannitol	99
Methyl red	91	Mannose	100
Voges-Proskauer	93	Melibiose	0
Gelatin hydrolysis	77	Rhamnose	0
Casein hydrolysis	74	Sorbitol	20
Tween® 20, 80,	82	Sucrose	0
hydrolysis		Trehalose	100

whole-cell serotypes of *Y. ruckeri* are recognized (Table 3). Stevenson and Daly[137] and Stevenson and Airdrie[144] divided the species into five serological groups (I and I' through IV). They concluded that the Australian isolate[145] should be included in Serotype I (I') and that two additional serotypes (V and VI) should be established. Daly et al.[146] detected five serotypes in Canada, most of which were in the serotype I and II classification.

4. EPIZOOTIOLOGY

Y. ruckeri is generally considered to be an obligate pathogen, but ERM is greatly influenced by the aquaculture environment because outbreaks are often closely associated with stress,[125,147] and the importance of stressful conditions to ERM epizootics can not be overemphasized. Hunter et al.[148] noted that asymptomatic *Y. ruckeri* carrier steelhead trout *(O. mykiss)* did not transmit the disease to noninfected fish unless the carriers were stressed. When the fish were stressed at 25°C the pathogen was transmitted to noninfected fish but no deaths resulted in the recipients. Disease outbreaks may also be precipitated by handling of apparently healthy trout and/or an increase in the ammonia and metabolic waste content in the water, which decreases oxygen levels.[124,135]

For all practical purposes infected trout and salmon are reservoirs of *Y. ruckeri*. Busch and Lingg[128] reported that within 45 d of rainbow trout surviving an ERM epizootic, *Y. ruckeri* became established in the lower intestine of 25% of the fish, thus becoming asymptomatic carriers. At 14.5°C a recurrent infection of ERM developed in these fish. It was further demonstrated that *Y. ruckeri* carrier trout shed the organism in higher numbers on 36- to 40-d cycles, which preceded clinical signs and mortality by 3 to 5 d. The periodicity of the shedding and ensuing infection is influenced by water temperature and other environmental factors and the full implication of the cyclic phenomenon is not fully understood.

Other research indicates that the incubation time between exposure to *Y. ruckeri* and clinical ERM will vary inversely with water temperature, overall health of the fish, and presence of additional stressors.[119,125] Generally, incubation time is 5 to 7 d between exposure and first deaths at 15°C; however, if fish have a history of prior exposure, the time may be reduced to 3 to 5 d before the first mortality. Onset of mortalities may also depend upon fish size and level of inoculum. Deaths may continue for 30 to 60 d.

ERM outbreaks usually occur only after fish have been exposed to large numbers of pathogens.[119,125] Rodgers[135] found that ERM occurs primarily between April and September, when fish range from 5 to 200 g each. Larger fish may also be affected, but incidence of infection coincides with a rise and fall

Table 3 **Serological groupings of *Yersinia ruckeri***

Serological group (serovar)	Designation	% of Isolates
I	Hagerman	59
I'	Salmonid blood spot	6
II	Oregon A	15
	Oregon B	8
III	Australian	6
IV	(Excluded)	
V	Colorado	3
VI	Ontario	3

in water temperature, overcrowding, poor water quality, and handling. Exposure to sublethal concentrations of copper (7 and 10 µg/l for 96 h) will also significantly increase susceptibility of steelhead trout to *Y. ruckeri*.[149]

Using a *Y. ruckeri* selective media (ROD), Rodgers[137] detected a very low number of cells in the intestines of carrier rainbow trout 6 to 8 weeks before kidney infections occurred. Also, significantly higher numbers of bacteria were present in the feces and the kidney of fish on a farm where the fish were more severely stressed by frequent handling and poor water exchange. Horizontal transmission occurs primarily through water from shedder fish to uninfected fish. Vertical transmission of *Y. ruckeri* has not been proven but isolation of the organism can be made from brood stock at time of spawning.

The effects of *Y. ruckeri* on a population of trout depend upon the age and size of fish, water temperature, stress level, and relative susceptibility.[132] Naturally infected hatchery salmonid populations suffer 30 to 70% mortality during the initial acute phase of disease and recurrent infections in survivor populations result in low, chronic mortality over an extended period of time. In laboratory infectivity studies, the mortality results depend upon the method of challenge. Intraperitoneal injection is more consistent but artificial, while waterborne exposure gives inconsistent results, with mortality resulting from these experimental exposures ranging from near 0 to 100%.[128,140]

Nonsalmonid fish may also carry *Y. ruckeri* and the appearance of the organism in imported bait fish in Europe tends to strongly incriminate these fish as the original source of the disease on that continent.[127] There could also be nonfish sources of *Y. ruckeri* because Willumsen[150] detected the organism in the intestinal content of sea gulls during an ERM outbreak and Stevenson and Daly[137] reported an isolate from the intestine of muskrat. *Y. ruckeri* may also survive in water for an extended length of time, and McDaniel[151] reported the adaptation of *Y. ruckeri* to a normal aquatic saprophytic state, under which conditions it could live for 2 months in mud.

5. PATHOLOGICAL MANIFESTATIONS

Due to the septicemia and hemorrhagic condition, packed red blood cell volume (hematocrits), hemoglobin content, and blood protein are reduced to about half of normal.[132,152] The histopathology of ERM is characterized by bacterial colonization of the capillaries of heavily vascularized tissues. Submucosal hemorrhage of the mouth, jaws, and under the head is most striking but not always present. Gills develop telangiectasis and petechial hemorrhage, congestion, and edema develops in gills, muscle, liver, kidney, spleen, and heart.[125] Inflammation of the reticuloendothelial tissue occurs, and necrotic foci develop in the liver with accumulation of mononuclear cells. Macrophages phagocytize bacterial cells throughout the vascular system and hematopoietic tissue of the kidney. The spleen becomes necrotic, resulting in a loss of lymphoid tissue that produces anemia. The digestive tract is characteristically hemorrhaged, edematous, and necrotic, with sloughing of the mucosa into the lumen.

6. SIGNIFICANCE

Enteric redmouth is a significant disease in coldwater aquaculture, particularly in rainbow trout. Its ability to cause up to 70% mortality in production-size trout, where approximately 75% of the investment costs lie, emphasizes the impact of this disease. In the Hagerman Valley alone, ERM caused 35% of the overall losses in the trout industry at an annual cost of over $2.5 million.[153] The detrimental effect that chronic *Y. ruckeri* infections have on reduced feeding, higher feed conversion, and retarded growth rates of trout is unknown, but it could be economically significant. Also, the fact that the geographic range of ERM has increased, either by dissemination of the organism or better and more intense surveillance and detection methods, has caused the disease to become an international problem. However, the common practice of vaccinating trout for ERM has significantly reduced its severity.

7. MANAGEMENT

ERM should be approached in three ways: prevention through management, chemotherapy, and vaccination. Sound management dictates that ponds and raceways be kept clean; culture unit loading limits should not be exceeded; the highest quality environmental conditions with adequate water flow must be maintained; and nets and other utensils used in the culture units should be sanitized and the units disinfected following fish removal. Also, removal of possible *Y. ruckeri* carrier fish from the water supply is essential. Prior to being handled, a suspect *Y. ruckeri* infected population can be fed a medicated feed containing Terramycin® for 3 consecutive d followed by 2 d of fasting. Theoretically, the chemoprophylaxis reduces stress on the intestine where the pathogen normally resides and inhibits the organism from becoming systemic. The disadvantage to this treatment is the enhancement of oxytetracycline resistant *Y. ruckeri,* and it is not an FDA-approved application of the drug. Also, trout eggs should be disinfected after fertilization and water hardening, or at the eyed stage when shipped (25 mg iodine per liter of water for 5 min) to prevent transmission of the organism on the egg surface.[154]

FDA-approved chemotherapy of infected populations includes oral application of Terramycin® at 50 mg/kg of body weight for 14 d.[46] The potentiated sulfonamide, Romet®, is also used in oral therapy and is administered at 50 mg/kg of body weight for 5 d followed by a 42-d withdrawal period, compared to a 21-d withdrawal period for Terramycin®, before trout can be marketed.[155]

ERM was one of the first fish diseases to be managed with a vaccine.[156] A commercial ERM vaccine was introduced in 1976 and has since become an integral part of controlling the disease in cultured salmonids in the U.S., Great Britain, Scotland, Scandinavia, and other parts of Europe.[157] Trout are vaccinated for ERM by immersion in killed bacterin.[158] Fish weighing less than 200 g should be vaccinated, however, 4- to 4.5-g fish are the most cost effective size for vaccination. Also, the larger the fish are when vaccinated, the longer the protection: 1.0-g fish (4 months), 2.0-g (6 months), and 4.0-g fish (12 months).[159] Fish should be immersed in vaccine for 30 s or the vaccine can be sprayed onto the fish. A secondary immune response develops following exposure to living *Y. ruckeri* for up to 7 months after initial vaccination and serves as a booster vaccination.[166]

Advantages of vaccination against ERM were noted by Tebbit et al.[161] in a study of *Y. ruckeri* vaccinates that showed an 84% reduction in mortalities, a 77% reduction in the need for medication, and a 13.7% lower food conversion rate. Horne and Robertson[157] confirmed these advantages by stating that vaccination of trout against ERM reduced losses, improved feed conversion and growth, significantly reduced the need for medication, and is overall cost effective. In a survey of trout farms in the U.K. by Rodgers,[162] about half of the farms that used vaccination for ERM felt that it failed to protect the fish, but the failure was usually attributed to the poor condition of the fish and low water temperature at the time of vaccination. A detrimental side effect of vaccinating trout is that a subclinical infection with IPN or IHN viruses can be exacerbated into a clinical state with the potential for substantial mortalities.[125] Emergence of these viruses following vaccination is theorized to be the result of stressful vaccination procedures.

E. BACTERIAL GILL DISEASE

Bacterial gill disease (BGD) was named by Davis[163b] when he observed large numbers of long filamentous bacteria on clubbed gills of brook and rainbow trout. The term has since been used to describe infections caused by several different species of bacteria that affect fish gills. The principle etiological agent of BGD is probably *Flavobacterium branchiophilum (branchiophila)* as described by Wakabayashi et al.[164-165] Originally the organism was *F. branchiophila,* but it was noted by von Graevenitz[166] that based on nomenclature rules the proper specific epithet should be *F. branchiophilum.* Also, Strohl and Tait[167] described *Cytophaga aquatilis* as a causative organism of BGD in trout and salmon. Apparently both organisms can actually cause the same clinical disease, with *F. branchiophilum* being more prevalent. Also, other undescribed bacteria may occasionally be involved. While recognizing that pathological gill changes can be caused by a variety of factors, including nutrition, toxicants, fungi, and bacteria,[168] only those gill infections of juvenile salmonids caused by members of the filamentous

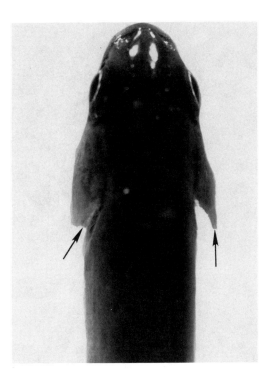

Figure 6 Bacterial gill disease of salmonids. Juvenile rainbow trout showing flared gill cover with some debris protruding from under one cover (arrows). (Photograph courtesy of H. Ferguson.)

bacteria *Flavobacterium* and *Cytophaga* will be discussed. Bacterial gill disease is the terminal result of gill injury usually caused by environmental factors, but bacteria plays a definite role in the death of infected fish.

1. GEOGRAPHICAL RANGE AND SPECIES SUSCEPTIBILITY

Since first being described in the northeastern U.S., BGD has been found in most parts of the world where fish are cultured and has essentially been confined to intensive, freshwater culture environments. *Flavobacterium branchiophilum* has been reported in Japan,[169] the U.S.,[164] Hungary, The Netherlands, and Canada.[170] *Cytophaga aquatilis*, described in Canada, is similar biochemically to gill isolates taken from numerous geographical locations suggesting that it also enjoys a wide geographical range.[167]

Juveniles of many species are susceptible to bacterial gill disease but salmonids are the most severely affected because of their mode of culture. BGD has also been reported in carp *(Cyprinus carpio)*, goldfish *(Carassius auratus)*, catfish *(Ictalurus punctatus)*, eels, fathead minnows, and other fish species.[171]

2. CLINICAL SIGNS AND FINDINGS

Fish affected by bacterial gill disease display inappetence, move toward water inflows, and if not disturbed will often position themselves equidistant from one another in the culture unit. Dark pigmentation, flared gill covers, debris protruding from beneath the operculum, and swollen, pale gills are characteristic of the disease (Figure 6). The upper part of the filament on yearling fish become whitish and swollen. There are no signs of systemic infection.

3. DIAGNOSIS AND BACTERIAL CHARACTERISTICS

Bacterial gill disease is clinically diagnosed by microscopic examination of gill tissue. Large numbers of filamentous bacteria, often in clumps, are present on the surface of the gills in wet mounts. Filaments begin to swell at the distal end and microscopic examination reveals tufts of bacteria between the lamellae at the filament tip. Swelling, or hyperplasia and hypertrophy of the epithelium produces a

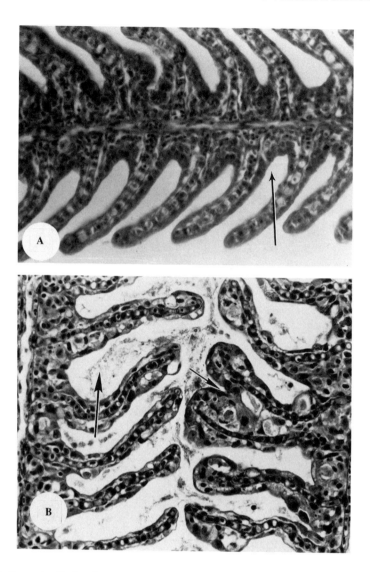

Figure 7   (A) Normal gill showing clear lamellar spaces (arrow); (B) gill of rainbow trout showing swollen lamellae, hyperplasia and hypertrophy of epithelial cells, fusion of lamellae (small arrow), and large numbers of the bacterial gill disease organism within the lamellar troughs (large arrows). (Photographs A and B courtesy of H. Ferguson.)

"clubbed gill" condition, which can become so acute that the filaments actually fuse (Figure 7). The bacteria are more easily seen in Gram-stained smears made from the gill lesions. Huh and Wakabayashi[172] compared light microscopy of wet mount material to IFAT for accuracy in detecting BGD and found that IFAT was more accurate and resulted in detection of a higher number of infected trout. Neither method detected *F. branchiophilum* on fish or in water unless clinical disease was in progress. The organism is confined to the gill epithelium but some bacteria may be found within and beneath degenerating and necrotic lamellar epithelial cells.[173]

   F. branchiophilum and *C. aquatilis* can be cultured aerobically on laboratory media where both form yellow pigmented colonies leading to their designation as "yellow pigmented bacteria" (YPB). These bacteria can be isolated from gill lesions but not from internal organs. *F. branchiophilum* is cultured on cytophaga agar where the light yellow colonies are round, transparent, smooth, and are about 0.5

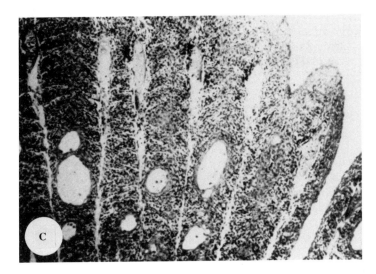

Figure 7 (C) Advanced bacterial gill disease with extensive gill filamentous hyperplasia and gill clubbing. (Photograph C courtesy of U.S. Fish and Wildlife Service.)

to 1 mm in diameter after incubation at 18°C for 5 d.[165] The organism does not grow on TSA agar unless the medium is diluted approximately 20-fold. *F. branchiophilum* grows at 10, 18, and 25°C on media containing up to 0.1% NaCl. Some strains will grow at 5 and 30°C, but not at 37°C. The cell, a Gram-negative rod that measures approximately 0.5 μm × 5 to 8 μm, is cytochrome oxidase positive, nonmotile, and does not glide or spread on agar. Most biochemical and biophysical characteristics are similar among all *F. branchiophilum* isolates (Table 4), however Ostland et al.[173] noted variability in carbohydrate utilization.

Cytophaga aquatilis (Table 4) was isolated on 2% tryptone agar plates.[167] Within 7 d, spreading colonies with pigmentation that varied from yellow to redish-orange appeared. The Gram-negative rods are 0.5 μm × 5 to 15 μm (average length is 8 μm), are motile by gliding with no flagella, and grow aerobically and anaerobically.

4. EPIZOOTIOLOGY

The epizootiology, and hence the etiological agent, of BGD has been confusing because different bacteria have been associated with it, including *Flexibacter* sp., *Cytophaga* sp., and *Flavobacterium* sp. Bullock[168] implicated *Cytophaga* sp. in BGD but did not identify the species and could not create infections at will without severe environmental stress. Strohl and Tait[167] described *C. aguatilis* from diseased gills of salmon and trout in Michigan, and Wakabayashi et al.[164] showed that the organism causing BGD in Japan, Oregon, and Hungary was a member of the genus *Flavobacterium*. *F. branchiophilum* strains from the three continents serologically possessed some identical antigens, but there were slight differences,[172] with strains from the U.S. and Hungary being most similar. *Flexibacter* sp. probably does not cause classic BGD although it does infect fish gills.

Apparently, *C. aguatilis* has not been experimentally transmitted under any conditions and, in general, experimental transmission of BGD is unpredictable. *F. branchiophilum* has been successfully transmitted in the laboratory; however, this often required extreme environmental stress.

It has even been suggested that bacterial gill disease may be incorrectly named because its development is directly dependent upon the conditions in the culture environment and any subtle injury that might occur to the fish's gill epithelium. It has been widely accepted that gills are usually injured by chemical or physical irritants in the water prior to bacterial colonization.[168] Gill injuries will occur if water exchange is inadequate, thus allowing elevated levels of ammonia and decreased oxygen concentrations, or if there is silt or excess feed present in the water. These injuries result in epithelial hyperplasia and make gills more susceptible to microbial invasion by *Flavobacterium, Cytophaga,* and occasionally other bacteria.

Table 4 **Characteristics of *Flavobacterium branchiophilum*[165] and *Cytophaga aquatilis*,[167] the two filamentous bacteria that have been associated with bacterial gill disease of trout**

Characteristics	*Flavobacterium branchiophilum*	*Cytophaga aquatilis*
Yellow to red pigment	+	+
Growth at 37°C	−	−
Motility	−	Gliding
Growth on nutrient agar	−	+
Hydrolysis of		
Gelatin	+	+
Casein	+	+
Chitin	−	+
Esculin	−	+
Starch	+	+
Indole from tryptone	−	−
Nitrate reduction	−	+
Congo red		−
Catalase production		+
H₂S production		(+)
ONPG		+
Growth in		
0% NaCl	+	+
0.5% NaCl	−	+
1.0% NaCl	−	+
2.0% NaCl	−	+
Acid from		
Glucose	+	+
Fructose	+	+
Lactose	−	−
Sucrose	+	−
Maltose	+	+
Trehalose	+	(+)
Cellobiose	(+)	+
Arabinose	−	+
Xylose	−	+
Raffinose	(+)	±
Mannitol	−	+

Note: + = positive; − = negative; (+) = weak reaction.

Studies by Speare et al.[174-175] presented a different scenario for development of BGD. After an extensive study of 23 separate outbreaks of BGD in rainbow trout they concluded that "no other disease conditions, no gross errors in management, nor recent exposure to chemotherapeutics" preceded BGD. Although the causative organism was not definitively identified, it is thought to have been *F. branchiophilum*. During a 5-month monitoring regime prior to the onset of natural outbreaks of the disease, gill morphology of examined fish remained unaltered and they proposed that *F. branchiophilum* can cause BGD without the environmental stress or gill injury. Nevertheless, most outbreaks of BGD are associated with some management factor such as excessive feeding, poor water quality, poor circulation, or inadequate water flow.

Wakabayashi and Iwado[176] reported that fish infected with *F. branchiophilum* were more susceptible to hypoxia because the bacterium impaired respiratory functions. Noninfected fish consumed 251 to 289 ml/kg/h O_2 while rates of oxygen consumption of infected fish at 2 and 5 d postinfection were 183 to 229 and 155 to 167 ml/kg/h O_2, respectively.

Historically, salmonid fry and fingerlings less than 5 cm in length are particularly susceptible to BGD but larger fish may also become infected. Adult rainbow and cutthroat trout and fall chinook salmon have suffered outbreaks of BGD.[177] Ferguson et al.[170] successfully transmitted *F. branchiophilum* to rainbow trout precipitating clinical infection within 24 h, and death in 48 h, in fish up to 3 years of age. In view of these reports the disease may be more of a problem in large trout than is realized.

Mortality due to BGD among small fish has the potential to become acute. In experimental BGD infection studies by Bullock[168] mortality reached 39 to 80% after 13 d. Speare et al.[175] found that morbidity could increase from approximately 5% when disease was first detected to over 80% within 24 to 48 h. During this time, mortality rates rose from 0 to 20% per day and then diminished by days 7 to 10, with few fish showing any clinical signs by days 10 to 14. BGD occurs at a wide range of temperature from 12 to 19°C.[178]

Viability of *F. branchiophilum* in water is not known but it is theorized that subclinically infected fish are the source of infection, however trout raised entirely in well water have become infected.[177] Heo et al.[178] detected *F. branchiophilum* in water immediately before and during an epizootic, but not after the disease abated. Ostland et al.[179] enumerated total bacteria and percentage of YPB on the gills of clinically healthy and BGD-infected rainbow trout. Healthy gills had 4.9×10^5 bacterial CFU/g of tissue, 15% of which were YPB. Severely BGD-infected fish had 3.9×10^6 CFU/g, of which 35% were filamentous YPB. These data suggest a strong association between severity of BGD and the presence of filamentous, yellow pigmented bacteria.

5. PATHOLOGICAL MANIFESTATIONS

Bacterial gill disease generally begins with hyperplasia of gill epithelium possibly caused by mild chronic toxins, or irritants present in the water. *F. branchiophilum* have large numbers of thin, fimbriae-like surface structures that bridge the space between the bacterium and gill.[174] These structures appear to be involved in attachment to the gill surface. Bacterial colonization and fusion of lamellae will begin on the distal end of filaments infected with BGD (Figure 7). In contrast, fish suffering from "nutritional gill disease" (due to insufficient dietary pantothenic acid) will have lamellar clubbing and filament fusion beginning at the proximal end of the filament and progressing distally. Speare et al.[174-175] confirmed and expanded earlier observations of bacterial colonization made by Kudo and Kimura[180] in stating that this event was preceded immediately by several gill changes that were detectable only at an ultrastructural level. The cause of these changes was not determined but environmental conditions were not considered a factor. Cytoplasmic blistering and degeneration of the microridges of superficial filament epithelium and slight irregularity of filament tips suggested mild hyperplasia. "Explosive" morbidity and acute mortality coincided with extensive bacterial proliferation on the lamellar surfaces, epithelial hydropic degeneration, necrosis, and edema (Figure 7). Lamellar fusion and epithelial hyperplasia were detected later in subacute (2 to 5 d) or chronic (7 to 14 d) changes.[174]

It has been suggested that bacteria physically occluded gill surfaces and inhibited respiratory exchange.[176] However, Speare et al.[175] challenged that suggestion because fish have large areas of underutilized gill surfaces and massive numbers of bacteria did not substantially cover the tissue-water interface. Why BGD-infected fish behave in a manner which indicates a lack of sufficient oxygen has still not been adequately explained. However, the oxygen consumption data of Wakabayashi and Iwado[176] supports the theory of increased opercular movement to increase water flow across gills to allow better oxygenation of blood.

BGD-infected trout develop a significant decrease in serum Na^+, Cl^-, and osmolality, resulting in hemoconcentration.[181] Fish with BGD exhibit increased respiration (tachybranchia) but are not hypoxic, therefore, the tachybranchia may not be a response to impaired oxygen exchange. It was suggested that fluctuations in blood components other than those that are acid-base related constitute fatal changes rather than hypoxia as the result of BGD infection. These circulatory disturbances result from the loss of blood electrolytes, which triggers a fluid shift from extracellular to intracellular and leads to death. Also, Wakabayashi and Iwado[176] concluded that a breakdown in gas exchange at the gills causes failure to provide enough oxygen to remove excess lactate from the muscle.

6. SIGNIFICANCE

Bacterial gill disease is one of the most common diseases of juvenile salmonids. Because BGD is often an acute condition with high mortality, it must be considered a potentially serious problem on trout and salmon culture facilities. The Fish Pathology Laboratory at Ontario Veterinary College, University of Guelph, Ontario, Canada, reported that BGD accounted for approximately 21% of all samples submitted for diagnosis from fish farms.[182]

7. MANAGEMENT

Prevention through proper sanitation and maintenance of high quality water and water flow, thus reducing environmental stress, is the best method of controlling bacterial gill disease.[183] The water supply should originate from a fish-free source, have a high oxygen concentration, and have low levels of accumulated metabolites such as ammonia. Insuring an adequate flow of clean water for the mass of fry or juvenile fish served is critical to maintaining high water quality. Feed with excessive "fines" should not be used for BGD-susceptible fish and excess feed should not be allowed to accumulate in tanks.

BGD therapy is effective when applied expediently because infected fish tend to recover rapidly after bacteria are reduced on, or removed from the gills. Disinfectants containing quaternary ammonia, such as benzalkonium chlorides in Hyamine® 1622 (98.8% active), Hyamine® 3500 (50% active), or Roccal® (several concentrations of active ingredient) are used as treatments. A 1- to 2-mg/l application of benzalkonium chloride for 1 h on 3 consecutive d is recommended, however caution is urged in soft water because this compound can be toxic to some species of trout in water with low hardness. The herbicide Diquat® has been used successfully in treating BGD at 2 to 4 mg/l of active ingredient. Potassium permanganate can also be used at 1 to 2 mg/l for 1 h, but the safety margin is very narrow. Chloramine-T applied at 8 to 10 mg/l for 1 h for 2 or 3 d is the most effective treatment for BGD.[177] Byrne et al.[181] suggested salt concentration in the water should be increased during BGD outbreaks or during chemotherapy treatment to compensate for loss of electrolytes. However, none of the compounds listed above are approved by the FDA for therapeutic use on fish with BGD that are destined for the table in the U.S.

F. BACTERIAL COLD-WATER DISEASE

Bacterial cold-water disease (BCWD) can cause low to moderate mortality among cultured salmonids. The disease, originally referred to as "peduncle disease" because of the location of the principal lesion, was first described in the eastern U.S. by Davis.[184] The etiological agent of BCWD is *Flexibacter psychrophilus,* which is synonymous with *Cytophaga psychrophila.* For clarification purposes the term *F. psychrophilus* will be used throughout the text.[185-186] *F. psychrophilus* is a member of the Cytophagaceae which includes the causative agents of "columnaris" in fresh *(F. columnaris)* and saltwater fish *(F. maritimus)* (Chapter IV).

1. GEOGRAPHICAL RANGE AND SPECIES SUSCEPTIBILITY

While bacterial cold-water disease occurs throughout most of the trout and salmon growing regions of North America, it most seriously affects salmon culture in the northwestern U.S. and western Canada.[187] The disease also occurs in Japan[188] and was recently reported in rainbow trout in France.[189]

All salmonid species are suspected to be susceptible to *F. psychrophilus,* however, rainbow, brook, lake, and cutthroat trout; and coho, sockeye, and chum salmon are known to be susceptible.[186,190-192] *F. psychrophilus* can occasionally cause disease on nonsalmonids; the pathogen has been detected in European eels *(Anguilla anguilla),* carp, tench *(Tinca tinca),* and crucian carp *(Carassius carassius)* in Germany.[193]

2. CLINICAL SIGNS AND FINDINGS

Clinical signs of BCWD vary with the size and age of the fish affected. When sac fry are infected the yolk skin becomes eroded,[194] while older fish exhibit lethargy and sometimes spiral swimming behavior.[195] The caudal peduncle appears whitish and the skin becomes necrotic, detached, and sloughs off, exposing underlying muscle (Figure 8). In fish up to 1 year of age, the caudal peduncle lesion is the most typical clinical sign of BCWD. Necrosis of the muscle may continue to the point that the vertebral column is exposed and the tail nearly separates from the body. Lesions may also appear dorsally, laterally, or on the isthmus. Hemorrhages on the fins and operculum and pale gills can occasionally be seen in infected fish. Darkly pigmented, moribund fish may be observed late in epizootics.[186]

Figure 8   Rainbow trout with typical lesion on the caudal peduncle (arrow) associated with bacterial cold-water disease *(Flexibacter psychrophila).*

Internally, petechiae may be present on the liver, pyloric cecae, adipose tissue, heart, swim bladder, and occasionally on the peritoneal lining. BCWD survivors may develop vertebrae deformities and/or nervous disorders which cause them to swim in a spiral.[195]

3. DIAGNOSIS AND BACTERIAL CHARACTERISTICS

Bacterial cold-water disease is diagnosed by recognition of characteristic clinical signs; detection of elongated vegetative cells in stained smears from lesions; and isolation of *F. psychrophilus* on appropriate media. Generally the organism can be isolated from muscle lesions or internal organs. On stained smears from lesions, the bacterium is a slender, Gram-negative bacillus with rounded ends that measures 0.3 µm × 2 to 2.5 µm. In broth culture cells, it can measure 0.3 to 0.75 µm × 2 to 7 µm but may be several times this length with branching of some cells.[186] *F. psychrophilus* is motile by gliding and does not produce fruiting bodies. *Cytophaga* media, either broth or agar, is the usual culture substrate,[196] however, it has been modified to improve the growth of *F. psychrophilus.*[197-198] On agar, *F. psychrophilus* may form a 1- to 5-mm-diameter yellow colony that is flat, spreading, and has a thin irregular edge; the colonies may be smooth, entire, and yellow; or cultures may be a mixture of the two types.[186] It is strictly aerobic with an optimum growth temperature of about 15°C, but will grow at temperatures from 4 to 25°C. Poor growth occurs at the extremes, and not at all at 30°C.[189,199] *F. psychrophilus* does not tolerate NaCl concentrations above 1%.

Biochemical properties do not vary greatly among the various strains of *F. psychrophilus* (Table 5).[185-186] Generally, the bacterium does not use carbohydrates but is proteolytic for casein, gelatin, albumin, and collagen. It is cytochrome oxidase positive by most accounts, weakly catalase positive, does not produce H_2S, and contains flexirubin pigments. These yellow pigments are of diagnostic value. Some strains degrade elastin.

While all strains of *F. psychrophilus* may not be identical, they are all serologically similar. Holt[186] reported that isolates of *F. psychrophilus* from New Hampshire, Michigan, and Alaska were antigenically similar to Oregon strains, but some differences were noted. Currently, serum agglutination is used for positive identification of *F. psychrophilus,*[186] however, Lorenzen and Karas[200] have developed a rapid diagnostic method using immunofluorescence on spleen imprints of diseased rainbow trout fry. A slight cross-reactivity with *F. columnaris* may occur, but this problem can be corrected by adsorption with the "columnaris" organism.

4. EPIZOOTIOLOGY

Normally, bacterial cold-water disease occurs in spring when water temperatures are from 4 to 10°C. Experimental infection was induced by Holt et al.[199] in coho and chinook salmon and rainbow trout at temperatures of 3 to 15°C, but as water temperatures rose past 15°C severity of disease decreased. The

Table 5   Biochemical characteristics of 28 isolates of *Flexibacter psychrophilus*[186]

Characteristic	Reaction	Characteristic	Reaction
Degradation of		Production of	
Cellulose	−	Ammonia	+
Starch	−	Hydrogen sulfide	−
Cascin	+	Catalase (slow)	+
Gelatin	+	Cytochrome oxidase	−
Albumin	+	Indole	−
Elastin	+[a]	Acetylmethyl carbinol	−
Collagen	+	Growth in broth	
Chitin	−	1% Tryptone	+
Tyrosine	+[a]	1% Casamino acid	−
Tributyrin	+	Growth on 0.01%	
Carbohydrate oxidation		sodium lauryl sulfate	−
Glucose	−	Methyl red	−
Lactose	−	Nitrate reduction	−
Sucrose	−		

[a] Variable reaction

shortest average of 2 d to death occurred at 15°C. Conversely, *F. psychrophilus* was recovered from 83% of steelhead trout that died at 22°C. Lehmann et al.[193] noted that *F. psychrophilus* infection in nonsalmonids occurs at water temperatures of 8 to 12°C.

Bacterial cold-water disease affects salmonids ranging from yolk-sac fry to yearling fish; the younger the fish the more severe the disease.[201] Mortality of yolk-sac fry generally ranges from 30 to 50%. As fish progress into a feeding mode, mortality is usually in the 20% range. Chronic infections may develop in yearling coho and other Pacific salmon during winter months when typical lesions of the peduncle can be seen and a low grade mortality ensues. Infected fish may also be anemic — a condition which is often compounded by the presence of erythrocytic inclusion body syndrome virus.[202] Experimentally, the virulence of *F. psychrophilus* varies with the bacterial strain because mortality of coho salmon infected with different strains varied from 0 to 100%.[186]

Two separate manifestations of bacterial cold-water disease may occasionally follow infections in juvenile coho salmon. Some fish have spinal deformities; a problem first reported by Wood[201] when he described infected fish which had compression of the vertebrae, lordosis, and scoliosis. Other fish may display nervous disorders such as spinning behavior, swelling in the posterior skull, and dark pigmentation on one side or the other. *F. psychrophilus* is readily isolated from brain tissue and less often from the internal organs of these fish.[203]

Although the natural reservoir of *F. psychrophilus* is not fully understood, adult coho and chinook salmon, and in all probability rainbow trout and other salmonids, serve as carriers and sources of *F. psychrophilus*.[204] Horizontal transmission is highly likely, although experimental transmission from fish to fish cannot be achieved unless the mucus layer and epithelium are injured. It is generally concluded that members of the Cytophagales (*Flexibacter* sp.) bacterial group are part of the normal microflora of the skin of trout and salmon, and if conditions are favorable can become pathogenic. The organism has been isolated from spleen, kidney, ovarian fluid, and milt of mature fish; therefore, transmission of the bacterium may occur on eggs during spawning because of its known presence in the reproductive fluids.[205] *F. psychrophilus* has been detected in 20 to 76% of sexually mature female Chinook and coho salmon in hatchery populations and in 0 to 66% of males on the same facility.[186,205] Whether or not salmon carry the organism through their marine maturation phase is not clear. Upon their return to fresh water these fish have a low incidence of *F. psychrophilus* carrier status, but the longer they stay in fresh water the higher the incidence of infection becomes. This would suggest that adults either become infected or reinfected upon their return to fresh water.

5. PATHOLOGICAL MANIFESTATIONS

Bacterial cold-water disease is a septicemia but the most obvious pathology is the necrotic, sloughing skin on the caudal peduncle. Although bacterial cells can be found throughout most organs and tissues of infected fish, death is generally attributed to severe heart lesions.[206] There is usually very little inflammation associated with BCWD; however, some researchers have noted mild mononuclear infiltration of macrophages in diseased tissues. Otis[207] has suggested that extracellular products of *F. psychrophilus* cause the characteristic epithelial lesions because they developed in fish that had been injected with either live bacteria or cell-free extracellular products. Lesions of internal organs were also found in fish injected with either live bacteria or crude cell-free substances.

6. SIGNIFICANCE

Bacterial cold-water disease is significant in the culture of salmonids, particularly of juvenile coho salmon in the northwestern region of North America where it is the most frequently isolated fish pathogen.[186]

7. MANAGEMENT

Increasing the water temperature above the optimum range is presently the only clear management approach for the prevention of BCWD; however, this is not practical in many hatcheries where the disease is endemic. Wood[201] also found that infections were less severe in coho salmon fry that were kept in shallow troughs rather than in deep ones. Excess water flow through vertical egg hatching apparatus tends to increase incidence of disease in yolk-sac fry. From these observations, it can be concluded that procedures which reduce skin and fin abrasions of juvenile fish can also reduce the incidence and severity of *F. psychrophilus*.

Most serious outbreaks of BCWD are difficult to treat because infection is systemic and affected fry will not feed, thus neutralizing any advantages of medicated feed.[186] Amend[208] reported that oxytetracycline and surface disinfectants applied in the water are ineffective treatments for BCWD of juvenile salmon, but conversely, Bullock and Snieszko[204] and Schachte[191] recommended bath treatments with water-soluble oxytetracycline at 10 to 50 mg/l or quaternary ammonium compounds at 2 mg/l while infections are confined to the skin. After the disease becomes systemic, oxytetracycline in the diet at 50 to 75 mg/l of fish daily for 10 d is effective.

Evidence indicates that *F. psychrophilus* can be transmitted on egg surfaces; therefore, it is recommended that eggs be disinfected with organic iodine compounds at 25 mg/l active iodine for 5 min,[191] but Holt[186] indicated that iodine treatment of eggs did not prevent the resulting hatched fry from contracting BCWD.

Vaccination of trout and salmon for BCWD is under investigation and has shown some promising results. Holt[186] demonstrated successful vaccination by immersion and injection of a formalin-killed bacterin of *F. psychrophilus*. Injection, with Freund's adjuvant, produced complete protection against challenge compared to 43% mortality in nonvaccinated controls. Immersion resulted in only 11% improved survival. Obach and Laurencin[210] reported that 40-day-old posthatch rainbow trout were not protected by immersion vaccination with a heat-inactivated preparation of *F. psychrophilus*. However, when immersion vaccination was applied to 50-day-old posthatch or older fish, significant protection was achieved. Vaccination of 2.2-g fish (90-day-old posthatch) by intraperitoneal injection elicited a very high degree of protection. They concluded that bath vaccination may prove practical and economical, but vaccination is perceived as beneficial only to 2-month-old to yearling fish, rather than to the fry.

G. BACTERIAL KIDNEY DISEASE

Bacterial kidney disease (BKD) is a chronic, rarely subacute, bacterial infection in salmonids that is caused by *Renibacterium salmoninarum*.[211] This disease was initially described in Scotland[212] in 1930 as "Dee" disease in Atlantic salmon and was reported in the U.S. in 1935.[213] Bacterial kidney disease has also been called corynebacterial kidney disease and salmonid kidney disease. High mortality is not usually associated with BKD.

1. GEOGRAPHICAL RANGE AND SPECIES SUSCEPTIBILITY

Bacterial kidney disease exists to some degree in most freshwater and marine environments of the world where trout and salmon occur, with the possible exception of Australia, New Zealand, and the former U.S.S.R.[214-215] Severe problems have been reported in Canada, Chile, England, France, Germany, Iceland, Italy, Japan, Spain, and the U.S. Most outbreaks in the U.S. occur in the northwest and upper midwest to the northeast.[216]

All salmonids are considered susceptible to *R. salmoninarum*, but brook trout and spring chinook salmon are considered to be the most severely affected species.[215,217] Natural outbreaks of BKD have occurred only in salmonids and few nonsalmonids are experimentally susceptible. However, experimental infections have been established in sable fish *(Anoplopoma fimbria)*,[218] Pacific herring *(Clupea harengus)*,[219] the shiner perch *(Cymatogaster aggregata)*,[217] common shiner *(Notropis cornutus)*, and fathead minnow.[220]

2. CLINICAL SIGNS AND FINDINGS

External clinical signs of BKD vary among species and are often obvious only in the terminal stages of disease.[215] The most common early signs of infection are dark pigmentation, exophthalmia, hemorrhages at the base of fins, abdominal distension — all associated with a bacteremia. Occasionally, superficial blisters and ulcerative abscesses occur on other body surfaces (Figure 9). Internally, a bloody ascitic fluid may be present in the coelomic cavity and petechia may appear in the muscle under the peritoneal lining of some fish. As BKD progresses, small grayish-white, granulomatous lesions appear on the surface of, or in, the kidney and to a lesser extent in the liver and/or spleen. These lesions may contain leukocytes, cellular debris, macrophages containing phagocytosed bacteria, and cell-free bacteria. As the disease progresses, the number and size of granulomatous lesions increase in all internal organs and the kidney becomes swollen with a "corrugated" surface (Figure 9). Although a systemic infection is a normal occurrence in BKD, infections may occasionally occur only in the eye or on the skin.[221-222]

3. DIAGNOSIS AND BACTERIAL CHARACTERISTICS

Clinical infections of *R. salmoninarum* are diagnosed by noting clinical signs and detection of Gram-positive diplobacillus bacteria in internal organs. Gram stain or fluorescent antibody technique (FAT) may be used on smears from infected kidney tissue to detect presence of the organisms.[223] Unless *R. salmoninarum* is present in abundance, Gram-stained smears give less precise results than FAT because pigment granules (melanin), which are naturally brownish in color, can be confused with the pathogen.

Detection of *R. salmoninarum* in subclinical carrier fish is somewhat more challenging, but due to improved FAT and the development of more sensitive enzyme-linked immunosorbent assays (ELISA)[224] more subclinical infections are being detected. Elliott and Barila[225] developed a membrane filtration-fluorescent antibody staining procedure that can detect less than 10^2 *R. salmoninarum* cells per milliliter of coelomic fluid. Lee[226] was also able to detect less than 10^2 cells per gram of enzyme-digested kidney using FAT and membrane filtrates.

Five methods for detecting subclinical *R. salmoninarum* in coho salmon were compared by Cipriano et al.[227] They found that Gram stain and direct fluorescent antibody were the least sensitive methods and counterimmunoelectrophoresis the most sensitive. Sakai et al.[228] modified an ELISA technique to a peroxidase-antiperoxidase procedure for detection of *R. salmoninarum* in coho salmon and found the modified method to be approximately ten times more sensitive than FAT. The procedure also gave excellent results in a BKD survey. Sakai et al.[229] compared eight different methods for detecting *R. salmoninarum* in supernatants from kidneys of rainbow trout. In their studies they found that Gram stain and immunodiffusion were the least sensitive, FAT was very sensitive, but the lowest bacterial cell concentration was detected by ELISA (direct and indirect) when used in conjunction with a dot blot assay. The latter method could detect from 10^2 to 10^3 cells per gram of kidney tissue.

An ELISA was used to detect soluble *R. salmoninarum* antigen (57 kDa protein) in tissues and body fluids by Pascho and Elliott.[230] Griffiths et al.[231] showed that detection of the soluble 57 kDa (p57 or "F" antigen) *R. salmoninarum* antigen by western blot was significantly more sensitive and reliable than detection of whole cells by direct FAT in tissue homogenates from actively infected Atlantic salmon. Whether or not the western blot is applicable to detection of carrier fish, where no active infection is ongoing, is not known. However, about one third of the fish surviving the epizootic were seropositive for *R. salmoninarum* by this method. Indications are that ELISA is the most sensitive method of detecting subclinical *R. salmoninarum*.

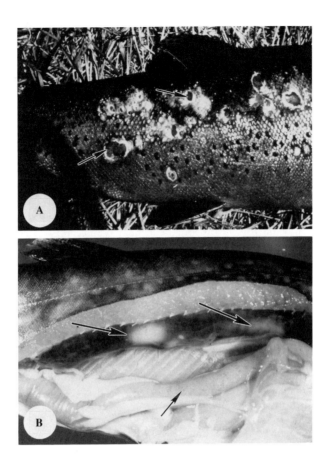

Figure 9   Bacterial kidney disease of salmonids. (A) Rainbow trout with epidermal, ulcerative lesions (arrow) typical of chronic *Renibacterium salmoninarum* infection. (B) Granulomatous lesions (large arrows) in the swollen, corrugated kidney of brook trout. Note the petechia on adipose tissue and intestine (small arrow). (Photograph courtesy of G. Camenisch.)

Isolation is the only current method of detecting viable *R. salmoninarum,* but because of its fastidious nature routine culture is difficult. The pathogen can be grown on bacterial culture media, but Evelyn[232] developed an improved medium — "kidney disease medium" (KDM2) — using peptone yeast extract agar that was supplemented with cysteine and serum. Evelyn et al.[233] improved the growth of *R. salmoninarum* on KDM2 by introducing "cross feeding" whereby a nonfastidious feeder, "nurse" organism, was placed next to the fastidious BKD organism. An additional growth improvement innovation was described by Evelyn et al.,[234] who supplemented KDM2 with a small amount of KDM2 broth in which *R. salmoninarum* had been grown (i.e., spent broth). At 15°C the organism requires up to 20 d for the appearance of nonpigmented, creamy, shiny, smooth, raised, entire colonies which measure approximately 2 mm in diameter. Because the organism grows so slowly, precautions must be taken to insure that the culture media remains moist during incubation.

R. salmoninarum is a Gram-positive bacillus (usually in pairs) that measures 0.6 to 1.0 μm × 0.3 to 0.5 μm.[235] It does not produce spores nor is it acid-fast, motile, or encapsulated. Smith[212] determined the optimum growth temperature to be 15°C with growth slowing as temperature was reduced to 5°C or elevated to 22°C, and no growth occurs at 37°C.

R. salmoninarum is biochemically homogeneous.[236] The bacterium is proteolytic in litmus milk, produces catalase, and does not liquefy gelatin. It produces acid and alkaline phosphatase, caprylate esterase, glucosidase, leucine arylamidase, α-mannosidase, and trypsinase. It degrades tributyrin, Tween® 40 and 60, but not Tween® 80. The organism grows at ph 7.8 on 0.0001% crystal violet, 0.00001% Nile blue, and poorly in 1% sodium chloride.[237-238]

R. salmoninarum isolates from various parts of the world are serologically homogeneous using polyclonal antisera.[239] However, Arakawa et al.[240] used monoclonal antibodies in an ELISA system to show that among nine different *R. salmoninarum* isolates from the U.S., Norway, and England, three serological groups were present. Supernate from only one clone (against an isolate from Norway) reacted with all isolates. By use of rocket electrophoresis and cross adsorption analysis, Getchell et al.[241] showed a common "F" (p57) antigen in seven isolates. This soluble, heat stable antigen is released into culture media and is considered to be a primary mediator of the bacteria's virulence.[242]

4. EPIZOOTIOLOGY

Bacterial kidney disease is primarily a disease of hatchery fish stocks but it also occurs in wild and anadromous trout and salmon. *R. salmoninarum* can cause an overt infection or can be present only in the carrier condition. In the proceedings of the National Workshop on Bacterial Kidney Disease,[216] it was noted that between 1986 and 1991 the pathogen was found at 63 locations in the U.S. where it was not causing disease and at 97 locations where overt disease was present: 9 states reported only the presence of the pathogen while 16 states reported either the presence of carrier populations or actual disease outbreaks, which would suggest that factors other than the mere presence of the pathogen are related to outbreaks of BKD.

Water quality may influence cumulative mortality and severity of the disease. A higher incidence of BKD has been noted at hatcheries with soft waters as opposed to those facilities using water with high total hardness.[243] Bacterial kidney disease has an adverse effect on survivability of salmon smolts as they move from fresh water to salt water. Over a 150-d period, coho salmon smolts experienced over 17% cumulative mortality in sea water compared to only 4% mortality in a group held in fresh water.[215] Paterson et al.[244] confirmed that heavy losses of Atlantic salmon smolts caused by BKD occurred as they became acclimated to sea water. Sanders et al.[245] found that 20% of hatchery-released and wild salmonids migrating down the Columbia River were infected with *R. salmoninarum*. However, the percent prevalence in fish held in fresh water was 9% compared to 46% prevalence in those held in salt water. Implications are that the disease becomes more severe when the young salmon reach the marine environment.

BKD occurs at a wide range of temperatures. Most BKD epizootics in chinook and sockeye salmon occur during fall (12 to 18°C) and winter (8 to 11°C), but Smith[212] observed that mortality accelerated in summer with rising water temperatures. Mortalities in cool water tend to be slow and chronic compared to a more subacute mortality pattern in warmer water. Water temperature also influences the time between exposure and death. At 11°C death occurs 30 to 35 d postexposure compared to 60 to 90 d at 7 to 10°C. Sanders et al.[246] presented conflicting experimental data concerning the relationship of temperature and mortality in BKD infections. They indicated that at 6.7 to 12°C, 78 to 100% mortality occurred; but as temperatures rose to 20°C, mortality declined to a low of 8 to 14°, indicating that factors other than temperature influence BKD-related mortality.

In addition to differential species susceptibility, various strains within a species of fish respond to BKD differently. Beacham and Evelyn[247] compared the susceptibility of three strains of chinook salmon to *R. salmoninarum* and found substantial differences in mortality rates among the strains. The susceptibility in the most affected strains was seven times higher than in the least affected strain. Breeding and genetic manipulation of fish stocks to increase resistance is an area of BKD management that should be further exploited.

Several studies indicate that dietary composition of feed will affect salmonid susceptibility to *R. salmoninarum*. Nutritional studies in Atlantic salmon indicate that insufficient amounts of dietary vitamin A, iron, zinc, iodine, and other minerals increased susceptibility to *R. salmoninarum*.[244] Woodall and Laroche[248] noted that salmon receiving insufficient amounts of iodine were more susceptible to BKD. The addition of 4.5 mg of iodine per kilogram of feed reduced the prevalence of BKD from 95 to 3%, and fluorine added at the same rate reduced the prevalence from 38 to 4%.[249]

According to Evelyn et al.,[250] *R. salmoninarum* was first reported in anadromous species in the U.S. early in the 1950s. Recent indications are that the disease is becoming more widespread, but this increase may be due to the aggressive use of more sensitive and accurate detection methods rather than due to actual increased geographical range. There has been a high incidence of the disease reported in adult Pacific and Atlantic salmon returning to spawn.[217] When salmon migrate from fresh water to sea water *R. salmoninarum* infections continue to develop with some fish succumbing to the disease while at sea.[251] Those that survive return to spawn and may transmit the pathogen to the next generation.

Bacterial kidney disease also exists in wild or feral trout and salmon populations that are confined to fresh water. Mitchum et al.[252] demonstrated the presence of BKD in feral populations of brook, brown, and rainbow trout in freshwater streams. Mitchum and Sherman[253] further showed that when noninfected hatchery-reared trout are stocked into these waters, the *R. salmoninarum* pathogen that is present in feral fish can be transmitted to newly stocked fish. *R. salmoninarum*-free rainbow trout became infected from the resident population and died in 9 months or less postexposure. Also, coho salmon in Lake Michigan have been extensively affected by BKD where it is believed to have a severe detrimental impact on the salmon fishery.[254] Souter et al.[255] found *R. salmoninarum*-infected Arctic char and lake trout in the Northwest Territories of Canada, an area that had received no stocking of hatchery fish. They postulated that this constituted a "natural" presence of the organism.

R. salmoninarum does not survive well outside of the host. Austin and Rayment[256] isolated the pathogen for 28 d from spiked filter-sterilized river water at 15°C but it is possible that it may survive for longer periods undetected. Fish appear to be the source of *R. salmoninarum* as it has been found only in clinically diseased or asymptomatic carrier fish.[215] Although Koch's postulates for BKD were fulfilled in 1956,[257] experimental horizontal transmission of the pathogen has been achieved with varying degrees of success. Wood and Wallis[258] demonstrated transmission of *R. salmoninarum* by feeding raw fish viscera to young salmon, but it can also be transmitted by cohabitation in either fresh water or sea water. In an experiment intended to demonstrate the relationship of nutritional quality to *R. salmoninarum* susceptibility in sockeye salmon, Bell et al.[259] successfully transmitted the pathogen from deliberately infected fish in cohabitation with the nutritional experimental fish. Murray et al.[260] described experimental induction of BKD to chinook salmon in sea water by immersion in a solution containing 10^4 to 10^6 cells/ml for 15 to 30 min, and to noninfected fish cohabitating with *R. salmoninarum*-injected fish. Both procedures required 5 to 6 months for clinical disease to appear. Any injury to the skin of fish enhances contraction of *R. salmoninarum* by immersion,[222] but the pathogen can be transmitted by injecting material from granulomatous lesions into noninfected fish.

Vertical transmission is an important aspect in the epizootiology of *R. salmoninarum* because the organism is uniquely adaptable to this method of transmission. It is small enough to enter the micropyle of the egg, which facilitates transmission; therefore, eggs may become infected during oogenesis or while they are surrounded by heavily contaminated coelomic fluid prior to, and during, the act of spawning.[261-263] During a long incubation period in fish it produces a chronic infection, and in the case of trout or Atlantic salmon the infection remains through multiple spawning periods or the entire life of the fish. Allison[264] first reported the presence of BKD in offspring from infected fish but actual transmission from adult to progeny was demonstrated by Bullock et al.[265] Evelyn et al.[266] reported the presence of *R. salmoninarum* within eggs of trout and salmon, further supporting vertical transmission.

5. PATHOLOGICAL MANIFESTATIONS

Austin and Austin[1] hypothesized that *R. salmoninarum* is a normal resident in the kidney and possibly the gastrointestinal tract of salmonids where it remains dormant in relatively low numbers. When the host is stressed, the pathogen multiplies in the kidney or breaks the epithelial barrier of the gut and becomes systemic. This theory is based on the organism's normally nonaggressive activity under most conditions. It has also been postulated that an unidentified substance may occur in the kidneys of rainbow trout that possibly inhibits growth of the pathogen, thus repressing its ability to cause disease.[267-268] The presence of very low numbers of *R. salmoninarum* in some brood rainbow trout, where clinical disease has not occurred, may be explained by the organism remaining dormant unless the host becomes stressed.

Infections of *R. salmoninarum* are usually systemic and are characterized by diffuse granulomatous inflammation. Granulomatous lesions develop primarily in the kidney (Figure 9) but may also occur in the liver, heart, spleen, and muscle.[269] The granulomas are often large, particularly in the kidney, with zones of caseation surrounded by epitheloid cells accompanied by lymphoid cell infiltration. These granulomas may or may not be encapsulated. Postorbital lesions of the eye and skin are also characterized by the presence of granulomas with chronic inflammation and infiltration by macrophages and leukocytes.[221-222]

According to Young and Chapman,[235] chronologically, *R. salmoninarum* develops a septicemia within 4 to 11 d of experimental infection. At about 18 d the bacteria had colonized on the outer surface of organs and by 25 d they had penetrated deeper into tissues where granulomas begin to form. These findings are congruous with the long incubation periods between exposure to *R. salmoninarum* and appearance of clinical disease and chronic mortality.

According to Bruno,[270] at 12°C rainbow trout and Atlantic salmon exhibited initial pathological signs at 10 d postexperimental infection, with both groups suffering 75 to 80% mortality by 35 d. In that study, the presence of *R. salmoninarum* in collecting ducts of the kidney may have affected the organ's function by causing sufficient damage to disrupt normal filtration processes. Death, therefore, was attributed to possible obliteration of normal kidney and liver structure due to proliferation of large granulomatous lesions. Death was further attributed to heart failure resulting from invasion of the myocardium by phagocytic cells containing *R. salmoninarum,* and release of hydrocytic and oxidizing enzymes from the disrupted macrophages. Sami et al.[271] described an extended pathogenesis in which glomerulonephritis was associated with chronic bacterial kidney disease in rainbow trout: 15 months after infection, pathology consisted of fibrosis adhesions in the Bowman's capsule, shrinkage or swelling of the capillary, and degeneration of proximal tubules.

R. salmoninarum produces an extracellular protein (57 kDa — p57 or F antigen) in naturally and experimentally infected fish which is presumed to be a significant factor in the pathogen's virulence.[270] Wiens and Kaattari,[272] using monoclonal antibody, characterized p57 and showed that the protein could agglutinate salmonid leukocytes *in vitro*. Whether or not p57 serves the same function *in vivo* is not known; however, it was suggested by Daly and Stevenson[273] that p57 may aid *R. salmoninarum*'s ability to attach to host cells and allow intracellular invasion.

6. SIGNIFICANCE

Due to the widespread existence of BKD in hatchery and feral stocks of trout and anadromous salmon it is one of the most serious and controversial diseases of salmonids. Though it seldom causes acute losses, cumulative losses can be significant. *R. salmoninarum*'s unique ability to be transmitted vertically within the egg, coupled with its intriguing epizootiology and absence of effective therapeutics, emphasizes its importance and economical impact on aquaculture.

7. MANAGEMENT

Because of its complex biological and epizootiological nature, BKD is one of the most difficult infectious diseases of fish to control and requires a variety of integrated management procedures. Management involves reducing vertical and horizontal transmission, chemotherapy, chemoprophylaxis, as well as environmental and population manipulation.[274]

Avoidance is the best method of preventing BKD but may not be the most practical. Due to improved sensitivity tests to detect the presence of *R. salmoninarum* in brood stock, "BKD-free certification" is now more feasible. Other methods to reduce the impact of BKD are to prevent the vertical transmission of *R. salmoninarum* from carrier adults to their offspring and, in facilities where the pathogen is endemic, employ some specific management practices.

Chemoprophylaxis is one method used to reduce vertical transmission of *R. salmoninarum*. Prior to spawning, adult female salmon are injected once or twice with 11 to 20 mg of erythromycin phosphate per kilogram of body weight, with the last injection being administered from a few to 30 d before spawning.[275-277] This improves survival of brood fish and increases spawning success.

Culling and segregation of brood fish and their eggs are management approaches to reduce the prevalence of BKD in hatchery-reared Pacific salmon.[274] Culling of brood stock involves the removal of those fish from a spawning population based on their probability of transmitting *R. salmoninarum*. Their ability to transmit the pathogen is based on the presence of clinical signs of BKD, or by FAT or ELISA detection methods. Culling and segregation involves the destruction of gametes from brood fish that are most likely to transmit *R. salmoninarum* to their offspring. Each brood fish is tested for the presence of *R. salmoninarum* at the time of spawning and the gametes are held separately under refrigeration, for delayed fertilization, until the BKD status of each individual fish is known. Gametes from BKD-positive fish can be destroyed while gametes of BKD-negative fish are used for fertilization. In the case of eggs already fertilized, those from BKD-negative pairings are incubated separately from BKD-positive pairings. Pascho et al.[230] used ELISA and FAT to detect *R. salmoninarum* in brood chinook salmon for the purpose of reducing the incidence of BKD in juvenile fish. They suggest that the two detection methods and resulting segregation of eggs and progeny of BKD-negative adults from progeny of BKD-positive adults will reduce the prevalence and levels of BKD in spring chinook salmon.

Chemotherapy and chemoprophylaxis have been practiced with limited success in managing BKD-infected populations. Erythromycin has been most widely used as an oral treatment at 4.5 g/45 kg (about 100 mg/kg) of body weight daily for 10 to 21 d.[274] Moffitt[278] found that erythromycin fed at

200 mg/kg of body weight daily for 21 d reduced mortality more than lower dosages for a shorter duration, but palatability of the medicated feed was a problem. Overall, chemotherapy of clinically infected BKD fish does not appear to be overly successful because drugs do not eliminate *R. salmoninarum* from all treated fish; therefore, relapses can be expected following application.[279]

The National Workshop on Bacterial Kidney Disease addressed the management aspect of BKD and arrived at the following recommendations for the U.S. Fish and Wildlife Service BKD management strategy:[216]

1. There is no reason to destroy stocks of fish that are infected with *R. salmoninarum*.
2. Approved methods to detect *R. salmoninarum* in fish include ELISA, FAT, and membrane filtration using FAT.
3. If *R. salmoninarum* is present at any broodstock facility, the stocks should be managed to reduce the severity of infection and if necessary the fish should be stocked in areas where the pathogen is present.
4. Eggs should not be taken from clinically diseased fish unless a specific program (i.e., endangered species) requires use of such eggs.
5. Asymptomatically infected fish should not be stocked in areas that are free of *R. salmoninarum*.
6. There is no need to treat BKD differently in one region of the U.S. from any other region.

In facilities where *R. salmoninarum* is endemic there are some strategies that can reduce the effect of the disease (Recommendation 3). The reduction of fish densities in raceways or ponds to half the normal density and keeping water temperatures below stressful levels may reduce the impact of BKD. Stocking of potentially BKD-infected fish (only into waters where the disease is endemic), prior to the season when the disease is most serious, will reduce losses. Utilization of more BKD-resistant strains or species of trout is also advisable.

Since the first reported experimental vaccination for BKD by Evelyn[280] numerous other attempts have been made to examine the potential and feasibility for vaccination.[281-283] Although some successful vaccinations have been reported, overall results have not been encouraging. Some problems encountered have been in delivery, longevity of immune response, and the ability of immunity to be protective.

REFERENCES

1. **Austin, B. and Austin, D. A.**, *Bacterial Fish Pathogens: Diseases in Farmed and Wild Fish*, Ellis Horwood, Chichester, 1987, 364.
2. **McGraw, B. M.**, Furunculosis of Fish, U.S. Fish and Wild. Serv., Spec. Sci. Rep. Fish. Ser. 84, 1952.
3. **Schubert, R. H. W.**, The taxonomy and nomenclature of the genus *Aeromonas* Kluyver and van Niel 1936. I. Suggestions on the taxonomy and nomenclature of the aerogenic *Aeromonas* species, *Int. J. Syst. Bacteriol.*, 17, 23, 1967.
4. **Shotts, E. B., Jr. and Bullock G. L.**, Bacterial diseases of fishes: procedures for Gram negative pathogens, *J. Fish. Res. Board Can.*, 32, 1243, 1975.
5. **McCarthy, D. H.**, Fish furunculosis, *J. Inst. Fish. Manage.*, 6, 13, 1975.
6. **Trust, T. J., Khouri, A. G., Austin, R. A., and Ashburner, L. D.**, First isolation in Australia of atypical *Aeromonas salmonicida*, *FEMS Microbiol. Lett.*, 9, 315, 1980.
7. **Herman, R. L.**, Fish furunculosis 1952–1966, *Trans. Am. Fish. Soc.*, 97, 221, 1968.
8. **Wichardt, U.-P., Johansson, N., and Ljunberg, O.**, Occurrence and distribution of *Aeromonas salmonicida* infections on Swedish fish farms, 1951–1987, *J. Aquat. Anim. Health*, 1, 187, 1989.
9. **Nomura, T., Yoshimizu, M., and Kimura, T.**, An epidemiological study of furunculosis in salmon propagation, in *Salmonid Diseases*, Kimura, T., Ed., Hokkaido University Press, Hakodate, Japan, 1992, 187.
10. **Beacham, T. D. and Evelyn, T. P. T.**, Population variation in resistance of pink salmon to vibriosis and furunculosis, *J. Aquat. Anim. Health*, 4, 168, 1992.
11. **Fryer J. L. and Rohovec, J. S.**, Principal bacterial diseases of cultured marine fish, *Helgol. Wiss. Meeresunters.*, 37, 533, 1984.
12. **Ostland, V. E., Hicks, B. D., and Daly, J. G.**, Furunculosis in baitfish and its transmission to salmonids, *Dis. Aquat. Org.*, 2, 163, 1987.

13. **Snieszko, S. F.,** Fish furunculosis, Fishery Leaflet 467, U.S. Fish and Wildlife Service, Washington, D.C., 1958, 4.

14. **Cipriano, R. C., Ford, L. A., Teska, J. D., and Hale, L. L.,** Detection of *Aeromonas salmonicida* in mucus of salmonid fishes, *J. Aquat. Anim. Health,* 4, 114, 1992.

15. **Munro, A. L. S. and Hastings, T. S.,** Furunculosis, in *Bacterial Diseases of Fish,* Inglis, V., Roberts, R. J., and Bromage, N. R., Eds., Blackwell Scientific, Oxford, 1993, 122.

16. **Udey, L. R. and Fryer, J. L.,** Immunization of fish with bacterins of *Aeromonas salmonicida, Mar. Fish. Rev.,* 40, 12, 1978.

17. **Paterson, W. D., Douey, D., and Desautels, D.,** Relationships between selected strains of typical and atypical *Aeromonas salmonicida, Aeromonas hydrophila* and *Haemophilus piscium, Can. J. Microbiol.,* 26, 588, 1980.

18. **Popoff, M.,** *Aeromonas* Kluyer and Van Niel 1936, 398, in *Bergey's Manual of Systematic Bacteriology,* Vol. 1, Krieg, N. R. and Holt, J. G., Eds., Williams & Wilkins, Baltimore, 1984.

19. **Austin, B., Bishop, I., Gray, C., Watt, B., and Dawes, J.,** Monoclonal antibody-based enzyme-linked immunosorbent assays for the rapid diagnosis of clinical cases of enteric redmouth and furunculosis in fish farms, *J. Fish Dis.,* 9, 469, 1986.

20. **Rabb, M., Cornick, J. W., and MacDermott, L.,** A macroscopic slide agglutination test for the presumptive diagnosis of furunculosis in fish, *Prog. Fish Cult.,* 26, 118, 1964.

21. **McCarthy, D. H. and Rawle, C. T.,** The rapid serological diagnosis of fish furunculosis caused by "smooth" and "rough" strains of *Aeromonas salmonicida, J. Gen. Microbiol.,* 86, 185, 1975.

22. **Rockey, D. D., Fryer, J. L., and Rohovec, J. S.,** Separation and *in vivo* analysis of two extracellular proteases and the T-hemolysin from *Aeromonas salmonicida, Dis. Aquat. Org.,* 5, 197, 1988.

23. **Michel, C. and Dubois-Darnaudpeys, A.,** Persistence of the virulence of *Aeromonas salmonicida* strains kept in river sediments, *Ann. Rech. Vet.,* 11, 375, 1980.

24. **Allen-Austin, D., Austin, B., and Colwell, R. R.,** Survival of *Aeromonas salmonicida* in river water, *FEMS Microbiol. Lett.,* 21, 142, 1984.

25. **Rose, A. S., Ellis, A. E., and Munro, A. L. S.,** The survival of *Aeromonas salmonicida* subsp. *salmonicida* in sea water, *J. Fish Dis.,* 13, 205, 1990.

26. **McCarthy, D. H.,** Some ecological aspects of the bacterial fish pathogen *Aeromonas salmonicida,* in *Aquatic Microbiology,* Symp. Soc. Appl. Bacteriol. No. 6, Skinner, E. A. and Roberts, D. H., Eds., Academic Press, London, 1977, 299.

27. **Cipriano, R. C.,** Furunculosis in brook trout: infection by contact, *Prog. Fish Cult.,* 44, 12, 1982.

28. **Bullock, G. L. and Stuckey, H. M.,** *Aeromonas salmonicida* detection of asymptomatically infected trout, *Prog. Fish Cult.,* 37, 237, 1975.

29. **Kingsbury, O. R.,** A possible control of furunculosis, *Prog. Fish Cult.,* 23, 136, 1961.

30. **Piper, R. G., McElwain, I. B., Orme, L. E., McCraren, J. P., Fowler, L. G., and Leonard, J. R.,** Fish Hatchery Management, U.S. Fish and Wildlife Service, Washington, D.C., 1982.

31. **McCarthy, D. H. and Roberts, R. J.,** Furunculosis of fish. The state of our knowledge, in *Advances in Aquatic Microbiology,* Droop. M. A. and Janasch, H. W., Eds., Academic Press, London, 1980, 193.

32. **Klontz, G. W., Yasutake, W. T., and Ross, A. J.,** Bacterial diseases of the Salmonidae in the Western United States: pathogenesis of furunculosis in rainbow trout, *Am. J. Vet. Res.,* 27, 1455, 1966.

33. **Anderson, D. P.,** Virulence and persistence of rough and smooth forms of *Aeromonas salmonicida* inoculated into coho salmon *(Oncorhynchus kisutch), J. Fish. Res. Board Can.,* 29, 204, 1972.

34. **Kay, W. W., Buckley, J. T., Ishiguro, E. E., Phipps, B. M., Monette, J. P. L., and Trust, T. J.,** Purification and disposition of a surface protein associated with virulence of *Aeromonas salmonicida, J. Bacteriol.,* 147, 1077, 1981.

35. **Evenberg, D. and Lugtenberg, B.,** Cell surface of the fish pathogenic bacterium *Aeromonas salmonicida.* II. Purification and characterization of a major cell envelope protein related to autoagglutination, adhesion and virulence, *Biochem. Biophys. Acta,* 684, 249, 1982.

36. **Trust, T. J., Eshiguro, E. E., Chart, H., and Kay, W. W.,** Virulence properties of *Aeromonas salmonicida, J. World Maricult. Soc.,* 14, 193, 1983.

37. **Munroe, A. L. S., Hastings, T. S., Ellis, A. E., and Livorsidge, J.,** Studies on an ichthyotoxic material produced extracellularly by the furunculosis bacterium *Aeromonas salmonicida,* in *Fish Diseases,* Ahne, W., Ed., Springer-Verlag, Berlin, 1980, 98.

38. **Titball, R. W. and Munn, C. B.**, Evidence for two haemolytic activities from *Aeromonas salmonicida, FEMS Microbiol. Lett.*, 12, 27, 1981.

39. **Cipriano, R. C., Griffin, B. R., and Lidgerding, B. C.**, *Aeromonas salmonicida:* relationship between extracellular growth products and isolate virulence, *Can. J. Fish Aquat. Sci.*, 38, 1322, 1981.

40. **Ellis, A. E., Hastings, T. S., and Munro, A. L. S.**, The role of *Aeromonas salmonicida* extracellular products in the pathology of furunculosis, *J. Fish Dis.*, 4, 41, 1981.

41. **Inglis, V. and Richards, R. H.**, Difficulties encountered in chemotherapy of furunculosis in Atlantic salmon (*Salmo salar* L.), in *Salmonid Diseases*, Kimura, T., Ed., Hokkaido University Press, Hakodate, Japan, 1992, 201.

42. **Herman, R. L.**, Prevention and control of fish diseases in hatcheries, in *A Symposium on Diseases of Fishes and Shellfishes*, Snieszko, S. F., Ed., Spec. Publ. No. 5, American Fisheries Society, Washington, D.C., 1970, 3.

43. **Ehlinger, N. F.**, Selective breeding of trout for resistance to furunculosis, *N.Y. Fish Game J.*, 24, 26, 1977.

44. **McFaddin, T. W.**, Effective disinfection of trout eggs to prevent egg transmission of *Aeromonas liquefaciens, J. Fish. Res. Board Can.*, 26, 2311, 1969.

45. **Amend, D. F.**, Comparative toxicity of iodophores to rainbow trout eggs, *Trans. Am. Fish. Soc.*, 103, 73, 1974.

46. **Schnick, R. A., Meyer, F. P., and Gray, D. L.**, A Guide to Approved Chemicals in Fish Production and Fishery Resources Management, U.S. Fish and Wildlife Service and Arkansas Coop. Ext. Serv., MP241-5M-3-89RV, Little Rock, AR, 1989.

47. **Hastings, T. S.**, Furunculosis vaccines, in *Fish Vaccination*, Ellis, A. E., Ed., Academic Press, London, 1988, 93.

48. **Rodgers, C. J.**, Immersion vaccination for control of fish furunculosis, *Dis. Aquat. Org.*, 8, 69, 1990.

49. **Paterson, W. D., Parker, W., Poy, M., and Horne, M. T.**, Prevention of furunculosis using orally applied vaccines, in *Salmonid Diseases*, Kimura, T., Ed. Hokkaido University Press, Hakodate, Japan, 1992, 225.

50. **Colwell, R. R. and Grimes, D. J.**, *Vibrio* diseases of marine fish populations, *Helgol. Wiss. Meeresunters.*, 37, 265, 1984.

51. **Egidius, E., Wiik, R., Andersen, K., Hoff, K. A., and Hjeltnes, B.**, *Vibrio salmonicida* sp. nov., a new fish pathogen, *Int. J. Syst. Bacteriol.*, 36, 518, 1986.

52. **Muroga, K., Lio-po, G., Pitogo, C., and Imada, R.**, *Vibrio* sp. isolated from milkfish *(Chanos chanos)* with opaque eyes, *Fish Pathol.*, 19, 81, 1984.

53. **Rucker, R. R.**, *Vibrio* infections among marine and fresh-water fish, *Prog. Fish Cult.*, 21, 22, 1959.

54. **Muroga, K.**, Studies on *Vibrio anguillarum* and *V. anguillarum* infection, *J. Fac. Fish. Anim. Husb. (Hiroshima University)*, 14, 101, 1975.

55. **Kitao, T., Aoki, T., Fukudome, M., Kawano, K., Wada, Y., and Mizuno, Y.**, Serotyping of *Vibrio anguillarum* isolated from diseased freshwater fish in Japan, *J. Fish Dis.*, 6, 175, 1983.

56. **Tajima, K., Ezura, Y., and Kimura, T.**, Studies on the taxonomy and serology of causative organisms of fish vibriosis, *Fish Pathol.*, 20, 131, 1985.

57. **Schiewe, M. H.**, Taxonomic status of marine vibrios pathogenic for salmonid fish, *Dev. Biol. Stand. Fish Biol. Serodiag. Vacc.*, 49, 149, 1981.

58. **Schiewe, M. H., Trust, T. J., and Crosa, J. H.**, *Vibrio ordalii* sp. nov.: a causative agent of vibriosis in fish, *Curr. Microbiol.*, 6, 343, 1981.

59. **Muroga, K.**, Vibriosis of cultured fishes in Japan, in *Salmonid Diseases*, Kimura, T., Ed., Hokkaido University Press, Hakodate, Japan, 1992, 165.

60. **Anderson, J. I. W. and Conroy, D. A.**, *Vibrio* disease in marine fishes, in *A Symposium on Diseases of Fishes and Shellfishes*, Snieszko, S. F., Ed., Spec. Publ. No. 5, American Fisheries Society, Washington, D.C., 1970, 266.

61. **Evelyn, T. P. T.**, First records of vibriosis in Pacific salmon cultured in Canada, and taxonomic status of the responsible bacterium, *Vibrio anguillarum, J. Fish. Res. Board Can.*, 28, 517, 1971.

62. **Novotny, A. J. and Harrell, L. W.**, Vibriosis — a common disease of Pacific salmon cultured in marine waters of Washington, College of Agriculture Ext. Bull. 663, Washington State University, Pullman, 1975, 8.

63. **Traxler, G. S. and Li, M. F.,** *Vibrio anguillarum* isolated from a nasal abscess of the cod fish *(Gadus morhua), J. Wildl. Dis.,* 8, 207, 1972.
64. **Tajima, K., Yoshimizu, M., Ezura, Y., and Kimura, T.,** Causative organisms of vibriosis among pen cultured coho salmon *(Oncorhynchus kisutch)* in Japan, *Bull. Jpn. Soc. Sci. Fish.,* 47, 35, 1981.
65. **Ransom, D. P., Lannan, C. N., Rohovec, J. S., and Fryer, J. L.,** Comparison of histopathology caused by *Vibrio anguillarum* and *Vibrio ordalii* and three species of Pacific salmon, *J. Fish Dis.,* 7, 107, 1984.
66. **West, P. A., Lee, J. V., and Bryant, T. N.,** A numerical taxonomic study of species of *Vibrio* isolated from the aquatic environment and birds in Kent, England, *J. Appl. Bacteriol.,* 55, 263, 1983.
67. **Schiewe, M. H., Crosa, J. H., and Ordal, E. J.,** Deoxyribonucleic acid relationships among marine vibrios pathogenic to fish, *Can. J. Microbiol.,* 23, 954, 1977.
68. **Pacha, R. E. and Kiehn, E. D.,** Characterization and relatedness of marine vibrios pathogenic to fish: physiology, serology and epidemiology, *J. Bacteriol.,* 100, 1242, 1969.
69. **Hastein, T. and Smith, J. E.,** A study of *Vibrio anguillarum* from farms and wild fish using principal components analysis, *J. Fish Biol.,* 11, 69, 1977.
70. **Chart, H. and Trust, T. J.,** Characterization of the surface antigens of the marine fish pathogens, *Vibrio anguillarum* and *Vibrio ordalii, Can. J. Microbiol.,* 30, 703, 1984.
71. **West, P. A. and Lee, J. V.,** Ecology of *Vibrio* species, including *Vibrio cholerae,* in natural waters of Kent, England, *J. Appl. Bacteriol.,* 52, 435, 1982.
72. **Larsen, J. L.,** *Vibrio anguillarum:* a comparative study of fish pathogenic, environmental, and reference strains, *Acta Vet. Scand.,* 24, 456, 1983.
73. **Sawyer, E. S., Strout, R. G., and Coutermarsh, B. A.,** Comparative susceptibility of Atlantic *(Salmo salar)* and coho *(Oncorhynchus kisutch)* salmon to three strains of *Vibrio anguillarum* from the Maine-New Hampshire coast, *J. Fish. Res. Board Can.,* 36, 280, 1979.
74. **Thorburn, M. A.,** Factors influencing seasonal vibriosis mortality rates in Swedish pen-reared rainbow trout, *Aquaculture,* 67, 79, 1987.
75. **Groberg, W. J., Jr., Rohovec, J. S., and Fryer, J. L.,** The effects of water temperature on infection and antibody formation induced by *Vibrio anguillarum* in juvenile coho salmon *(Oncorhynchus kisutch), J. World Maricult. Soc.,* 14, 240, 1983.
76. **Harrell, L. W.,** Vibriosis and current salmon vaccination procedures in Puget Sound, Washington, *Mar. Fish. Rev.,* 40(3), 24, 1978.
77. **Watkins, W. D., Wolke, R. E., and Cabelli, V. J.,** Pathogenicity of *Vibrio anguillarum* for juvenile winter flounder, *Pseudopleuronectes americanus, Can. J. Fish. Aquat. Sci.,* 38, 1045, 1981.
78. **Muroga, K., Yamanoi, H., Hironaka, Y., Yamamoto, S., Tatani, M., Jo, Y., Takahashi, S., and Hanada, H.,** Detection of *Vibrio anguillarum* from wild fingerlings of ayu *Plecoglossus altivelis, Bull. Jpn. Soc. Sci. Fish.,* 50, 591, 1984.
79. **Harbell, S. C., Hodgins, H. O., and Schiewe, M. H.,** Studies on the pathogenesis of vibriosis in coho salmon *Oncorhynchus kisutch* (Walbaum), *J. Fish Dis.,* 2, 391, 1979.
80. **Miyazaki, T. and Kubota, S. S.,** Histopathological studies on vibriosis of the Japanese eel *(Anguilla japonica).* II. Responses of fishes pretreated with bacterial sonicate against challenge with viable bacteria, *Fish Pathol.,* 16, 101, 1981.
81. **Ransom, D. P.,** Bacteriologic, Immunologic and Pathologic Studies of *Vibrio* sp., Pathogenic to Salmonids, Ph.D. dissertation, Oregon State University, Corvallis, 1978.
82. **Umbreit, T. H. and Tripp, M. R.,** Characterization of the factors responsible for death of fish infected with *Vibrio anguillarum, Can. J. Microbiol.,* 21, 1272, 1975.
83. **Austin, B., Morgan, D. A., and Alderman, D. J.,** Comparison of antimicrobial agents for control of vibriosis in marine fish, *Aquaculture,* 26, 1, 1981.
84. **Rohovec, J. S., Garrison, R. L., and Fryer, J. L.,** Immunization of fish from the control of vibriosis, Third U.S.-Japan Meeting on Aquaculture, Tokyo, October 15 to 16, 1975, 105.
85. **Amend, D. F. and Johnson, K. A.,** Current status and future needs of *Vibrio anguillarum* bacterins, *Dev. Biol. Stand. Fish Biol. Serodiag. Vacc.,* 49, 403, 1981.
86. **Song, Y.-L., Chen, S.-N., Kou, G.-H., Lin, C.-I., and Ting, Y.-Y.,** Evaluation of Hivax *Vibrio anguillarum* bacterin in the vaccination of milkfish *(Chanos chanos)* fingerlings, *CAPD Fish. Ser. No. 3, Rep. Fish Dis. Res.,* III, 103, 1980.
87. **Kawano, K., Aoki, T., and Kitao, T.,** Duration of protection against vibriosis in ayu *Plecoglossus altivelis* vaccinated by immersion and oral administration with *Vibrio anguillarum, Bull. Jpn. Soc. Sci. Fish.,* 50, 771, 1984.

88. **Tiecco, G., Sebastio, C., Francioso, E., Tantillo, G., and Corbari, L.,** Vaccination trials against "red plague" in eels, *Dis. Aquat. Org.,* 4, 105, 1988.

89. **Rogers, W. A. and Xu, D.,** Protective immunity induced by a commercial *Vibrio* vaccine in hybrid striped bass, *J. Aquat. Anim. Health,* 4, 303, 1992.

90. **Horne, M. T., Tatner, M., McDerment, S., and Agius, C.,** Vaccination of rainbow trout, *Salmo gairdneri* Richardsons, at low temperatures and the long-term persistence of protection, *J. Fish Dis.,* 5, 343, 1982.

91. **Håstein, T., Hallingstad, F., Refsti, T., and Roald, S. O.,** Recent experience of field vaccination trials against vibriosis in rainbow trout *(Salmo gairdneri),* in *Fish Diseases,* Ahne, W., Ed., 3rd COPRAQ session, Springer-Verlag, Berlin, 1980, 53.

92. **Thornburn, M. A., Carpenter, T. E., and Plant, R. E.,** Perceived vibriosis risk by Swedish rainbow trout net-pen farmers: its effect on purchasing patterns and willingness-to-pay for vaccination, *Prev. Vet. Med.,* 4, 419, 1987.

93. **Hjeltnes, B. and Roberts, R. J.,** Vibriosis, in *Bacterial Diseases of Fish,* Inglis, V., Roberts, R. J., and Bromage, N., Eds., Blackwell Scientific, Oxford, 1993, 109.

94. **Egidius, E., Andersen, K., Clausen, E., and Raa, J.,** Cold-water vibriosis or "Hitra disease" in Norwegian salmonid farming, *J. Fish Dis.,* 4, 353, 1981.

95. **Holm, K. O., Strøm, E., Stensvåg, K., Raa, J., and Jørgensen, T.,** Characteristics of a *Vibrio* sp. associated with the Hitra disease of Atlantic salmon in Norwegian fish farms, *Fish Pathol.,* 20, 125, 1985.

96. **Egidius, E., Wiik, R., Andersen, K., Hoff, K. A., and Hjeltnes, B.,** *Vibrio salmonicida* sp. nov., a new fish pathogen, *Int. J. Syst. Bacteriol.,* 36, 518, 1986.

97. **Fjølstad, M. and Heyeraas, A. L.,** Muscular and myocardial degeneration in cultured Atlantic salmon, *Salmo salar* L., suffering from "Hitra disease", *J. Fish Dis.,* 8, 367, 1985.

98. **Poppe, T. T., Håstein, T., and Salte, R.,** "Hitra disease" (haemorrhagic syndrome) in Norwegian salmon farming: present status, in *Fish and Shellfish Pathology,* Ellis, A. E., Ed., Academic Press, New York, 1985, 22.

99. **Poppe, T. T., Håstein, T., Frøslie, A., Koppang, N., and Norheim, G.,** Nutritional aspects of haemorrhagic syndrome ("Hitra disease") in farmed Atlantic salmon *Salmo salar, Dis. Aquat. Org.,* 1, 155, 1986.

100. **Bruno, D. W., Hastings, T. S., and Ellis, A. E.,** Histopathology, bacteriology and experimental transmission of cold-water vibriosis in Atlantic salmon *Salmo salar, Dis. Aquat. Org.,* 1, 163, 1986.

101. **Hjeltnes, B. Andersen, K., Ellingsen, H. M., and Egidius, E.,** Experimental studies on the pathogenicity of a *Vibrio* sp. isolated from Atlantic salmon, *Salmo salar* L., suffering from Hitra disease, *J. Fish Dis.,* 10, 21, 1987.

102. **Dalsgaard, I., Jurgens, O., and Mortensen, A.,** *Vibrio salmonicida* isolated from farmed Atlantic salmon in the Faroe Islands, *Bull. Eur. Assoc. Fish Pathol.,* 8, 53, 1988.

103. **Mitchell, H.,** Connors Aquaculture, Inc., Eastport, Maine, personal communication, 1993.

104. **Egidius, E., Soleim, O., and Andersen, K.,** Further observations on cold-water vibriosis of Hitra disease, *J. Fish Dis.,* 10, 85, 1984.

105. **Jørgensen, T., Midling, K., Espelid, S., Nilsen, R., and Stensvåg, K.,** *Vibrio salmonicida,* a pathogen in salmonids, also causes mortality in net-pen captured cod *(Gadus morhua), Bull. Eur. Assoc. Fish Pathol.,* 9, 42, 1989.

106. **Schroder, M. B., Espelid, S., and Jørgensen, T. O.,** Two serotypes of *Vibrio salmonicida* isolated from diseased cod *(Gadus morhua* L.); virulence, immunological studies and vaccination experiments, *Fish Shellfish Immunol.,* 2, 211, 1992.

107. **Espelid, S., Holm, K. O., Hjelmeland, K., and Jørgensen, T.,** Monoclonal antibodies against *Vibrio salmonicida:* the causative agent of coldwater vibriosis ("Hitra disease") in Atlantic salmon, *Salmo salar* L., *J. Fish Dis.,* 11, 207, 1988.

108. **Evensen, O., Espelid, S., and Håstein, T.,** Immunohistochemical identification of *Vibrio salmonicida* in stored tissue of Atlantic salmon *Salmo salar* from the first known outbreak of cold-water vibriosis ("Hitra disease"), *Dis. Aquat. Org.,* 10, 185, 1991.

109. **Wiik, R., Andersen, K., Daae, F. L., and Hoff, F. A.,** Virulence studies based on plasmid profiles of the fish pathogen *Vibrio salmonicida, Appl. Environ. Microbiol.,* 55, 819, 1989.

110. **Hjeltnes, B., Andersen, K., and Egidius, K.,** Multiple antibiotic resistance in *Vibrio salmonicida, Bull. Eur. Assoc. Fish Pathol.,* 7, 85, 1987.

111. **Hjeltnes, B. and Julshamn, K.**, Concentrations of iron, copper, zinc and selenium in liver of Atlantic salmon *Salmo salar* infected with *Vibrio salmonicida, Dis. Aquat. Org.,* 12, 147, 1992.

112. **Hjeltnes, B.**, Institute of Marine Research, Bergen, Norway, personal communication, 1993.

113. **Totland, G. K., Nylund, A., and Holm, K. O.**, An ultrastructural study of morphological changes in Atlantic salmon, *Salmo salar* L., during the development of cold water vibriosis, *J. Fish Dis.,* 11, 1, 1988.

114. **Holm, K. O. and Jørgensen, T.**, A successful vaccination of Atlantic salmon, *Salmo salar* L., against "Hitra disease" or coldwater vibriosis, *J. Fish Dis.,* 10, 85, 1987.

115. **Lillehaug, A. A.**, A field trial of vaccination against cold-water vibriosis in Atlantic salmon (*Salmo salar* L.), *Aquaculture,* 84, 1, 1990.

116. **Lillehaug, A., Sorum, R. H., and Ramstad, A.**, Cross-protection after immunization of Atlantic salmon, *Salmo salar* L., against different strains of *Vibrio salmonicida, J. Fish Dis.,* 13, 519, 1990.

117. **Hjeltnes, B., Andersen, K., and Ellingsen, H. M.**, Vaccination against *Vibrio salmonicida:* the effect of different routes of administration and revaccination, *Aquaculture,* 83, 1, 1989.

118. **Lillehaug, A.**, Vaccination of Atlantic salmon (*Salmo salar* L.) against coldwater vibriosis — duration of protection and effect on growth rate, *Aquaculture,* 92, 99, 1991.

119. **Ross, A. J., Rucker, R. R., and Ewing, W. H.**, Description of a bacterium associated with redmouth disease of rainbow trout (*Salmo gairdneri*), *Can. J. Microbiol.,* 19, 763, 1966.

120. **Bullock, G. L., Stuckey, H. M., and Shotts, E. B., Jr.**, Enteric redmouth bacterium: comparison of isolates from different geographic areas, *J. Fish Dis.,* 1, 351, 1978.

121. **Davies, R. L. and Frerichs, G. N.**, Morphological and biochemical differences among isolates of *Yersinia ruckeri* obtained from wide geographical areas, *J. Fish Dis.,* 12, 357, 1989.

122. **Stevenson, R., Flett, D., and Raymond, B. T.**, Enteric redmouth (ERM) and other enterobacterial infections of fish, in *Bacterial Diseases of Fish,* Inglis, V., Roberts, R. J., and Bromage, N. R., Eds., Blackwell Scientific, Oxford, 1993, 80.

123. **Kusuda, R. and Kawai, K.**, Bacterial infections of fish in Japan, in *Salmonid Diseases,* Kimura, T., Ed., Hokkaido University Press, Hakodate, Japan, 1992, 137.

124. **Bullock, G. L. and Snieszko, S. F.**, Enteric Redmouth Disease of Salmonids, Fish Disease Leaflet No. 57, U.S. Fish and Wildlife Service, Washington, D.C., 1979.

125. **Busch, R. A.**, Enteric redmouth disease (*Yersinia ruckeri*), in *Les Antigenes des Micro-organismes Pathogenes des Poissons,* Anderson, D. P., Dorson, M., and Dubourget, Ph., Eds., Collection Fondation Marcel Meriux, Lyon, France, 1983, 201.

126. **McAardle, J. F. and Dooley-Martyn, C.**, Isolation of *Yersinia ruckeri* Type I (Hagerman strain) from goldfish *Carassius auratus* (L), *Bull. Eur. Assoc. Fish Pathol.,* 5, 10, 1985.

127. **Michel, C., Faivre, B., and de Kinkelin, P.**, A clinical case of enteric redmouth in minnows (*Pimephales promelas*) imported in Europe as bait-fish, *Bull. Eur. Assoc. Fish Pathol.,* 6, 97, 1986.

128. **Busch, R. A. and Lingg, A. J.**, Establishment of an asymptomatic carrier state infection of enteric redmouth disease in rainbow trout (*Salmo gairdneri*), *J. Fish. Res. Board Can.,* 32, 2429, 1975.

129. **Frerichs, G. N., Stewart, J. A., and Collins, R. O.**, Atypical infection of rainbow trout, *Salmo gairdneri* Richardson, with *Yersinia ruckeri, J. Fish Dis.,* 8, 383, 1985.

130. **Meier, W.**, Enteric redmouth disease: outbreak in rainbow trout in Switzerland, *Dis. Aquat. Org.,* 2, 81, 1986.

131. **Fuhrmann, H., Bohm, K. H., and Schlotfeldt, H. J.**, On the importance of enteric bacteria in the bacteriology of freshwater fish, *Bull. Eur. Assoc. Fish Pathol.,* 4, 42, 1984.

132. **Wobeser, G.**, An outbreak of redmouth disease in rainbow trout (*Salmo gairdneri*) in Saskatchewan, *J. Fish. Res. Board Can.,* 30, 571, 1973.

133. **Waltman, W. D. and Shotts, E. B., Jr.**, A medium for the isolation and differentiation of *Yersinia ruckeri, Can. J. Fish. Aquat. Sci.,* 41, 804, 1984.

134. **Hastings, T. S. and Bruno, D. W.**, Enteric redmouth disease: survey in Scotland and evaluation of a new medium, Shotts-Waltman, for differentiating *Yersinia ruckeri, Bull. Eur. Assoc. Fish Pathol.,* 5(2), 32, 1985.

135. **Rodgers, C. J.**, Development of a selective-differential medium for the isolation of *Yersinia ruckeri* and its application in epidemiological studies, *J. Fish Dis.,* 15, 243, 1992.

136. **Ewing, W. H., Ross, A. J., Brenner, D. J., and Fanning, G. R.**, *Yersinia ruckeri* sp. nov., the redmouth (RM) bacterium, *Int. J. Syst. Bacteriol.,* 28, 37, 1978.

137. **Stevenson, R. M. W. and Daly, J. G.**, Biochemical and serological characteristics of Ontario isolates of *Yersinia ruckeri, Can. J. Fish. Aquat. Sci.,* 39, 870, 1982.

138. **Cipriano, R. C. and Pyle, J. B.**, Development of a culture medium for determination of sorbitol utilization among strains of *Yersinia ruckeri*, *Microbios Lett.*, 28, 79, 1985.

139. **Valtonen, E. T., Rintamaki, P., and Koskivaara, M.**, Occurrence and pathogenicity of *Yersinia ruckeri* at fish farms in northern and central Finland, *J. Fish Dis.*, 15, 163, 1992.

140. **de Grandis, S. A., Krell, P. J., Flett, D. E., and Stevenson, R. M. W.**, Deoxyribonucleic acid relatedness of serovars of *Yersinia ruckeri*, the enteric redmouth bacterium, *Int. J. Syst. Bacteriol.*, 38, 49, 1988.

141. **Johnson, K. A., Wobeser, G., and Rouse, B. T.**, Indirect fluorescent antibody technique for detection of RM bacterium of rainbow trout *(Salmo gairdneri)*, *J. Fish. Res. Board Can.*, 31, 1975, 1974.

142. **Hansen, C. B. and Lingg, A. J.**, Inert particle agglutination tests for detection of antibody to enteric redmouth bacterium, *J. Fish. Res. Board Can.*, 33, 2857, 1976.

143. **Cipriano, R. C., Schill, W. B., Pyle, S. W., and Horner, R.**, An epizootic in chinook salmon *(Oncorhynchus tshwaytscha)* caused by a sorbitol-positive serovar II strain of *Yersinia ruckeri*, *J. Wildl. Dis.*, 22, 488, 1986.

144. **Stevenson, R. M. W. and Airdrie, D. W.**, Serological variation among *Yersinia ruckeri* strains, *J. Fish Dis.*, 7, 247, 1984.

145. **Bullock, G. L., Stuckey, H. M., and Shotts, E. B., Jr.**, Early records of North American and Australian outbreaks of enteric redmouth disease, *Fish Health News*, 6(2), 96, 1977.

146. **Daly, J. G., Lindvik, B., and Stevenson, R. M. W.**, Serological heterogeneity of recent isolates of *Yersenia ruckeri* from Ontario and British Columbia, *Dis. Aquat. Org.*, 1, 151, 1986.

147. **Hester, F. E.**, Fish health: a nationwide survey of problems and needs, *Prog. Fish Cult.*, 35, 11, 1973.

148. **Hunter, V. A., Knittel, M. D., and Fryer, J. L.**, Stress-induced transmission of *Yersinia ruckeri* infection from carriers to recipient steelhead trout *Salmo gairdneri* Richardson, *J. Fish Dis.*, 3, 467, 1980.

149. **Knittel, M. D.**, Susceptibility of steelhead trout *Salmo gairdneri* Richardson to redmouth infection *Yersinia ruckeri* following exposure to copper, *J. Fish Dis.*, 4, 33, 1981.

150. **Willumsen, B.**, Birds and wild fish as potential vectors of *Yersinia ruckeri*, *J. Fish Dis.*, 12, 275, 1989.

151. **McDaniel, D. W.**, Hatchery Biologist Quarterly Report, First Quarter, 1972, U.S. Fish and Wildlife Service, Washington, D.C., 1972.

152. **Quentel, C. and Aldrin, J. F.**, Blood changes in catheterized rainbow trout *(Salmo gairdneri)* intraperitoneally inoculated with *Yersinia ruckeri*, *Aquaculture*, 53, 169, 1986.

153. **Busch, R. A.**, Enteric red mouth disease (Hagerman strain), MFR Pap. 1296, *Mar. Fish. Rev.*, 40(3), 42, 1978.

154. **Ross, A. J. and Smith, C. A.**, Effect of two iodophors on bacterial and fungal fish pathogens, *J. Fish. Res. Board Can.*, 29, 1359, 1972.

155. **Bullock, G. L., Maestrone, G., Starliper, C., and Schill, B.**, Potentiated sulfonamide therapy of enteric redmouth disease, *Can. J. Fish. Aquat. Sci.*, 40, 101, 1983.

156. **Ross, A. J. and Klontz, G. W.**, Oral immunization of rainbow trout *(Salmo gairdneri)* against the etiologic agent of "redmouth disease", *J. Fish. Res. Board Can.*, 22, 713, 1965.

157. **Horne, M. T. and Robertson, D. A.**, Economics of vaccination against enteric redmouth disease of salmonids, *Aquacult. Fish. Manage.*, 18, 131, 1987.

158. **Johnson, K. A., Flynn, J. K., and Amend, D. F.**, Onset of immunity in salmonid fry vaccinated by direct immersion in *Vibrio anguillarum* and *Yersinia ruckeri* bacterins, *J. Fish Dis.*, 5, 197, 1982.

159. **Johnson, K. A., Flynn, J. K., and Amend, D. F.**, Duration of immunity in salmonids vaccinated by direct immersion with *Yersinia ruckeri* and *Vibrio anguillarum* bacterins, *J. Fish Dis.*, 5, 206, 1982.

160. **Lamers, C. H. J. and Muiswinkel, W. B.**, Primary and secondary immune responses in carp *(Cyprinus carpio)* after administration of *Yersinia ruckeri* O-antigen, in *Acuigrup Fish Diseases*, Editora ATP, Madrid, 1984, 119.

161. **Tebbit, G. L., Erickson, J. D., and VandeWater, R. B.**, Development and use of *Yersinia ruckeri* bacterins to control enteric redmouth disease, *Dev. Biol. Stand.*, 49, 395, 1981.

162. **Rodgers, C. J.**, The use of vaccination and antimicrobial agents for control of *Yersinia ruckeri*, *J. Fish Dis.*, 14, 291, 1991.

163a. **Anderson, D. P., Dixon, O. W., and Robertson, B. S.,** Kinetics of the primary immune response in rainbow trout after flush exposure to *Yersinia ruckeri* O-antigen, *Dev. Comp. Immunol.,* 3, 739, 1979.

163b. **Davis, H. S.,** A new gill disease of trout, *Trans. Am. Fish. Soc.,* 56, 156, 1926.

164. **Wakabayashi, H., Egusa, S., and Fryer, J. L.,** Characteristics of filamentous bacteria isolated from a gill disease of salmonids, *Can. J. Fish. Aquat. Sci.,* 37, 1499, 1980.

165. **Wakabayashi, H., Huh, G. J., and Kimura, N.,** *Flavobacterium branchiophila* sp. nov., a causative agent of bacterial gill disease of freshwater fishes, *Int. J. Syst. Bacteriol.,* 39, 213, 1989.

166. **von Graevenitz, A.,** Revised nomenclature of *Campylobacter laridis, Enterobacter intermedium* and *"Flavobacterium branchiophila", Int. J. Syst. Bacteriol.,* 40, 211, 1990.

167. **Strohl, W. R. and Tait, L. R.,** *Cytophaga aquatilis* sp. nov., a facultative anaerobe isolated from the gill of freshwater fish, *Int. J. Syst. Bacteriol.,* 28, 293, 1978.

168. **Bullock, G. L.,** Studies on selected myxobacteria pathogenic for fishes and on bacterial gill disease in hatchery-reared salmonids, *Bur. Sport Fish. Wildl. Res. Rep.,* 60, 30, 1972.

169. **Kimura, N., Wakabayashi, H., and Kudo, S.,** Studies on bacterial gill disease in salmonids. I. Selection of bacterium transmitting gill disease, *Fish Pathol.,* 12, 233, 1978.

170. **Ferguson, H. W., Ostland, V. E., Byrne, P., and Lumsden, J. S.,** Experimental production of bacterial gill disease in trout by horizontal transmission and by bath challenge, *J. Aquat. Anim. Health,* 3, 118, 1991.

171. **Ostland, V. E., Ferguson, H. W., and Stevenson, R. M. V.,** Case report: bacterial gill disease in goldfish *Carassias auratus, Dis. Aquat. Org.,* 6, 179, 1989.

172. **Huh, G.-J. and Wakabayashi, H.,** Serological characteristics of *Flavobacterium branchiophila* isolated from gill diseases of freshwater fishes in Japan, USA, and Hungary, *J. Aquat. Anim. Health,* 1, 142, 1989.

173. **Ostland, V. E., Lumsden, J. S., MacPhee, D. D., and Ferguson, H. W.,** Characteristics of *Flavobacterium branchiophilum,* the cause of salmonid bacterial gill disease in Ontario, Canada, *J. Aquat. Anim. Health,* 6(1), in press.

174. **Speare, D. J., Ferguson, H. W., Beamish, F. W. M., Yager, J. A., and Yamashiro, S.,** Pathology of bacterial gill disease: ultrastructure of branchial lesions, *J. Fish Dis.,* 14, 1, 1991.

175. **Speare, D. J., Ferguson, H. W., Beamish, F. W. M., Yager, J. A., and Yamashiro, S.,** Pathology of bacterial gill disease: sequential development of lesions during natural outbreaks of disease, *J. Fish Dis.,* 14, 21, 1991.

176. **Wakabayashi, H. and Iwado, T.,** Changes in glycogen, pyruvate and lactate in rainbow trout with bacterial gill disease, *Fish Pathol.,* 20, 161, 1985.

177. **Holt, R.,** Department of Microbiology, Oregon State University, Corvallis, personal communication, 1992.

178. **Heo, G.-J., Kasai, K., and Wakabayashi, H.,** Occurrence of *Flavobacterium branchiophila* associated with bacterial gill disease at a trout hatchery, *Fish Pathol.,* 25, 99, 1990.

179. **Ostland, V. E., Ferguson, H. W., Prescott, J. F., Stevenson, R. M. W., and Barker, I. K.,** Bacterial gill disease of salmonids; relationship between the severity of gill lesions and bacterial recovery, *Dis. Aquat. Org.,* 9, 5, 1990.

180. **Kudo, S. and Kimura, N.,** Scanning electron microscopic studies on bacterial gill disease in rainbow trout fingerlings, *Jpn. J. Ichthyol.,* 30(4), 393, 1984.

181. **Byrne, P., Ferguson, H. W., Lumsden, J. S., and Ostland, V. E.,** Blood chemistry of bacterial gill disease in brook trout *Salvelinus fontinalis, Dis. Aquat. Org.,* 10, 1, 1991.

182. **Speare, D. J. and Ferguson, H. W.,** Clinical features of bacterial gill disease of salmonids in Ontario, *Can. Vet. J.,* 30, 882, 1989.

183. **Snieszko, S. F.,** Bacterial gill disease of freshwater fishes, Fish Disease Leaflet No. 6, U.S. Fish and Wildlife Service, Washington, D.C., 1981.

184. **Davis, H. S.,** Care and Diseases of Trout, Res. Rep. No 12, U.S. Department of the Interior, Washington, D.C., 1946.

185. **Bernardet, J. F. and Grimont, P. A. D.,** Deoxyribonucleic acid relatedness and phenotypic characterization of *Flexibacter columnaris* sp., nov., nom. rev., *Flexibacter maritimus* Wakabyashi, Hikida and Masumura 1986, *Int. J. Syst. Bacteriol.,* 39, 346, 1989.

186. **Holt, R. A.,** Bacterial cold-water disease, in *Bacterial Diseases of Fish,* Inglis, V., Roberts, R. J., and Bromage, N. R., Eds., Blackwell Scientific, Oxford, 1993, 3.

187. **Anderson, J. I. W. and Conroy, D. A.**, The pathogenic myxobacteria with special reference to fish disease, *J. Appl. Bacteriol.*, 32, 30, 1969.

188. **Wakabayashi, H., Horiuchi, M., Bunya, T., and Hoshiai, G.**, Outbreaks of cold-water disease in coho salmon in Japan, *Fish Pathol.*, 26, 211, 1992.

189. **Bernardet, J. F. and Kerouault, B.**, Phenotypic and genomic studies of *Cytophaga psychrophila* isolated from diseased rainbow trout *(Oncorhynchus mykiss)* in France, *Appl. Environ. Microbiol.*, 55, 1795, 1989.

190. **Rucker, R. R., Earp. B. J., and Ordal, E. J.**, Infectious diseases of Pacific salmon, *Trans. Am. Fish. Soc.*, 83, 297, 1953.

191. **Schachte, J. H.**, Coldwater disease, in *A Guide to Integrated Fish Health Management in the Great Lakes Basin*, Spec. Publ. No. 83-2, Meyer, F. P., Warren, J. W., and Carey, T. G., Eds., Great Lakes Fisheries Commission, Ann Arbor, MI, 1983, 193.

192. **Amos, K., Ed.**, *Procedures For the Detection and Identification of Certain Fish Pathogens*, 3rd ed., Fish Health Section, American Fisheries Society, Corvallis, OR, 1985, 57.

193. **Lehmann, J., Mock, D., Sturenberg, F. J., and Bernard, J. F.**, First isolation of *Cytophaga psychrophilus* from a systemic disease in eel and cyprinids, *Dis. Aquat. Org.*, 10, 217, 1991.

194. **Wood, J. W.**, *Diseases of Pacific Salmon: Their Prevention and Treatment*, 2nd ed., Department of Fisheries, Hatchery Division, Washington State, Olympia, 1974, 22.

195. **Kent, M. L., Groff, J. M., Morrison, J. K., Yasutake, W. T., and Holt, R. A.**, Spiral swimming behavior due to cranial and vertebral lesions associated with *Cytophaga psychrophila* infections in salmonid fishes, *Dis. Aquat. Org.*, 6, 11, 1989.

196. **Anacker, R. L. and Ordal, E. J.**, Studies on the myxobacterium *Chondrococcus columnaris*. I. Serological typing, *J. Bacteriol.*, 78, 25, 1959.

197. **Wakabayashi, H. and Egusa, S.**, Characteristics of myxobacteria associated with some freshwater fish diseases in Japan, *Bull. Jpn. Soc. Sci. Fish.*, 40, 751, 1974.

198. **Bullock, G. L., Hsu, T. C., and Shotts, E. B.**, Columnaris Disease of Fishes, Fish Disease Leaf. No. 72, U.S. Fish and Wildlife Service, Washington, D.C., 1974, 9.

199. **Holt, R. A., Amandi, A., Rohovec., J. S., and Fryer, J. L.**, Relation of water temperature to bacterial cold-water disease in coho salmon, chinook salmon and rainbow trout, *J. Aquat. Anim. Health*, 1, 97, 1989.

200. **Lorenzen, E. and Karas, N.**, Detection of *Flexibacter psychrophilus* by immunofluorescence in fish suffering from fry mortality syndrome: a rapid diagnostic method, *Dis. Aquat. Org.*, 13, 231, 1992.

201. **Wood, J. W.**, Diseases of Pacific Salmon: Their Prevention and Treatment, 2nd ed., Department of Fisheries, Washington State, Olympia, 1974.

202. **Leek, S. L.**, Viral erythrocytic inclusion body syndrome (EIBS) occurring in juvenile spring chinook salmon *(Oncorhynchus tshwytscha)* reared in fresh water, *Can. J. Fish. Aquat. Sci.*, 44, 685, 1987.

203. **Meyers, T. R.**, Apparent chronic bacterial myeloencephalitis in hatchery-reared juvenile coho salmon *Oncorhynchus kisutch* in Alaska, *Dis. Aquat. Org.*, 6, 217, 1989.

204. **Bullock, G. L. and Snieszko, S. F.**, Fin Rot, Coldwater Disease, and Peduncle Disease of Salmonid Fishes, Fish. Leaf. No. 462, Department of the Interior, Washington, D.C., 1970.

205. **Holt, R. A.**, Characterization and Control of *Cytophaga psychrophila* (Borg), the Causative Agent of Low Temperature Disease in Young Coho Salmon *(Oncorhynchus kisutch)*, M.S. thesis, Oregon State University, Corvallis, 1972.

206. **Wood, E. M. and Yasutake, W. T.**, Histopathology of fish. III. Peduncle ("cold-water") disease, *Prog. Fish Cult.*, 18, 58, 1956.

207. **Otis, E. J.**, Lesions of Coldwater Disease in Steelhead Trout *(Salmo gairdneri):* The Role of *Cytophaga psychrophila* Extracellular Products, M.S. Thesis, University of Rhode Island, Kingston, 1984; as cited in Holt, R. A., in *Bacterial Diseases of Fish*, Blackwell Scientific, Oxford, 1993, 3.

208. **Amend, D. F.**, Myxobacterial infections of salmonids: prevention and treatment, in *A Symposium on Diseases of Fishes and Shellfishes*, Snieszko, S. F., Ed., Spec. Publ. No. 5, American Fisheries Society, Washington, D.C., 1970, 258.

209. **Amend, D. F.**, Comparative toxicity of two iodophors to rainbow trout eggs, *Trans. Am. Fish. Soc.*, 103, 763, 1974.

210. **Obach, A. and Laurencin, F. B.**, Vaccination of rainbow trout *Oncorhynchus mykiss* against the visceral form of coldwater disease, *Dis. Aquat. Org.*, 12, 13, 1991.

211. **Sanders, J. E. and Fryer, J. L.,** *Renibacterium salmoninarum* gen. nov., sp. nov., the causative agent of bacterial kidney disease in salmonid fishes, *Int. J. Syst. Bacteriol.,* 30, 496, 1980.

212. **Smith, I. W.,** The occurrence of pathology of Dee disease, *U.S. Dep. Agric. Fish. Scot. Freshwa. Sal. Fish. Res.,* 34, 1, 1964.

213. **Belding, D. L. and Merill, B.,** A preliminary report upon a hatchery disease of Salmonidae, *Trans. Am. Fish. Soc.,* 65, 76, 1935.

214. **Evelyn, T. P. T. and Prosperi-Porta, L.,** A new medium for growing the kidney disease bacterium: its performance relative to that of other currently used media, in *Salmonid Diseases,* Kimura, T., Ed., Hokkaido University Press, Hakodate, Japan, 1992, 143.

215. **Fryer, J. L. and Sanders, J. L.,** Bacterial kidney disease of salmonid fish, *Annu. Rev. Microbiol.,* 35, 273, 1981.

216. **Mangin, S., Ed.,** National Workshop on Bacterial Kidney Disease, U.S. Fish and Wildlife Service, Phoenix, AZ, Nov. 20 to 21, 1991.

217. **Evelyn, T. P. T.,** Bacterial kidney disease — BKD, in *Bacterial Diseases of Fish,* Inglis, V., Roberts, R. J., and Bromage, N. R., Eds., Blackwell Scientific, Oxford, 1993, 177.

218. **Bell, G. R., Hoffmann, R. W., and Brown, L. L.,** Pathology of experimental infections of the sablefish, *Anoplopoma fimbria* (Pallas), with *Renibacterium salmoninarum,* the agent of bacterial kidney disease in salmonids, *J. Fish Dis.,* 13, 355, 1990.

219. **Traxler, G. S. and Bell, G. R.,** Pathogens associated with impounded Pacific herring *Clupea harengus pallasi,* with emphasis on viral erythrocytic necrosis (VEN) and atypical *Aeromonas salmonicida, Dis. Aquat. Org.,* 5, 93, 1988.

220. **Hicks, B. D., Daly, J. G., and Ostland, V. E.,** Experimental infection of minnows with the bacterial kidney disease bacterium *Renibacterium salmoninarum,* 3rd Annu. Meet. Aquacult. Assoc. Can., Guelph, Ontario, 1986 (Abstr.).

221. **Hendricks, J. D. and Leek, S. L.,** Kidney disease postorbital lesions in spring chinook salmon *(Oncorhynchus tshawytscha), Trans. Am. Fish. Soc.,* 104, 805, 1975.

222. **Hoffman, R., Popp, W., and Van der Graaff, S.,** Atypical BKD predominantly causing ocular and skin lesions, *Bull. Eur. Assoc. Fish Pathol.,* 4, 7, 1984.

223. **Bullock, G. L., Griffin, B. R., and Stuckey, H. M.,** Detection of *Corynebacterium salmoninus* by direct fluorescent antibody test, *Can. J. Fish. Aquat. Sci.,* 37, 719, 1980.

224. **Pascho, R. J. and Mulcahy, D.,** Enzyme-linked immunosorbent assay for a soluble antigen of *Renibacterium salmoninarum,* the causative agent of bacterial kidney disease, *Can. J. Fish. Aquat. Sci.,* 44, 183, 1987.

225. **Elliott, D. G. and Barila, T. Y.,** Membrane filtration-fluorescent antibody staining procedure for detecting and quantifying *Renibacterium salmoninarum* in coelomic fluid of chinook salmon *(Oncorhynchus tshawytscha), Can. J. Aquat. Sci.,* 44, 206, 1987.

226. **Lee, E. G.-H.,** Technique for enumeration of *Renibacterium salmoninarum* in fish kidney tissues, *J. Aquat. Anim. Dis.,* 1, 25, 1989.

227. **Cipriano, R. C., Starliper, C. E., and Schachte, J. H.,** Comparative sensitivities of diagnostic procedures used to detect bacterial kidney disease in salmonid fishes, *J. Wildl. Dis.,* 21, 144, 1985.

228. **Sakai, M., Koyama, G., Atsuta, S., and Kobayashi, M.,** Detection of *Renibacterium salmoninarum* by a modified peroxidase-antiperoxidase (PAP) procedure, *Fish Pathol.,* 22, 1, 1987.

229. **Sakai, M., Atsuta, S., and Kobayashi, M.,** Comparison of methods used to detect *Renibacterium salmoninarum,* the causative agent of bacterial kidney disease, *J. Aquat. Anim. Health,* 1, 21, 1989.

230. **Pascho, R. J., Elliott, D. G., and Streufert, J. M.,** Brood stock segregation of spring chinook salmon *Oncorhynchus tshawytscha* by use of the enzyme-linked immunosorbent assay (ELISA) and the fluorescent antibody technique (FAT) affects the prevalence and levels of *Renibacterium salmoninarum* infection in progeny, *Dis. Aquat. Org.,* 12, 25, 1991.

231. **Griffiths, S. G., Olivier, G., Fildes, J., and Lynch, W. H.,** Comparison of western blot, direct fluorescent antibody and drop-plate culture methods for the detection of *Renibacterium salmoninarum* in Atlantic salmon *(Salmo salar), Aquaculture,* 89, 117, 1991.

232. **Evelyn, T. P. T.,** An improved growth medium for the kidney bacterium and some notes in using the medium, *Bull. Off. Int. Epizool.,* 87, 511, 1977.

233. **Evelyn, T. P. T., Bell, G. R., Prosperi-Porta, L., and Ketcheson, J. E.,** A simple technique for accelerating the growth of the kidney disease bacterium *Renibacterium salmoninarum* on a commonly used culture medium (KDM1), *Dis. Aquat. Org.,* 7, 231, 1989.

234. **Evelyn, T. P. T., Prosperi-Porta, L., and Ketcheson, J. E.,** Two new techniques for obtaining consistent results when growing *Renibacterium salmoninarum* on KDM2 culture medium, *Dis. Aquat. Org.,* 9, 209, 1990.
235. **Young, C. L. and Chapman, G. B.,** Ultrastructural aspects of the causative agent and renal histopathology of bacterial kidney disease in brook trout *(Salvelinus fontinalis), J. Fish. Res. Board Can.,* 35, 1234, 1978.
236. **Bruno, D. W. and Munro, A. L. S.,** Uniformity in the biochemical properties of *Renibacterium salmoninarum* isolates obtained from several sources, *FEMS Microbiol. Lett.,* 33, 247, 1986.
237. **Embley, T. M., Goodfellow, M., and Austin, B.,** A semi-defined growth medium for *Renibacterium salmoninarum, FEMS Microbiol. Lett.,* 14, 299, 1982.
238. **Goodfellow, M., Embley, T. M., and Austin, B.,** Numerical taxonomy and emended description of *Renibacterium salmoninarum, J. Gen. Microbiol.,* 131, 2739, 1985.
239. **Bullock, G. L., Stuckey, H. M., and Chen, P. K.,** Corynebacterial kidney disease of salmonids: growth and serological studies on the causative bacterium, *Appl. Microbiol.,* 28, 811, 1974.
240. **Arakawa, C. K., Sanders, J. E., and Fryer, J. L.,** Production of monoclonal antibodies against *Renibacterium salmoninarum, J. Fish Dis.,* 10, 249, 1987.
241. **Getchell, R. G., Rohovec, J. S., and Fryer, J. L.,** Comparison of *Renibacterium salmoninarum* isolates by antigenic analysis, *Fish Pathol.,* 29, 149, 1985.
242. **Bruno, D. W.,** Presence of a saline extractable protein associated with virulent strains of the fish pathogen *Renibacterium salmoninarum, Bull. Eur. Assoc. Fish Pathol.,* 136, 949, 1990.
243. **Warren, J. W.,** Kidney disease of salmonid fishes and the analysis of hatchery waters, *Prog. Fish Cult.* 25, 121, 1963.
244. **Paterson, W. D., Lall, S. P., and Desautels, D.,** Studies on bacterial kidney disease in Atlantic salmon *(Salmo salar)* in Canada, *Fish Pathol.,* 15, 283, 1981.
245. **Sanders, J. E., Long, J. J., Arakawa, C. K., Bartholomew, J. L., and Rohovec, J. S.,** Prevalence of *Renibacterium salmoninarum* among downstream migrating salmonids in the Columbia River, *J. Aquat. Anim. Health,* 4, 72, 1992.
246. **Sanders, J. E., Pilcher, K. S., and Fryer, J. L.,** Relation of water temperature to bacterial kidney disease in coho salmon *(Oncorhynchus kisutch),* sockey salmon *(O. nerka)* and steelhead trout *(Salmo gairdneri), J. Fish. Res. Board Can.,* 35, 8, 1978.
247. **Beacham, T. D. and Evelyn, T. P. T.,** Population and genetic variation in resistance of chinook salmon to vibriosis, furunculosis, and bacterial kidney disease, *J. Aquat. Anim. Health,* 4, 153, 1992.
248. **Woodall, A. N. and Laroche, G.,** Nutrition of salmonid fishes. XI. Iodide requirements of chinook salmon, *J. Nutr.,* 824, 475, 1964.
249. **Lall, S. P., Paterson, W. D., Hines, J. A., and Adams, N. J.,** Control of bacterial kidney disease in Atlantic salmon, *Salmo salar* L., by dietary modification, *J. Fish Dis.,* 8, 113, 1985.
250. **Evelyn, T. P. T., Hoskins, G. E., and Bell G. R.,** First record of bacterial kidney disease in an apparently wild salmonid in British Columbia, *J. Fish. Res. Board Can.,* 30, 1578, 1973.
251. **Banner, C. R., Long, J. J., Fryer, J. L., and Rohovec, J. S.,** Occurrence of salmonid fish infected with *Renibacterium salmoninarum* in the Pacific Ocean, *J. Fish Dis.,* 9, 273, 1986.
252. **Mitchum, D. L., Sherman, L. E., and Baxter, G. T.,** Bacterial kidney disease in feral populations of brook trout *(Salvelinus fontinalis),* brown trout *(Salmo trutta),* and rainbow trout *(Salmo gairdneri), J. Fish. Res. Board Can.,* 36, 1370, 1979.
253. **Mitchum, D. L. and Sherman, L. E.,** Transmission of bacterial kidney disease from wild to stocked hatchery trout, *Can. J. Fish. Aquat. Sci.,* 38, 547, 1981.
254. **Hnath, J. G.,** Bacterial kidney disease in Michigan, Workshop Bacterial Kidney Disease, Phoenix, AZ, 1991, 46, (Abstr.).
255. **Souter, B. W., Dwilow, A. G., and Knight, K.,** *Renibacterium salmoninarum* in wild Arctic char *Salvelinus alpinus* and lake trout *S. namaycush* from the Northwest Territories, Canada, *Dis. Aquat. Org.,* 3, 151, 1987.
256. **Austin, B. and Rayment, J.,** Epizootiology of *Renibacterium salmoninarum,* the causal agent of bacterial kidney disease in salmonid fish, *J. Fish Dis.,* 8, 505, 1985.
257. **Ordal, E. J. and Earp, B. J.,** Cultivation and transmission of etiological agent of kidney disease in salmonid fishes, *Proc. Soc. Exp. Biol. Med.,* 92, 85, 1956.
258. **Wood, J. W. and Wallis, J.,** Kidney disease in adult chinook salmon and its transmission by feeding young chinook salmon, *Fish. Com. Oregon Res. Br.,* 6, 32, 1955.

259. **Bell, G. R., Higgs, D. A., and Traxler, G. S.,** The effect of dietary ascorbate, zinc, and manganese on the development of experimentally induced bacterial kidney disease in sockeye salmon *(Oncorhynchus nerka), Aquaculture,* 36, 293, 1984.

260. **Murray, C. B., Evelyn T. P. T., Beacham, T. D., Barner, L. W., Ketcheson, J. E., and Prosperi-Porta, L.,** Experimental induction of bacterial kidney disease in chinook salmon by immersion and cohabitation challenges, *Dis. Aquat. Org.,* 12, 91, 1992.

261. **Bruno, D. W. and Munro, A. L. S.,** Observations on *Renibacterium salmoninarum* and the salmonid egg, *Dis. Aquat. Org.,* 1, 83, 1986.

262. **Lee, E. G. H. and Gordon, M. R.,** Immunofluorescence screening of *Renibacterium salmoninarum* in the tissues and eggs of farmed chinook salmon spawners, *Aquaculture,* 65, 7, 1987.

263. **Lee, E. G. H. and Evelyn, T. P. T.,** Effect of *Renibacterium salmoninarum* levels in the ovarian fluid of spawning chinook salmon on the prevalence of the pathogen in their eggs and progeny, *Dis. Aquat. Org.,* 7, 179, 1989.

264. **Allison, L. N.,** Multiple sulfa therapy of kidney disease among brook trout, *Prog. Fish Cult.,* 20, 66, 1958.

265. **Bullock, G. L., Stuckey, H. M., and Mulcahy, D.,** Corynebacterial kidney disease: egg transmission following iodophore disinfection, *Fish Health News,* 76, 51, 1978.

266. **Evelyn, T. P. T., Ketcheson, J. E., and Prosperi-Porta, L.,** Further evidence for the presence of *Renibacterium salmoninarum* in salmonid eggs and for the failure of povidine-iodine to reduce the intra-ovum infection in water-hardened eggs, *J. Fish Dis.,* 7, 173, 1984.

267. **Evelyn, T. P. T., Ketcheson, J. E., and Prosperi-Porta, L.,** The clinical significance of immunofluorescence-based diagnosis of the bacterial kidney disease carrier, *Fish Pathol.,* 15, 293, 1981.

268. **Daly, J. G. and Stevenson, R. M. W.,** Inhibitory effects of salmonid tissue on the growth of *Renibacterium salmoninarum, Dis. Aquat. Org.,* 4, 169, 1988.

269. **Bell, G. R.,** Two epidemics of apparent kidney disease in cultured pink salmon *(Oncorhynchus gorbuscha), J. Fish. Res. Board Can.,* 18, 559, 1961.

270. **Bruno, D. W.,** Histopathology of bacterial kidney disease in laboratory infected rainbow trout, *Salmo gairdneri,* Richardson, and Atlantic salmon, *Salmo salar* L. with reference to naturally infected fish, *J. Fish Dis.,* 9, 523, 1986.

271. **Sami, S., Fischer-Scherl, T., Hoffman, R. W., and Pfeil-Putzien, C.,** Immune complex-mediated glomerulonephritis associated with bacterial kidney disease in the rainbow trout *(Oncorhynchus mykiss), Vet. Pathol.,* 29, 169, 1992.

272. **Wiens, G. D. and Kaattari, S. L.,** Monoclonal antibody characterization of a leukoagglutinin produced by *Renibacterium salmoninarum, Infect. Immunol.,* 59, 631, 1991.

273. **Daly, J. G. and Stevenson, R. M.,** Characterization of the *Renibacterium salmoninarum* haemagglutinin, *J. Gen. Microbiol.,* 136, 949, 1990.

274. **Elliott, D. G., Pascho, R. J., and Bullock, G. L.,** Developments in the control of bacterial kidney disease of salmonid fishes, *Dis. Aquat. Org.,* 6, 201, 1989.

275. **Groman, D. B. and Klontz, G. W.,** Chemotherapy and prophylaxis of bacterial kidney disease with erythromycin, *J. World Maricult. Soc.,* 14, 226, 1983.

276. **Evelyn, T. P. T., Ketcheson, J. E., and Prosperi-Porta, L.,** Use of erythromycin as a means of preventing vertical transmission of *Renibacterium salmoninarum, Dis. Aquat. Org.,* 2, 7, 1986.

277. **Brown, L. L., Albright, L. J., and Evelyn, T. P. T.,** Control of vertical transmission of *Renibacterium salmoninarum* by injection of antibiotics into maturing female coho salmon *Oncorhynchus kisutch, Dis. Aquat. Org.,* 9, 127, 1990.

278. **Moffitt, C. M.,** Survival of juvenile chinook salmon challenged with *Renibacterium salmoninarum* and administered oral doses of erythromycin thiocyanate for different durations, *J. Aquat. Anim. Health,* 4, 119, 1992.

279. **Austin, B.,** Evaluation of antimicrobial compounds for the control of bacterial kidney disease in rainbow trout, *Salmo gairdneri* Richardson, *J. Fish Dis.,* 8, 209, 1985.

280. **Evelyn, T. P. T.,** Agglutinin response in sockeye salmon vaccinated intraperitoneally with a heat-killed preparation of the bacterium responsible for salmonid kidney disease, *J. Wildl. Dis.,* 7, 328, 1971.

281. **Paterson, W. D., Desautel, D., and Wever, J. M.,** The immune response of Atlantic salmon, *Salmo salar* L., to the causative agent of bacterial kidney disease, *Renibacterium salmoninarum, J. Fish Dis.,* 4, 99, 1981.

282. **McCarthy, D. H., Croy, T. R., and Amend, D. F.,** Immunization of rainbow trout, *Salmo gairdneri* Richardson, against bacterial kidney disease: preliminary efficacy evaluation, *J. Fish Dis.,* 7, 65, 1984.

283. **Bruno, D. W.,** Serum agglutinating titers against *Renibacterium salmoninarum* the causative agent of bacterial kidney disease, in rainbow trout, *Salmo gairdneri* Richardson and Atlantic salmon, *Salmo salar* L., *J. Fish Biol.,* 30, 327, 1987.

Chapter XVI

Miscellaneous Bacterial Diseases

A. ACID-FAST STAINING BACTERIA (*MYCOBACTERIUM* AND *NOCARDIA*)

Acid-fast staining bacteria were first discovered as fish pathogens in carp in Europe during the latter part of the 19th century.[1] The disease was first known as "fish tuberculosis" because of the taxonomic and overt similarity of the organisms that cause infections in fish and tuberculosis in humans. Members of two genera of acid-fast bacteria cause infections in fish: *Mycobacterium* (Mycobacteriosis) and *Nocardia* (Nocardiosis).

1. MYCOBACTERIOSIS

Mycobacteriosis is an infrequently encountered disease of fishes that is usually chronic and terminal. It was suggested by Parisot and Wood[2] that fish "tuberculosis" should be more correctly called fish "mycobacteriosis" because other than the organism's acid-fast staining characteristic and taxonomic classification, there is very little similarity between the infection that occurs in fish and tuberculosis in humans. Fish mycobacteriosis is caused by *M. marinum, M. fortuitum,*[3] or *M. chelonei.*[4]

a. Geographical Range and Species Susceptibility

Mycobacteriosis occurs throughout the world in marine environments and with increasing regularity in freshwater fishes. The disease is most severe in cultured or aquarium fish but it is found in wild populations where consequences are mild. Nigrelli and Vogel[5] listed over 150 marine and freshwater fish species, including salmonids and ornamentals, in which *M. fortuitum* and/or *M. marinum* had been documented, however, all teleosts should be considered as possible hosts. Under certain conditions, cultured striped bass (*Morone saxatilis*) and hybrids of striped bass and white bass (*M. chrysops*) are particularly susceptible to *M. marinum. M. chelonei* occurs in salmonids in Japan.[4] *M. fortuitum* and *M. marinum* are pathogenic to humans and other homeotherms as well as frogs, snakes, and lizards.[3,6-7]

b. Clinical Signs and Findings

Gross external clinical signs of mycobacteriosis vary depending on the species of fish.[3] Fish may lose their appetite, become emaciated, develop sunken abdomens; exhibit grayish irregular ulcerations and hemorrhaging in the skin (Figure 1); have frayed fins, deformed vertebrae and mandibles; become exophthalmic; and lose one or both eyes. Scales develop lepidorthosis before being lost. Nodular tubercles which develop in the muscle appear externally as diffuse, light brown spots or swollen areas that may rupture. White streaks parallel to the cartilage support of the filaments occur in the gills. Secondary sexual characteristics (hooked jaw and color changes) fail to develop in adult Pacific salmon and they may be smaller than normal, are more darkly colored, and have undeveloped gonads. Diseased ornamental fish usually lose their bright coloration.

Internal gross pathology is more consistent between species. The spleen and head kidney become extremely enlarged and develop a granular appearance. Small to large grayish tubercles or nodules develop on the liver, spleen, kidney, and throughout the visceral cavity (Figure 1). The liver may contain dark pigmented spots but these granulomatous nodules, which show up in more chronic infections, contain masses of acid-fast bacteria early in their development. Advanced mycobacterial lesions of internal organs are void of acid-fast bacteria but resemble those associated with bacterial kidney disease of salmonids.

c. Diagnosis and Bacterial Characteristics

Mycobacteriosis is routinely diagnosed by detection of acid-fast staining (red) bacteria in smears from nodules or histological sections of the tubercles (Figure 2) but it is difficult to differentiate *Mycobacterium* from *Nocardia* in histological sections. *Mycobacterium* spp. are Gram-positive, strongly acid-fast (Ziehl-Neelsen), nonmotile, pleomorphic rods that measure 0.25 to 0.35 μm $\times$ 1.5 to 2.0 μm.[8-10]

Special media, such as Lowenstein-Jensen, provide the best growth of *Mycobacterium* spp., but *M. marinum, M. fortuitum,* and *M. chelonei* can be isolated on blood agar if the plates are sealed to retain moisture. Their growth rates differ depending upon species. *M. fortuitum* and *M. chelonei* will form

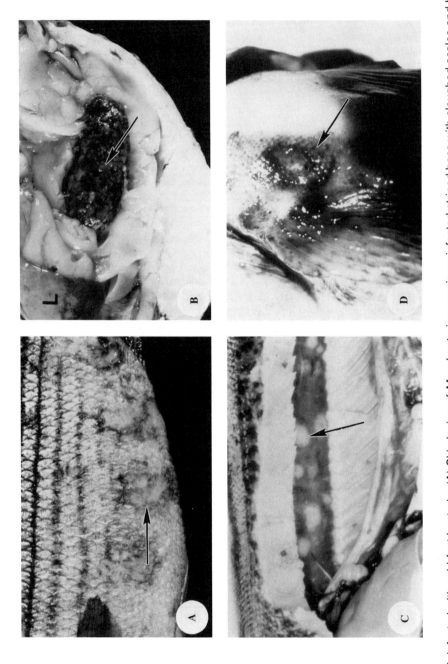

Figure 1 Fish infected with acid-fast bacteria. (A) Skin lesion of *Mycobacterium marinum*-infected striped bass with sloughed scales and hemorrhage on the margins (arrow); (B) viscera of *M. marinum*-infected striped bass with pale and mottled liver (L) and swollen spleen with granulomas (arrow); (C) kidney of *M. marinum*-infected rainbow trout with swollen kidney and large granulomas (arrow); (D) hemorrhagic lesion (arrow) at the base of the pectoral fin of rainbow trout with *Nocardia* sp. infection. (Photographs A and B by J. Hawke; photographs C and D by L. Ashburner.)

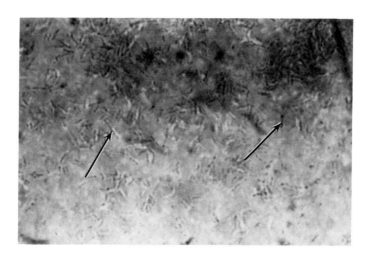

Figure 2 Acid-fast *Mycobacterium marinum* (arrow) in a granuloma in the kidney of hybrid striped bass × white bass. (Photograph courtesy of J. Newton.)

colonies in about 7 d when incubated at 25°C. *M. marinum* grows more slowly whereby it forms colonies in 2 to 3 weeks at an incubation temperature of 25°C. Developing colonies may be smooth, rough, moist, dry, raised, or flat, depending upon the type of media and age of the culture. Material from skin and other tissues containing mixed bacterial species must be treated with 0.3% Sepheran® prior to primary culture and then pure cultures can be maintained on Lowenstein-Jensen media. *M. marinum, M. fortuitum,* and *M. chelonei* can be differentiated based on several biophysical and biochemical properties (Table 1). *M. marinum* produces nicotinamidase and pyrazinamidase but *M. fortuitum* does not. *M. fortuitum* is positive for nitrate reductase while *M. marinum* and *M. chelonei* are negative. *M. marinum* produces a yellow-orange pigment when exposed to light while *M. fortuitum* lacks the pigment production.[11] The optimum growth temperature of *M. marinum* is 25 to 30°C. It normally does not grow at 33°C or above, but may adapt to the higher temperature in the laboratory. *M. fortuitum* grows at 19 to 42°C with an optimum range of 30 to 37°C.[3]

d. Epizootiology

Clark and Shepard[12] demonstrated experimentally that infected fish do shed acid-fast bacteria into the water but little is actually known about the epizootiology of mycobacteriosis in wild fish. It is likely that a low level of infection in wild fish serves as a reservoir for the bacterium and is, therefore, often the source of infection in cultured fish. However, months or possibly years pass between natural exposure to the bacterium and the presence of clinical disease. During an epidemiological study of the mackerel *(Scomber scombrus)* from the northeastern Atlantic ocean, fish over 2 years old showed evidence of mycobacterial infection.[13] The increase of *Mycobacterium* sp. infection with age indicated a chronic disease in these wild fish; however, generally mycobacteriosis can occur in any age or size fish.

Pathogen ingestion via raw contaminated fish viscera was the probable source of mycobacteriosis in salmon in the northwestern U.S. in the 1950s when inclusion of raw fish in the diet was a common practice.[14] Fish free of the acid-fast bacilli were seen only in hatcheries where raw fish was not fed, and when a raw fish diet was discontinued the mycobacterial disease problems disappeared. Chinabut et al.[15] successfully transmitted mycobacteria in snakehead *(Chana striatus)* by feeding raw offal to naive fish. However, Ashburner[16] reported *Mycobacterium* in freshwater cultured chinook salmon *(Oncorhynchus tshawytscha)* in Australia where feed was eliminated as a source of infection. He presented evidence that the pathogen was passed to the F_1 generation during spawning, probably via ovarian fluid.

Mycobacteriosis was not diagnosed in Oregon salmon hatcheries between 1964 and 1981. Fryer and Sanders,[17] however, suggest that the organism is still present and occurs at various levels of prevalence throughout the life cycle of anadromous salmon. Arakawa and Fryer[4] found the prevalence of *Mycobacterium* to be 0 to 26% in wild juvenile coho *(O. kisutch)* compared to 1.4 and 4.0% in chinook salmon.

Table 1   Selected differentiating characteristics between *Mycobacterium marinum*, *Mycobacterium fortuitum*, and *Mycobacterium chelonei*[4,9]

Characteristics	*Mycobacterium marinum*	*Mycobacterium fortuitum*	*Mycobacterium chelonei*
Colony color	Yellow-orange	White	White, smooth
Cell size	2 × 0.25–0.35 μm		1–4 × 0.3–0.6 μm
Growth temperature			
10°C	?	?	+
30°C	+	+	+
37°C	−	+	−
Nitrate reduced	−	+	−
Degrade Tween® 80	+	±	?
Nicotinamidase production	+	−	−
Pyrazinamidase production	+	−	−

Mycobacteriosis is a chronic disease that results in low morbidity and moderate to high cumulative mortality. The disease usually develops more rapidly in warm water. At 18°C juvenile rainbow trout *(O. mykiss)* experimentally infected with *M. chelonei* suffered 20 to 52% mortality, whereas 98% of juvenile chinook salmon died within 10 d postinfection.[4] Mortality of other farmed fish have varied from 35% in naturally infected three-spot gouramy *(Macropodus opercularis)* in Columbia to 100% in experimentally infected pejerrey *(Odonthestes banariensis)*.[18] Hedrick et al.[19] reported that 50% of the population of yearling striped bass infected with *M. marinum* died within months of stocking into an intensive culture system, and the greater the intensification of the aquaculture environment the more serious the disease becomes. In diagnostic cases submitted to the Fish Disease Diagnostic Laboratory at Louisiana State University, it was found that in closed recirculating systems, where hybrid striped bass were experiencing chronic mortalities, 30 to 50% of the fish randomly sampled from an affected population had characteristic mycobacterial granulomas present in the internal organs.[20]

In the U.S. *M. marinum* is becoming such a serious disease problem in certain, intensive striped bass/white bass hybrid recirculating culture systems that one facility terminated the production of these fish.[21] The source of the bacterium in these recirculating systems could not be determined, but no *Mycobacterium* disease problems have been reported in pond-cultured hybrid striped bass.

When humans contract mycobacteria from fish the disease usually manifests itself as superficial nodules that may become crusted or ulcerated granulomatous tissue on the fingers, hands, wrists, and forearms.[22] Fish isolates are able to adapt somewhat to the higher temperature of humans, but apparently the lower skin temperature of the extremities is more conducive to establishment of the infection.[12] Strains of the *Mycobacterium* spp. isolated from humans develop on culture media in 7 to 10 d at an optimum temperature of 30 to 33°C. Optimum culture temperature for mycobacteria from fish is 25 to 30°C. Although mycobacteria usually do not spread to the lungs or other internal organs of humans, as does the human tuberculosis organism *(M. tuberculum)*, *M. fortuitum* may be found infecting the lungs, lymph nodes, and internal organs because of its higher temperature tolerance.[3] Occasionally, *M. marinum* has been known to be more invasive, causing infection of the tendon sheaths, joints, and bone.[23] Workers who clean marine fish for a living, handle certain cultured fish (especially marine fish), or work with saltwater ornamental fish, are at risk of contracting the disease. Also, the presence of open scratches or wounds on the skin most likely enhances infection. Immunocompromised individuals, such as those suffering from human immune deficiency virus infections, may be particularly susceptible to these so-called "atypical mycobacterial infections".[24] *M. marinum* is also the causative agent of "swimming pool granuloma", a disease which appears as a cutaneous granuloma on the elbows, fingers, knees, feet, and toes of humans. *M. fortuitum* is an opportunistic pathogen in humans following superficial trauma or surgery.

e. Pathological Manifestations

Fish react to mycobacterial infection with proliferation of connective tissue but exhibit little inflammatory response other than granulomatous inflammation.[3] The pathology of mycobacteriosis has little resemblance to leprosy and only superficial similarity to tuberculosis of humans. In teleosts, mycobacteriosis is considered less cellular than tuberculosis in higher animals because some workers refute the presence of Langerhans giant cells which are characteristic of the mammalian tubercle.[5,22] However, Timur et al.[25] showed caseation, typical Langerhans cell production, and cell-mediated immunity in *M. marinum* infections in plaice *(Pleuranectus platessa)*. Large masses of bacteria were found in the visceral adipose tissue, hematopoietic tissue of the kidney, spleen, and liver of young fish. The disease in adult fish appeared to be merely an extension of that observed in the juveniles. Foci of bacteria surrounding the intestinal tract of young fish disappeared in older fish, leaving large areas of caseous necrosis. Spleen, liver, and kidney had severe lesions characterized by massive concentrations of acid-fast bacteria. Caseous necrosis also formed in the kidney.

In infected fish, bacterial metabolism is inhibited by the walling off of the pathogen by connective tissue, leading to death of the organism.[3] The resulting tubercle becomes necrotic and mineralization is the final process leading to cavitation. Acid-fast bacteria are usually seen in young nodules but are generally absent in older granulomas. Necrotic lesions may merge into conglomerations, thus forming larger nodules. Caseation of the nodules may appear in the center and be localized or widespread. Tubercles may also contain black pigmentation.

f. Significance

Because raw fish is no longer fed to salmon, mycobacteriosis is seldom a problem in cultured salmon today. Currently mycobacteriosis has the greatest impact on some hybrid striped-white bass in recirculating culture systems in the southern and eastern U.S., where *M. marinum* has become established. Mycobacteriosis continues to be a chronic disease problem in ornamental fish in home and large public aquaria. Also, any time an organism can be transmitted from a lower vertebrate to humans, as mycobacteria can, the significance of that disease is elevated.

g. Management

Elimination of raw fish products in diets of cultured fish will usually disrupt the cycle of mycobacteriosis infection. Once fish become infected with the disease it is extremely difficult to treat, and it has been suggested that infected fish populations be destroyed, buried, and the facility sterilized.[26] It is recommended that fish be removed from contaminated facilities and the system cleaned with an acid wash to remove as much organic material from the water lines as possible. The system may then be disinfected with 200 mg/l of HTH, chlorine dioxide, or phenolic compounds. If destruction and disinfection is not acceptable, or possible, oral application of sulfisoxazole at 2 mg/g of feed is recommended by Van Duijn.[3] Recent efforts to control mycobacterial infections in striped bass culture populations has shown that some isolates are sensitive to Terramycin®, however, other isolates are resistant to most antibiotics.[21] The long-term antibiotic therapy required to keep the disease in remission in a fish population is not economically feasible.

Recirculating aquaculture systems should be equipped with water sterilization equipment such as UV and/or ozone units; however, the water treatment rates and actual value of these units have not been studied in detail with mycobacterial infections in fish.

2. NOCARDIOSIS

Infections of *Nocardia*, "nocardiosis", in fish are difficult to distinguish from mycobacteriosis because both are acid-fast staining organisms and the resulting lesions are overtly very similar. Two bacterial species are known to cause nocardiosis in fish: *N. asteroides* and *N. kampachi*.[27]

a. Geographical Range and Species Susceptibility

Nocardia infections in fish have been reported specifically from the U.S., Argentina, Germany,[28] Japan,[29] and Taiwan,[30] but in all likelihood it occurs worldwide in saltwater and, on occasion, in freshwater fish.

To date, *Nocardia* infections have been documented in rainbow trout, brook trout *(Salvelinus fontinalis)*, neon tetra *(Paracheirodon innesi)*, yellowtail *(Seriola guingueradiata)*, Formosa snakehead *(C. maculata)*, giant gorami *(Osphronemus goramy)*, and largemouth bass *(Micropterus salmoides)*.[31-33]

b. Clinical Signs and Findings

Clinical signs of *Nocardia* infections are similar to those of mycobacteriosis[1] (Figure 1). Fish swim in a rapid tail-chasing mode or are sluggish and become anorexic, emaciated, and show abdominal distension. Fish may also lose scales, frequently exhibit exophthalmia, and have opaque eyes. Blood pools under the epithelium of the oral and opercular cavities, and multiple yellowish white nodules, varying in size from 0.5 to 2.0 cm in diameter are scattered throughout the muscle, gill, heart, liver, spleen, ovary, and the mesenteries.

c. Diagnosis and Bacterial Characteristics

Nocardia spp. are Gram-positive, weakly acid-fast, nonmotile, long and branching bacilli that can be detected either in sections or smears from the granulomatous nodules.[30] Differentiation between *Nocardia* and *Mycobacterium* is impossible without culturing, but *Nocardia* is more easily isolated than is *Mycobacterium* on BHI, blood agar, or Ogawa medium. *Nocardia* spp. are longer rods than *Mycobacterium* spp. and are less fastidious while growing more rapidly than *M. marinum*. There are other morphological and biochemical characteristics that differentiate the two genera. Colonies are irregular and rough; white, pinkish, orange or yellow in color; and require up to 21 d at 18 to 37°C for growth.[10]

d. Epizootiology

Like mycobacteriosis, nocardiosis is a slowly developing chronic infection but little is known about its epizootiology. The organism may be a normal inhabitant of the environment (i.e., soil and water) or carrier fish may serve as a reservoir for the pathogen.

There is little information on mortality of *Nocardia*-infected fish. However, in a documented case of *N. asteroides*-infected Formosa snakehead cultured in a freshwater pond in Taiwan, 20% of 30,000 8- to 9-month-old (20- to 30-cm) fish were killed in 2 weeks.[29] Successful experimental transmission has been inconsistent. Snieszko et al.[33] were unable to transmit *N. asteroides* from rainbow trout to other trout by feeding, but were somewhat successful in transmission by injection, requiring 1 to 3 months post-injection for disease to develop. Chen[31] transmitted *N. asteroides* from largemouth bass to other individuals of the same species by intramuscular injection that resulted in 100% mortality, and characteristic granular nodules developed in the visceral organs. Kusuda[34] had the greatest success transmitting the disease in yellowtail by injection or by smearing surface wounds with *N. seriolae* (*N. kampachi*). These experiments suggest that the route of infection is more likely through epidermal injury than orally.

e. Pathological Manifestation

The pathology is similar to that of mycobacteriosis. However, in contrast to that of mycobacteriosis, according to Chen,[31] in the early stages of infection of largemouth bass there was acute, serous inflammation resulting in production of an exudate containing cellular and bacterial debris. As infection progressed these lesions became granulomatous. The nodules consist of necrotizing foci surrounded by some epitheloid cells and fibroblasts or fibrous encapsulation. The most characteristic structures in the tubercular nodules are bacillary masses within small cavities that are surrounded by concentrical layers of fibrous tissue. The long, branching, filamentous, weakly acid-fast bacteria lie within the nodule.

f. Significance

Nocardiosis does not appear to be a particularly significant disease because generally only a few fish have the disease at any given time, and outbreaks have been sporadic.

g. Management

There is no known therapy for nocardiosis. The most practical management approach is removal and proper disposal of infected fish, and disinfection of facilities and equipment.

B. PASTEURELLOSIS

Pasteurellosis, also known as "pseudotuberculosis", is a chronic to acute systemic infection of wild and cultured fish caused by *Pasteurella piscicida*.[35]

Table 2   **Characteristics of *Pasteurella piscicida*[39,41,43]**

Characteristic	Reaction	Characteristic	Reaction
Gram stain	− (bipolar)	Degradation of	
Production of		Arginine	+
Catalase	+	Casein	−
Oxidase	+	Gelatin	−
Gluconate oxidase	−	Starch	+
Phenylalanine		Tween® 80	+
deaminase	−	Urea	−
Arginine		Growth at	
dihydrolase	+	0.5–3.0% NaCl	+
β-Galactosidase	−	30°C	+
H₂S	−	37°C	−
Indole	−	Growth on McConkey	−
Lysine		On potassium	
decarboxylase	−	cyanamide broth	−
Ornithine		Sensitive to 0/129	+
decarboxylase	−	Utilize citrate	−
2,3,-Butanediol		Acid from	
dehydrogenase	+	Fructose	+
Nitrate reduction	−	Galactose	+
Methyl red	+	Glucose	+
Voges-Proskauer	+	Mannose	+

1. GEOGRAPHICAL RANGE AND SPECIES SUSCEPTIBILITY

P. piscicida was initially isolated from white perch *(M. americana)* and striped bass from the Chesapeake Bay of Virginia and Maryland in 1963.[36] Its known range has been extended to Japan, Norway, Taiwan, and the Gulf of Mexico coast of the U.S.[37-39] Kitao[40] lists eight species of fish, primarily marine, in Japan that have been infected with *P. piscicida*. It is also infrequently found in freshwater fishes. In addition to the two species previously mentioned, the known susceptible fish to *P. piscicida* include striped bass, yellowtail *(Seriola quinqueradiate)*, red sea bream *(Pagrus major)*, black sea bream *(Mylio macrocephalus)*, Formosa snakehead *(C. maculata)*, and striped jack *(Pseudocaranx dentex)*.[36,39-42]

2. CLINICAL SIGNS AND FINDINGS

Few clinical signs are associated with pasteurellosis other than loss of mobility and sinking in the water column.[41] The disease is characterized in most fish by darkening of pigmentation and the presence of discrete, mild petechiae at the base of fins and on the operculum. Internal lesions associated with the disease are enlargement of the spleen and kidney and the presence of multiple whitish circumscribed areas (granulomatous-like) throughout these organs and the liver.

3. DIAGNOSIS AND BACTERIAL CHARACTERISTICS

P. piscicida can be isolated on ordinary bacteriological media (BHI) to which 0.5 to 3.0% NaCl has been added.[41] Colonies of *P. piscicida* are 1 to 2 mm in diameter and are regular, convex, viscous, and opaque to translucent when incubated at 25°C for 48 to 72 h. The organism is a Gram-negative, bipolar staining, nonmotile bacillus that is oxidase and catalase positive.[35-36] The cell measures 0.5 to 0.8 μm wide and 0.7 to 2.6 μm in length. Pleomorphism of the coccoid to long rods may be evident in older cultures. *P. piscicida* has an optimum growth temperature of 23 to 27°C, and will grow at 10 to 30°C but not at 37°C. Phenotypically *P. piscicida* is generally considered to be a homogeneous species[43] (Table 2).

Kimura and Kitao[37] were unable to separate the Japanese isolates from the American isolates serologically, but Kusuda et al.[44] differentiated these isolates by immunoelectrophoresis, implying that more than one serological strain of *P. piscicida* may be involved. Also, Kitao and Kimura[45] were able to identify *P. piscicida* 100% of the time utilizing a direct fluorescent antibody technique on impression smears of organs that exhibited the characteristic white lesions. Mori et al.[46] detected incipient infection in the spleens and/or kidneys of yellowtail by using fluorescent antibody and suggested that this technique could be used for detection of subclinical infection.

4. EPIZOOTIOLOGY

The initial epizootic of *P. piscicida* occurred in Chesapeake Bay and killed millions of white perch and thousands of striped bass.[36] The epizootic began in the lower Potomac River in June and spread into the Bay proper during July. At the time of the epizootic the white perch population was high while the Bay and its tributaries were heavily polluted with organic material — conditions thought to contribute to the outbreak.[47] The following year commercial harvest of white perch in the Bay was reduced by almost half, leading to speculation that *P. piscicida* had killed approximately 50% of the population. No other episode of *P. piscicida* in a wild fish population has been as devastating as was the epizootic in Chesapeake Bay.

Mortality from *P. piscicida* may vary in cultured fish, but generally younger fish are more susceptible. Higher mortalities seem to occur in young cultured yellowtail and black sea bream, but only a 30% loss was noted in cultured snakehead in Taiwan with the feeding of an antibiotic.[38] Hawke et al.[39] reported a mortality of about 80% in a cultured juvenile striped bass population in the U.S. It has been estimated that a 30 to 80% mortality occurred in hybrid striped bass cultured on coastal Louisiana fish farms due to *P. piscicida* from the fall of 1990 through the fall of 1992, during which time water temperatures ranged from 20 to 30°C.[21] Nakai et al.[42] reported that *P. piscicida* was responsible for the loss of 34% of 10,000 juvenile striped jack being raised in sea pens in Japan. Environmental conditions probably play a major role in determining the seriousness of an epizootic resulting from pasteurellosis. Matsusato[48] reported that the disease incidence in yellowtail rose during the rainy season when salinity dropped below 30 ppt and when an optimum water temperature of approximately 25°C existed.

The route of transmission of *P. piscicida* is unknown but its short-lived nature in brackish water has led to the conclusion that transmission is from fish to fish, even though a carrier or latent state has not been proven. In support of the fish reservoir theory, Toranzo et al.[49] demonstrated that *P. piscicida* survived for less than 2 d in fresh water and less than 5 d in sea water.

P. piscicida can be highly pathogenic, but this varies from species to species. An injectible LD_{50} of $10^{1.9}$ CFU/ml was established in Formosa snakehead by Tung et al.,[38] but in other experiments greater numbers of organisms were required to kill fish. Nakai et al.[42] established an LD_{50} of 10^3 CFU/fish in striped jack by injection and an LD_{50} of 10^7 CFU/fish in red sea bream.

5. PATHOLOGICAL MANIFESTATIONS

Few pathological changes can be noted in fish with acute pasteurellosis, whereas a chronic infection is characterized by miliary lesions in the kidney and spleen.[41] In a chronic form of the disease, necrotic lymphoid and peripheral blood cells collect in the spleen and focal areas of necrotic hepatocytes occur in the liver. No inflammatory response occurs in infected white perch, but spleens of experimentally infected juvenile striped bass developed multiple foci of bacterial colonies and reduced cellular densities. Live bacteria were present in phagocytes of yellowtail and these phagocytes became swollen to the point that they blocked capillary blood flow.[50]

In naturally *P. piscicida*-infected striped bass, there was extensive acute multifocal necrosis of splenic lymphoid tissue, characterized by a loss of cells, coagulation necrosis, karyorrhexis, and the presence of large colonies of bacteria.[39] Similar histopathology was seen in the liver but to a lesser degree. Rod-shaped bacteria were observed in the sinusoids and within the hepatic vessels, and large areas of the liver exhibited hyperplasia of reticuloendothelial cells lining the hepatic sinusoids. The pathogenesis of *P. piscicida* is unknown, however, Nakai et al.[42] found that extracellular products of *P. piscicida* were as pathogenic to striped jack and red sea bream as the bacteria themselves, suggesting pathology is the result of extracellular toxins.

6. SIGNIFICANCE

Pasteurellosis is not a common disease but under certain conditions it can be highly significant, especially in the Japanese yellowtail culture industry.[51] It has also become a significant problem in brackish water striped bass culture in the U.S.[21]

7. MANAGEMENT

According to Kitao,[40] avoidance of overcrowding and good management may help prevent pasteurellosis. Application of medicated feed when the disease occurs has been successful. Ampicillin fed at 10 and 100 mg/kg of body weight reduced mortality to 30 and 0%, respectively, compared to 100% mortality of nonmedicated yellowtail.[52] Hawke et al.[39] fed oxytetracycline at 50 to 150 mg/kg of body weight of striped bass per day with only slightly reduced mortality. However, Nakai et al.[42] controlled a *P. piscicida* infection in striped jack with oxytetracycline and oxolinic acid as a feed additive. Since antibiotics have been used, however, R plasmids have been discovered which confer antibiotic resistance.[53]

Vaccination has been shown to be an effective method of *Pasteurella* prevention, especially in Japan, and this is expected to be a part of managing the disease in the future. *Pasteurella* vaccines can be effectively delivered by intraperitoneal injection, immersion, spray, or orally in the feed with formalin-killed whole-cell bacterins,[54] but Fukuda and Kusuda[55] found that LPS preparations delivered by immersion or spray provided better protection than whole-cell preparations. However, Kusuda et al.[56] also demonstrated a high degree of protection with an injectable ribosomal vaccine prepared from *P. piscicida* and vaccination by immersion with a live attenuated preparation, but commercial immunogenic preparations have not been developed.

C. *STREPTOCOCCUS* AND *ENTEROCOCCUS* SEPTICEMIA

Streptococcus septicemia affects a variety of fish species. The disease may be acute but more often manifests itself as a chronic condition. Although Austin and Austin[1] listed seven distinct species of *Streptococcus* (*S. agalactiae, S. dysgalactiae, S. equi, S. equisimilis, S. faecium, S. pyogenes,* and *S. zooepidemicus*) that are at various times pathogenic to fish, many infections are caused by unspeciated streptococci that differ from other known members of the genus. However, several isolates from Japan appear to be similar to *S. faecalis* or *S. faecium*[57] while an additional, previously unspeciated, isolate from Japan has been compared to *S. iniae.*[58] Some previously identified *Streptococcus* from fish in Japan have also recently been described and named as *Enterococcus seriolicida.*

1. GEOGRAPHICAL RANGE AND SPECIES SUSCEPTIBILITY

Streptococcus infections have been reported in Japan, the U.S., South Africa, Israel, Great Britain, and Norway.[60-62] Members of *Enterococcus* have been described in Japan and the U.S.

Kitao[57] lists 22 species of fish that are naturally susceptible to *Streptococcus* infection, but the most severely affected cultured species are yellowtail, eels (*Anguilla* spp.), ayu (*Plecoglossus altivelis*), tilapia (*Tilapia* spp.), rainbow trout, and striped bass or their hybrids. Other known susceptible species include cultured channel catfish and golden shiners. Plumb et al.[63] isolated *Streptococcus* from about ten species of wild marine fish in the estuaries of the northern Gulf of Mexico including menhaden (*Brevortia patronus*), silver trout (*Cynoscion nothus*), pinfish (*Lagodon rhomboides*), croaker (*Micropogon undulatus*), and sea catfish (*Arius felis*). The *E. seriolicida* occurs in yellowtail and eels in Japan.

2. CLINICAL SIGNS AND FINDINGS

Clinical signs of moribund fish afflicted by streptococci are not particularly specific but generally are darkly pigmented and exhibit a lethargic, erratic, spiraling swimming pattern and a curved body.[57,63] Infected fish are exophthalmic and exhibit abdominal distension, hemorrhaging in the eye and corneal opacity, and diffused hemorrhage in the operculum, base of the fins, and skin (Figure 3). These epidermal lesions are more superficial than those associated with the aeromonads or vibriosis. Bloody mucoid fluid in the lower intestine exudes from the anus. Internal findings include a bloody exudate in the abdominal cavity, pale livers, hyperemic digestive tract, and a greatly enlarged nearly black spleen. The lower gut is flaccid and hyperemic. *Enterococcus* infections produce similar host responses.

3. DIAGNOSIS AND BACTERIAL CHARACTERISTICS

Actual isolation of the bacterium is preferable, but presumptive diagnosis may be accomplished by detecting the Gram-positive coccus (sometimes ovoid) in histological sections or smears from infected

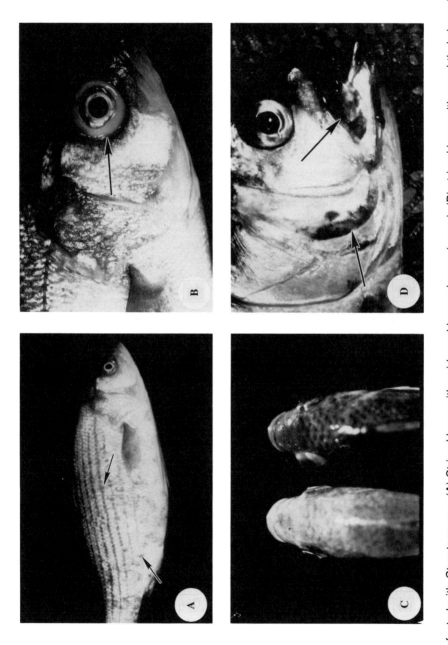

Figure 3 Fish infected with *Streptococcus*. (A) Striped bass with epidermal hemorrhages (arrow); (B) striped bass with exophthalmia and fluid in the eye (arrow); (C) *Tilapia nilotica* with inflamed and exophthalmic eye; (D) menhaden with hemorrhage of the mandible and operculum (arrows). (Photographs A and B by J. Hawke; Photograph C by P.-H. Chang.)

Table 3 **Characteristics of principal groups of**
Streptococcus **spp. and** *Enterococcus seriolicida*
isolated from diseased fish[57,59,63,65]

Characteristic	Streptococcus spp.			Enterococcus seriolicida
Lancefield group	B	D[a]	D[b]	Not D
Hemolysis	None	Alpha	Beta	Alpha
Growth at				
10°C	−	+	+	+
45°C	−	+	−	+
6.5% NaCl	−	+	−	+
pH 9.5	−	+	−	+
0.1% MB	−	+	−	+
40% Bile	±	+	−	+
Hydrolysis of				
Gelatin	−	−	−	?
Starch	−	−	+	?
Hippurate	+	−	−	−
Esculin	−	+	+	?
Arginine	±	+	+	+
Arabinose	−	−	−	−
Glycerol	−	−	±	?
Inulin	?	?	−	?
Lactose	−	−	±	−
Salicin	±	+	+	?
Sorbitol	−	+	−	?
Sucrose	±	−	+	−
Trahalose	+	+	+	?

[a]Characteristics similar to *S. faecalis* and *S. faecium*. [b]Serologically similar to *S. iniae* by fluorescent antibody.[58]

tissues. *Streptococcus* and *Enterococcus* are not acid fast, motile, capsulated, or spore formers. The bacteria may be single or paired but rarely form chains in infected fish. The organism can be isolated on Todd-Hewitt media, tryptose agar with blood,[64] nutrient agar supplemented with rabbit blood,[65] or BHI agar with or without the addition of blood. Inoculated media should be incubated at 20 to 30°C, where yellowish to gray, translucent, rounded, slightly raised colonies that are approximately 0.5 to 1.0 mm in diameter are visible at 24 to 48 h. When the streptococci are grown on media (agar or broth) they have more of a tendency to form chains of up to seven or eight cells.

Streptococcus isolated from fish may be divided into four major groups (Table 3): (1) group B nonhemolytic, (2) group D alpha and group D beta hemolytic, (3) alpha hemolytic strains that do not react with any recognized Lancefield antisera, and (4) other streptococci from freshwater and marine fish.[21] Group B nonhemolytic streptococci are pathogens of freshwater and marine fish in the U.S. They were first reported by Robinson and Meyer[61] but have since been found in other epizootics including wild fishes and cultured striped bass. Group B *Streptococcus* may be presumptively identified by a positive CAMP test on sheep blood agar and are similar to strains of human and bovine origin identified as *S. agalactiae*.

Kawahara and Kusuda[58] proposed that the beta hemolytic streptococci isolated in Japan be considered identical to *S. iniae*. Group D alpha and beta hemolytic streptococci conform to the enterococcus group in that they grow at 45°C, in 40% bile, in 6.5% NaCl, and at pH 9.6. The cells are more ovoid than round but do occur in chains. They were also isolated from rainbow trout in South Africa by Boomker et al.[64] and hybrid striped bass in freshwater systems in the U.S.[21] Isolates from rainbow trout, yellowtail, and eel in Japan in the 1970s appear to fit into this group. An alpha hemolytic strain isolated from yellowtail in Japan does not react with the Lancefield Group D antisera, but have characteristics similar to the *Enterococcus;* therefore, Kusuda et al.[59] proposed that they be named *Enterococcus seriolicida*.[66]

Whether or not the enterococcus described in Japan is the same as the isolate from striped bass in the U.S. is not known.

Other streptococci have been infrequently reported from freshwater and marine fish that do not fit into any known species or the previously described groups.[21] Obviously, streptococci that can infect fish are varied and poorly understood and require additional taxonomic clarification.

4. EPIZOOTIOLOGY

Streptococcus infections are more likely to occur in marine and brackish water fish than in freshwater fish and, generally, infections are more severe in saltwater fish. *In vitro* salinity tolerance studies of streptococcus support this phenomenon.[67-68] However, most isolates from freshwater fishes do not grow at salinity above 3% NaCl.[65] Streptococci isolates from various ecosystems have adapted to those salinity levels and they are generally less virulent in a different ecosystem.

It has been shown in Japan that *Streptococcus* sp. remain in sea water and mud in the vicinity of fish rearing facilities throughout the year, with higher numbers being present during the summer. However, Kitao et al.[69] found that isolation was easier from mud during autumn and winter.

In naturally occurring infections, transmission of streptococcus is thought to be by contact; however, experimental transmission is accomplished by immersion, injection, or cohabitation. Transmission is enhanced by injury to the epithelium or by stressful environmental conditions. Plumb et al.[63] reported that over 90% of fish killed by streptococcus in estuaries were menhaden, while numerous other species of fish that died in the epizootic were observed feeding on infected menhaden carcasses; therefore, it was theorized that these scavenger fish were infected with *Streptococcus* through ingestion. Minami[70] also isolated *Streptococcus* from rough fish used in the feed of cultured yellowtail in Japan; therefore, sources of infection could be the water, mud, fish in the diet, or from carrier fish in the environment.

Epizootics in wild fish populations may also be associated with environmental conditions. Plumb et al.[63] noted that fish kills along the Gulf of Mexico coast were confined to bays which had restricted flow to the open Gulf. This restricted flow was thought to interfere with flushing of fresh water into the bays and created the potential for poor water quality conditions which were likely stressful to the fish. A similar phenomenon was also noted in the Chesapeake Bay when fish were killed by *Streptococcus*.[68] In the U.S. most epizootics occur during the summer, but in cultured yellowtail in Japan most epizootics occur during August to November.[57]

5. PATHOLOGICAL MANIFESTATIONS

Streptococcus sp. affects the spleen, liver, eye, and in some cases the kidney of infected fish. Rasheed et al.[71] found that the spleens of infected Gulf killifish *(Fundulus grandis)* were about ten times larger than normal and the splenic pulp in the enlarged organ was severely congested and some cells were necrotic. It appeared that phagocytosed bacteria were multiplying. Livers of these fish were edematous, and sinusoids were dilated and possessed atrophied hepatocytes. Focal necrosis was also apparent. A general bacteremia occurred in tilapia with the eye showing severe granulomatous inflammation.[72] Also, infiltration of bacterial-laden macrophages were noted in these fish and granulomas were observed in infected lesions of the epicardium, capsules of the liver and spleen, peritoneum, stomach, intestine, brain, ovary, and testes.

Pathogenesis of *Streptococcus* is thought to be facilitated by exotoxins. Kimura and Kusuda[73] reported that exposure of yellowtail by injection to cell-free culture media of *Streptococcus* increased susceptibility during subsequent exposure to the bacterium. They also found that following *Streptococcus* infection by injection the kidney and spleen were more conducive to high numbers of bacteria than were the blood, liver, or intestines.

6. SIGNIFICANCE

Streptococcus sp. is generally not a major disease in North American aquaculture, but it does cause serious problems in striped bass/white bass culture populations, particularly along coastal areas. As culture of these fish becomes more extensive, *Streptococcus* could become a limiting factor. The disease is potentially serious in Japanese mariculture, particularly in yellowtail, eel, and tilapia. In 1989 there were 484 reported incidences of *Streptococcus* in cultured yellowtail in Japan.[57]

7. MANAGEMENT

Available information on control of *Streptococcus* in fish is sketchy but avoidance of stress due to poor water quality, overcrowding, overfeeding, or unnecessary handling during critical periods should be practiced. Also, infected fish and carcasses should be promptly removed from the culture unit. Antibiotics have been shown to be somewhat effective in treating *Streptococcus* infections in fish. Robinson and Meyer[61] used 12 mg/l of oxytetracycline in aquaria to successfully prevent the spread of *Streptococcus* among golden shiners. Although there are presently no drugs approved by the FDA for use on fish infected with *Streptococcus* in the U.S., Group B *Streptococcus* isolated from striped bass in Louisiana were sensitive to Terramycin® but resistant to Romet®. *Enterococcus,* from the same region, was sensitive to Erythromycin® but resistant to Terramycin®, Romet®, and most other drugs tested.[21] Oral application of Erythromycin® at a rate of 25 to 50 mg/kg of body weight per day for 4 to 7 d was effective in controlling *Streptococcus* in yellowtail in Japan.[74] However, Aoki et al.[75] showed that about 17% of *Streptococcus* sp. isolated from yellowtail exhibited drug resistance to lincomycin, tetracycline, and chloramphenicol. Vaccines have been used experimentally to prevent streptococcus infections in Japan.[76]

REFERENCES

1. **Austin, B. and Austin, D. A.,** *Bacterial Fish Pathogens: Diseases in Farmed and Wild Fish,* Ellis Horwood, Chichester, 1987, 364.
2. **Parisot, T. J. and Wood, E. M.,** A comparative study of the causative agents of a mycobacterial disease of salmonid fishes, *Am. Rev. Respir. Dis.,* 82, 212, 1960.
3. **Van Duijn, C.,** Tuberculosis in fishes, *J. Small Anim. Pract.,* 22, 391, 1981.
4. **Arakawa, C. K. and Fryer, J. L.,** Isolation and characterization of a new subspecies of *Mycobacterium chelonei* infectious for salmonid fish, *Helgol. Wiss. Meeresunters.,* 37, 329, 1984.
5. **Nigrelli, R. F. and Vogel, H.,** Spontaneous tuberculosis in fishes and in other cold-blooded vertebrates with special reference to *Mycobacterium fortuitum* Cruz from fish and human lesions, *Zoologica,* 48, 130, 1963.
6. **Goodfellow, M. and Wayne, L. G.,** Taxonomy and nomenclature, in *The Biology of the Mycobacteria,* Ratledge, C. and Stanfor, J. L., Eds., Academic Press, New York, 1982, 471.
7. **Jansen, W. A.,** Fish as potential vectors of human bacterial diseases, in *A Symposium on Diseases of Fishes and Shellfishes,* Snieszko, S. F., Ed., Spec. Publ. No. 5, American Fisheries Society, Washington, D.C., 1970, 284.
8. **Dulin, M. P.,** A review of tuberculosis (mycobacteriosis) in fish, *Vet. Med. Small Anim. Clin.,* May, 735, 1979.
9. **Wayne, L. G. and Kubica, S.,** Genus *Mycobacterium,* in *Bergey's Manual of Systematic Bacteriology,* Vol. 2, Sneath, P. H. A., Mair, N. S., Sharpe, M. E., and Holt, J. G., Eds., Williams & Wilkins, Baltimore, 1986, 1436.
10. **Frerichs, G. N.,** Mycobacteriosis: Nocardiosis, in *Bacterial Diseases of Fish,* Inglis, V., Roberts, R. J., and Bromage, N. R., Eds., Blackwell Scientific, Oxford, 1993, 219.
11. **Wheeler, A. P. and Graham, B. S.,** Saturday conference: atypical mycobacterial infections, *South. Med. J.,* 82, 1250, 1989.
12. **Clark, H. F. and Shepard, C. C.,** Effect of environmental temperatures on infection with *Mycobacterium marinum* (Balnei) of mice and a number of poikilothermic species, *J. Bacteriol.,* 86, 1057, 1963.
13. **MacKenzie, K.,** Presumptive mycobacteriosis in northeast Atlantic mackerel, *Scomber scombrus, J. Fish Biol.,* 32, 263, 1988.
14. **Ross, A. J.,** Mycobacteriosis among salmonid fishes, in *A Symposium on Diseases of Fishes and Shellfishes,* Snieszko, S. F., Ed., Spec. Publ. No. 5, American Fisheries Society, Washington, D.C., 1970, 279.
15. **Chinabut, S., Limsuwan, C., and Chanaratchakool, P.,** Mycobacteriosis in snakehead, *Chana striatus* (Fowler), *J. Fish Dis.,* 13, 531, 1990.

16. **Ashburner, L. D.,** Mycobacteriosis in hatchery-confined chinook salmon (*Oncorhynchus tshawytscha* Walbaum) in Australia, *J. Fish Biol.,* 10, 523, 1977.

17. **Fryer, J. L. and Sanders, J. E.,** Bacterial kidney disease in salmonid fish, *Annu. Rev. Microbiol.,* 35, 273, 1981.

18. **Hatai, K., Lawhavinit, O., Kubota, S., Toda, K., and Suzuki, N.,** Pathogenicity of *Mycobacterium* sp. isolated from pejerrey *Odonthestes banariensis, Fish Pathol.,* 23, 155, 1988.

19. **Hedrick, R. P., McDowell, T., and Groff, J.,** Mycobacteriosis in cultured striped bass from California, *J. Wildl. Dis.,* 23, 391, 1987.

20. **Newton, J.,** College of Veterinary Medicine, Louisiana State University, Baton Rouge, personal communication, 1992.

21. **Hawke, J.,** College of Veterinary Medicine, Louisiana State University, Baton Rouge, personal communication, 1992.

22. **Giavenni, R.,** Alcuni aspetti zoonosici delle micobatteriosi di origine Ittica, *Riv. Ital. Pisci. Ittiopat.,* 14, 123, 1979.

23. **Wolinski, E.,** Mycobacterial diseases other than tuberculosis, *Clin. Infect. Dis.,* 15, 1, 1992.

24. **Frerichs, G. N. and Roberts, R. J.,** The bacteriology of teleosts, in *Fish Pathology,* 2nd ed., Roberts, R. J., Ed., Balliere Tindall, London, 1989, 289.

25. **Timur, G., Roberts, R. J., and McQueen, A.,** The experimental pathogenesis of focal tuberculosis in the plaice *(Pleuronectes platessa), J. Comp. Anat.,* 87, 83, 1977.

26. **Lebovitz, L.,** Fish tuberculosis (Mycobacteriosis), *J. Am. Vet. Med. Assoc.,* 176, 415, 1980.

27. **Conroy, D. A.,** Notes on the incidence of piscine tuberculosis in Argentina, *Prog. Fish Cult.,* 26, 89, 1964.

28. **Post, G.,** *Textbook of Fish Diseases,* T. F. H. Publication, Neptune, NJ, 1987.

29. **Kariya, T., Kubota, S., Nakamura, Y., and Kira, K.,** Nocardial infection in cultured yellowtail (*Seriola quinquiradiata* and *S. purpurascens*). I. Bacteriological study, *Fish Pathol.,* 3, 16, 1968.

30. **Chen, S.-C., Tung, M.-C., and Tsai, W.-C.,** An epizootic in Formosa snake-head fish (*Chana maculata* Lacepede), caused by *Nocardia asteroides* in fresh water pond in southern Taiwan, *COA Fish. Ser. No. 15, Fish Dis. Res. (IX),* 6, 42, 1989.

31. **Chen, S.-C.,** The study on the pathogenicity of *Nocardia asteroides* to largemouth bass *Micropterus salmoides* Lacepede, *Fish Pathol.,* 27, 1, 1992.

32. **Kitao, T., Ruangpan, L., and Fukudome, M.,** Isolation and classification of a *Nocardia* species from diseased giant gorami *Osphronemus goramy, J. Aquat. Anim. Health,* 1, 154, 1989.

33. **Snieszko, S. F., Bullock, G. L., Dunbar, C. E., and Pettijohn, L. L.,** Nocardial infection in hatchery-reared fingerling rainbow trout *(Salmo gairdneri), J. Bacteriol.,* 88, 1809, 1964.

34. **Kusuda, R.,** Nocardial Infection in Cultured Yellowtails, Spec. Publ., Fish. Agency, Jpn. Sea Reg. Fish Res. Lab., 1975, 63.

35. **Jansen, W. A. and Surgalla, M. J.,** Morphology, physiology, and serology of *Pasteurella* species pathogenic for white perch, *J. Bacteriol.,* 96, 1606, 1968.

36. **Snieszko, S. F., Bullock, G. L., Hollis, E., and Boone, J. G.,** *Pasteurella* sp. from an epizootic of white perch *(Roccus americanus)* in Chesapeake Bay tidewater areas, *J. Bacteriol.,* 88, 1814, 1964.

37. **Kimura, M. and Kitao, T.,** On the etiological agent of ''bacterial tuberculosis'' of *Seriola, Fish Pathol.,* 6, 8, 1971.

38. **Tung, M.-C., Tsai, S.-S., Ho, L.-F., Huang, S.-H., and Chen, S.-C.,** An acute septicemic infection of *Pasteurella* organism in pond-cultured Formosa snake-head fish (*Channa maculata* Lacepede) in Taiwan, *Fish Pathol.,* 20, 143, 1985.

39. **Hawke, J. P., Plakas, S. M., Minton, R. V., McPhearson, R. M., Snider, T. G., and Guarino, A. M.,** Fish pasteurellosis of cultured striped bass *(Morone saxatilis)* in coastal Alabama, *Aquaculture,* 65, 193, 1987.

40. **Kitao, T.,** Pasteurellosis, in *Bacterial Diseases of Fish,* Inglis, V., Roberts, R. J., and Bromage, N. R., Eds., Blackwell Scientific, Oxford, 1993, 159.

41. **Robohm, R. A.,** *Pasteurella piscicida,* in *Antigens of Fish Pathogens,* Anderson, D. P., Dorson, M., and Dubourget, Ph., Eds., Collection Fondation Marcel Merieux, Lyon, France, 1983, 161.

42. **Nakai, T., Fujiie, N., Muroga, K., Arimoto, M., Mizuto, Y., and Matsuoka, S.,** *Pasteurella piscicida* infection in hatchery-reared juvenile striped jack, *Fish Pathol.,* 27, 103, 1992.

43. **Kusuda, R. and Yamoaka, M.**, Etiological studies on bacterial pseudotuberculosis in cultured yellowtail with *Pasteurella piscicida* as the causative agent. I. On morphological and biochemical properties, *Nippon Suisan Gakkaishi*, 38, 1325, 1972.

44. **Kusuda, R., Kawai, K., and Masui, T.**, Etiological studies on bacterial pseudotuberculosis in cultured yellowtails with *Pasteurella piscicida* as the causative agent. II. On the serological properties, *Fish Pathol.*, 13, 179, 1978.

45. **Kitao, T. and Kimura, M.**, Rapid diagnosis of pseudotuberculosis in yellowtail by means of fluorescent antibody technique, *Bull. Jpn. Sci. Fish.*, 40, 889, 1974.

46. **Mori, M., Kitao, T., and Kimura, M.**, A field survey by means of the direct fluorescent antibody technique for diagnosis of pseudotuberculosis in yellowtail, *Fish Pathol.*, 11, 11, 1976.

47. **Sinderman, C. J.**, *Principal Diseases of Marine Fish and Shellfish*, Academic Press, New York, 1970, 312.

48. **Matsusato, T.**, Bacterial Tuberculoidosis of Cultured Yellow Tail, Spec. Publ., Fish. Agency Jpn. Sea Reg. Fish. Res. Lab., 1975, 115.

49. **Toranzo, A. E., Barja, J. L., and Hetrick, F. M.**, Survival of *Vibrio anguillarum* and *Pasteurella piscicida* in estuarine and fresh waters, *Bull. Eur. Assoc. Fish Pathol.*, 3, 43, 1982.

50. **Kubota, S., Kimura, M., and Egusa, S.**, Studies of a bacterial tuberculoidosis of the yellowtail. III. Findings on nodules and bacterial colonies in tissues, *Fish Pathol.*, 6, 69, 1972.

51. **Egusa, S.**, Disease problems in Japanese yellowtail *Seriola guinguiradiata* culture: a review, in *Diseases of Commercially Important Marine Fish and Shellfish*, Stewart, J. E., Ed., l'Exploration de la Mer, Copenhagen, 1983, 10.

52. **Kusuda, R. and Inoue, K.**, Studies on the application of ampicillin for pseudotuberculosis of cultured yellowtails. I. *In vitro* studies on sensitivity, development of drug-resistance, and reversion of acquired drug-resistance of *Pasteurella piscicida*, *Bull. Jpn. Soc. Sci. Fish.*, 42, 9969, 1976.

53. **Toranzo, A. E., Barja, J. L., Colwell, R. R., and Hetrick, F. M.**, Characterization of plasmids in bacterial fish pathogens, *Infect. Immunol.*, 39, 184, 1983.

54. **Fukuda, Y. and Kusuda, R.**, Efficacy of vaccination for pseudotuberculosis in cultured yellowtail by various routes of administration, *Bull. Jpn. Soc. Sci. Fish.*, 47, 147, 1981.

55. **Fukuda, Y. and Kusuda, R.**, Vaccination of yellowtail against pseudotuberculosis, *Fish Pathol.*, 20, 421, 1985.

56. **Kusuda, R., Ninomiya, M., Hamacuchi, M., and Muroaka, A.**, The efficacy of ribosomal vaccine prepared from *Pasteurella piscicida* against pseudotuberculosis in cultured yellowtail, *Fish Pathol.*, 23, 191, 1988.

57. **Kitao, T.**, Streptococcal infections, in *Bacterial Diseases of Fish*, Inglis, V., Roberts, R. J., and Bromage, N. R., Eds., Blackwell Scientific, Oxford, 1993, 196.

58. **Kawahara, E. and Kusuda, R.**, Direct fluorescent antibody technique for differentiation between alpha- and beta-hemolytic *Streptococcus* spp., *Fish Pathol.*, 22, 77, 1987.

59. **Kusuda, R., Kawai, K., Salati, F., Banner, C. R., and Fryer, J. L.**, *Enterococcus seriolicida* sp. nov., a fish pathogen, *Int. J. Syst. Bacteriol.* 41, 406, 1991.

60. **Hoshina, T., Sano, T., and Morimoto, Y.**, A streptococcus pathogenic to fish, *J. Tokyo Coll. Fish.*, 44, 57, 1958.

61. **Robinson, J. A. and Meyer, F. P.**, Streptococcal fish pathogen, *J. Bacteriol.*, 92, 512, 1966.

62. **Barham, W. T., Schoonbee, H., and Smit, G. L.**, The occurrence of *Aeromonas* and *Streptococcus* in rainbow trout *(Salmo gairdneri)*, *J. Fish Biol.*, 15, 457, 1979.

63. **Plumb, J. A., Schachte, J. H., Gaines, J. L., Peltier, W., and Carroll, B.**, *Streptococcus* sp. from marine fishes along the Alabama and northwest Florida coast of the Gulf of Mexico, *Trans. Am. Fish. Soc.*, 103, 358, 1974.

64. **Boomker, J., Imes, G. D., Jr., Cameron, C. M., Nause, T. W., and Schoonbee, H. J.**, Trout mortalities as a result of *Streptococcus* infection, *Onderstepoort J. Vet. Res.*, 46, 71, 1979.

65. **Kitao, T., Aoki, T., and Sakoh, R.**, Epizootic caused by B-haemolytic *Streptococcus* species in cultured freshwater fish, *Fish Pathol.*, 15, 301, 1981.

66. **Kitao, T.**, The method for detection of *Streptococcus* sp., causative bacteria of streptococcal disease of cultured yellowtail *(Seriola guingueradiata)*. Especially their cultural, biochemical and serological properties, *Fish Pathol.*, 17, 17, 1982.

67. **Rasheed, V. M.**, *Streptococcus* sp. Infection in Bullminnows *(Fundulus grandis)*, Ph.D. dissertation, Auburn University, Alabama, 1983, 61.

68. **Baya, A. M., Lupiani, B., Hetrick, F. M., Roberson, B. S., Lukacovic, R., May, E., and Poukish, C.,** Association of a *Streptococcus* sp. with fish mortalities in the Chesapeake Bay and its tributaries, *J. Fish Dis.,* 13, 251, 1990.

69. **Kitao, T., Aoki, T., and Iwata, K.,** Epidemiological study on streptococciosis of cultured yellowtail *(Seriola guingueradiata).* I. Distribution of *Streptococcus* sp. in seawater and muds around yellowtail farms, *Bull. Jpn. Soc. Sci. Fish.,* 45, 567, 1979.

70. **Minami, T.,** *Streptococcus* sp., pathogenic to cultured yellowtail, isolated from fishes for diets, *Fish Pathol.,* 14, 15, 1979.

71. **Rasheed, V., Limsuwan, C., and Plumb, J.,** Histopathology of bullminnows, *Fundulus grandis* Baird and Girard, infected with a non-haemolytic group B *Streptococcus* sp., *J. Fish Dis.,* 8, 65, 1985.

72. **Miyazaki, T., Kubota, S. S., Kaige, N., and Miyashita, T.,** A histopathological study of streptococcal disease in tilapia, *Fish Pathol.,* 19, 167, 1984.

73. **Kimura, H. and Kusuda, R.,** Studies on the pathogenesis of streptococcal infection in cultured yellowtails *Seriola* spp.: effect of the cell free culture on experimental streptococcal infection, *J. Fish Dis.,* 2, 501, 1979.

74. **Katao, H.,** Erythromycin — the application to streptococcal infections in yellowtails, *Fish Pathol.,* 17, 77, 1982.

75. **Aoki, T., Takami, K., and Kitao, T.,** Drug resistance in a non-hemolytic *Streptococcus* sp. isolated from cultured yellowtail *Seriola guingueradiata, Dis. Aquat. Org.,* 8, 171, 1990.

76. **Sakai, M., Kubota, R., Atsuta, S., and Kobayashi, M.,** Vaccination of rainbow trout, *Salmo gairdneri* against beta-hemolytic streptococcal disease, *Bull Jpn. Soc. Sci. Fish.,* 53, 1373, 1987.

INDEX

A

Acanthopagrus schleyeli, see Black sea bream
Acid-fast staining bacteria, 223–228, see also
    Mycobacteriosis; Nocardiosis; specific types
Acid-sulfate soils, 12
Acipenser
  fulvescens, see Lake sturgeon
  transmontanus, see White sturgeon
Acipenseridae, see Sturgeon
Acute disease, 34, see also specific types
Acyclovir quanine, 101
Adenoviruses, 113, see also specific types
Aeration, 10, 11
Aeromonas
  caviae, 149, 150, 153
  hydrophila, 6, 7, 15, 21, 31, 34, 41, 133, see also
    Motile *Aeromonas* septicemia
    ANAI and, 163
    antibiotic resistance of, 154
    in carp, 165
    in catfish, 149
    description of, 144
    Edwardsiella tarda and, 169
    ESC and, 143
    EVE and, 69
    Flexibacter columnaris and, 138
    in humans, 153
    identification of, 150
    in other aquatic animals, 153
    significance of, 152, 154
    susceptibility to, 151
    SVC and, 57
    virulence of, 154
    Yersinia ruckeri and, 191
  salmonicida achromogenes, 57, 161, 162, 164, 178
  salmonicida masoucida, 178
  salmonicida salmonicida, 17, 177, see also
    Furunculosis
  sobria, 149, 150, 153
  spp., 5, 15, 31, 138, 148, 150
    atypical nonmotile, see Atypical nonmotile
      Aeromonas infection
    motile, see Motile *Aeromonas* septicemia
    septicemia of, see Motile *Aeromonas* septicemia
    susceptibility to, 151
Aflatoxins, 19
Agasi carp, 61
Akita Prefecture, 97
Alewife, 126
Alosa pseudoharangus, see Alewife
Amago, 177
Amberjack, 50
Ameirus melas, see Black bullheads

American eel, 69, 167
Ammonia, 6, 21
Ampicillin, 154, 231
ANAI, see Atypical nonmotile *Aeromonas* infection
Anemia, 38
Aneurysm, 39
Angelfish, 123
Anguilla
  anguilla, see European eel
  japonica, see Japanese eel
  rostrata, see American eel
Anguillidae, see Eels
Anoplopoma fimbria, see Sable fish
Antibiotic resistance, 154
Aquareovirus spp., 64
Aquazine, 10
Arctic char, 88, 177
Argulus spp., 60
Aristichthys nobilis, see Bighead carp
Arius felis, see Sea catfish
Ascites, 38
Aspergillus spp., 19
Atlantic cod, 88, 126, 127, 129, 162, 186
Atlantic herring, 126, 129
Atlantic menhaden, 16, 84
Atlantic salmon
  ANAI in, 161, 164
  BKD in, 206
  edwardsiellosis in, 167
  EED in, 101
  furunculosis in, 177
  herpesvirus in, 96
  IHNV in, 77
  IPNV in, 83
  red spot disease in, 171
  vibriosis in, 186, 189, 190
Atrophy, 40
Atypical mycobacterial infections, 226, see also specific
    types
Atypical nonmotile *Aeromonas* infection (ANAI),
    161–165
Ayu, 171, 182, 231

B

Bacteria, 16, 20, 31, 34, 133–134, 223–235, see also
    specific types
  acid-fast staining, 223–228
  in carp, 161–165
  in catfish, 135, see also specific types
  cell morphology of, 133
  distribution of, 32
  in eels, 167–174
  in goldfish, 161

H

I